# Evolutionary Economics and Social Complexity Science

Volume 9

The Japanese Association for Evolutionary Economics (JAFEE) always has adhered to its original aim of taking an explicit "integrated" approach. This path has been followed steadfastly since the Association's establishment in 1997 and, as well, since the inauguration of our international journal in 2004. We have deployed an agenda encompassing a contemporary array of subjects including but not limited to: foundations of institutional and evolutionary economics, criticism of mainstream views in the social sciences, knowledge and learning in socio-economic life, development and innovation of technologies, transformation of industrial organizations and economic systems, experimental studies in economics, agent-based modeling of socio-economic systems, evolution of the governance structure of firms and other organizations, comparison of dynamically changing institutions of the world, and policy proposals in the transformational process of economic life. In short, our starting point is an "integrative science" of evolutionary and institutional views. Furthermore, we always endeavor to stay abreast of newly established methods such as agent-based modeling, socio/econo-physics, and network analysis as part of our integrative links.

More fundamentally, "evolution" in social science is interpreted as an essential key word, i.e., an integrative and /or communicative link to understand and re-domain various preceding dichotomies in the sciences: ontological or epistemological, subjective or objective, homogeneous or heterogeneous, natural or artificial, selfish or altruistic, individualistic or collective, rational or irrational, axiomatic or psychological-based, causal nexus or cyclic networked, optimal or adaptive, micro- or macroscopic, deterministic or stochastic, historical or theoretical, mathematical or computational, experimental or empirical, agentbased or socio/econo-physical, institutional or evolutionary, regional or global, and so on. The conventional meanings adhering to various traditional dichotomies may be more or less obsolete, to be replaced with more current ones vis-á-vis contemporary academic trends. Thus we are strongly encouraged to integrate some of the conventional dichotomies.

These attempts are not limited to the field of economic sciences, including management sciences, but also include social science in general. In that way, understanding the social profiles of complex science may then be within our reach. In the meantime, contemporary society appears to be evolving into a newly emerging phase, chiefly characterized by an information and communication technology (ICT) mode of production and a service network system replacing the earlier established factory system with a new one that is suited to actual observations. In the face of these changes we are urgently compelled to explore a set of new properties for a new socio/economic system by implementing new ideas. We thus are keen to look for "integrated principles" common to the above-mentioned dichotomies throughout our serial compilation of publications. We are also encouraged to create a new, broader spectrum for establishing a specific method positively integrated in our own original way.

More information about this series at http://www.springer.com/series/11930

Yuji Aruka • Alan Kirman
Editors

# Economic Foundations for Social Complexity Science

## Theory, Sentiments, and Empirical Laws

Springer

*Editors*
Yuji Aruka
Faculty of Commerce
Chuo University
Hachioji, Tokyo, Japan

Alan Kirman
Directeur d'études à l'EHESS, Paris
Professeur Emerite Aix-Marseille Université
Aix-en-Provence, France

ISSN 2198-4204 ISSN 2198-4212 (electronic)
Evolutionary Economics and Social Complexity Science
ISBN 978-981-10-5704-5 ISBN 978-981-10-5705-2 (eBook)
DOI 10.1007/978-981-10-5705-2

Library of Congress Control Number: 2017952057



Printed on acid-free paper

This Springer imprint is published by Springer Nature
The registered company is Springer Nature Singapore Pte Ltd.
The registered company address is: 152 Beach Road, #21-01/04 Gateway East, Singapore 189721, Singapore

*Dedicated to the memory of Dr. Jun-ichi Inoue, the late associate professor of the faculty of the Graduate School of Information Science and Technology, Hokkaido University.*

# Preface

This book focuses on how important massive information is and how sensitive outcomes are to information. In this century, humans now are coming up against the massive utilisation of information in various contexts. The advent of super intelligence is drastically accelerating the evolution of the socio-economic system. Our traditional analytic approach must therefore be radically reformed in order to adapt to an information-sensitive framework, which means giving up myopic purification and the elimination of all considerations of massive information. In this book, authors who have shared and exchanged their ideas over the last 20 years offer thorough examinations of the theoretical–ontological basis of complex economic interaction, econophysics and agent-based modelling during the last several decades. This book thus provides the indispensable philosophical–scientific foundations for this new approach and then moves on to empirical–epistemological studies concerning changes in sentiments and other movements in financial markets.

The book was principally motivated by the workshop titled the International Conference on Socio-economic Systems with ICT and Networks, 26–27 March 2016, Tokyo, Japan. This conference was sponsored by JSPS grant no. 26282089 entitled "A study on resilience from systemic risks in the socio-economic system". Due to the success of this conference, we were provided with an excellent opportunity for our JSPS project members to exchange with and profit from interactions with the conference participants, in particular, with the guest speakers of the conference. Thus, just after the conference, the interactive process of discussions naturally around our subjects began to attain the collection of the essays in this volume.

Professor Alan Kirman, the coeditor of this volume, has not only promoted to advance an intelligent integration of this volume but also given the leading introductory perspective to this book. Our book readers will, we hope, easily understand the spirit of our project and to what extent our aims and scope are attained.

Project leader, A study on resilience from systemic risks in the socio-economic system (JSPS Grant no. 26282089) — Yuji Aruka
May 12, 2017

# Contents

# Contributors

**Yuji Aruka** Faculty of Commerce, Chuo University, Higashinakano Hachioji-shi, Tokyo, Japan

**Anindya S. Chakrabarti** Economics Area, Indian Institute of Management, Ahmedabad/Vastrapur, Gujarat, India

**Bikas K. Chakrabarti** Condensed Matter Physics Division, Saha Institute of Nuclear Physics, Kolkata, India

Economic Research Unit, Indian Statistical Institute, Kolkata, India

**Anirban Chakraborti** School of Computational and Integrative Sciences, Jawaharlal Nehru University, New Delhi, India

**Arnab Chatterjee** Condensed Matter Physics Division, Saha Institute of Nuclear Physics, Kolkata, India

TCS Innovation Lab, Delhi, India

**Shu-Heng Chen** Department of Economics, AI-ECON Research Center, National Chengchi University, Taipei, Taiwan

**Bertram Düring** Department of Mathematics, University of Sussex, Brighton, UK

**Kunika Fukuda** Department of Applied Mathematics and Physics, Graduate School of Informatics, Kyoto University, Kyoto, Japan

**Nicos Georgiou** Department of Mathematics, University of Sussex, Brighton, UK

**Asim Ghosh** Department of Computer Science, Aalto University School of Science, Aalto, Finland

**Tomoki Ito** Department of Systems Innovation, School of Engineering, The University of Tokyo, Tokyo, Japan

**Kiyoshi Izumi** Department of Systems Innovation, School of Engineering, The University of Tokyo, Tokyo, Japan

**Taisei Kaizoji** The Graduate School of Arts and Sciences, International Christian University, Tokyo, Japan

**Alan Kirman** CAMS, EHESS, Paris, Aix-Marseille University, Aix en Provence, France

**Masao Kubo** National Defense Academy of Japan, Yokosuka, Kanagawa, Japan

**Enda Liu** Department of Systems Innovation, School of Engineering, The University of Tokyo, Tokyo, Japan

**Jürgen Mimkes** Physics Department, Paderborn University, Paderborn, Germany

**Michiko Miyano** The Graduate School of Arts and Sciences, International Christian University, Tokyo, Japan

**J. Barkley Rosser Jr.** Department of Economics, James Madison University, Harrisonburg, VA, USA

**Marina V. Rosser** Department of Economics, James Madison University, Harrisonburg, VA, USA

**Aki-Hiro Sato** Department of Applied Mathematics and Physics, Graduate School of Informatics, Kyoto University, Kyoto, Japan

**Hiroshi Sato** National Defense Academy of Japan, Yokosuka, Kanagawa, Japan

**Enrico Scalas** Department of Mathematics, University of Sussex, Brighton, UK

**Shreyansh Shah** Indian Institute of Technology, Banaras Hindu University, Varanasi, India

**Kiran Sharma** School of Computational and Integrative Sciences, Jawaharlal Nehru University, New Delhi, India

**Hiroshi Suzuki** Department of Systems Innovation, School of Engineering, University of Tokyo, Tokyo, Japan

**Fujio Toriumi** Department of Systems Innovation, School of Engineering, University of Tokyo, Tokyo, Japan

**Kota Tsubouchi** Yahoo Japan Corporation, Tokyo, Japan

**Ragupathy Venkatachalam** Institute of Management Studies, Goldsmiths, University of London, London, UK

**Akihiro Yamaguchi** Fukuoka Institute of Technology, Fukuoka, Japan

**Tatsuo Yamashita** Yahoo Japan Corporation, Tokyo, Japan

# Chapter 1
# The Economy as a Complex System

**Alan Kirman**

It is paradoxical that economists wish to consider the economy as a system which can be studied almost independently of the fact that it is embedded in a much larger socio-economic framework. Our difficulties in analysing and diagnosing economic problems lie, in effect, precisely in the fact that the social system constantly generates feedbacks into the economy and that this is at the root of much of the instability that the overall system exhibits.

If we think, for a moment, of the framework that is supposed to underlie the economic functioning of our society, liberalism, it is based on an explanation that is not justified but rather assumed. What is the basic account that has been current since Adam Smith (1776) gave his account of how the independent actions of self-interested individuals would "automatically" further the common good? The answer is that the economy is a system made up of individuals each of whom pursue their own selfish goals and that such a system will naturally self-organise into a stable and socially satisfactory state. To be fair to Adam Smith, his account of the system's tendency to achieve this was much more nuanced than modern economic theory would lead us to believe. Yet this idea is now so firmly rooted that few contest its validity.

It is worth recalling that Walras and the founders of the "Marginal Revolution", who wished to formalise Smith's basic idea, based much of their analysis on physics using it as an example of the sort of science which economics could and would become. How did they think of the evolution of the economic system? Essentially as a physical system which would tend to an "equilibrium" that is, a state from which the system had no tendency to move. The physical analysis on which they based

A. Kirman (✉)
CAMS, EHESS, Paris, France
e-mail: alan.kirman@ehess.fr



Y. Aruka, A. Kirman (eds.), *Economic Foundations for Social Complexity Science*,
Evolutionary Economics and Social Complexity Science 9,
DOI 10.1007/978-981-10-5705-2_1

their understanding of the economy was drawn from classical mechanics.[1] The equilibrium in question is a situation in which, at given prices, what is demanded of each good is exactly what is supplied of that good. This is far from how physicists in the twentieth and twenty-first centuries would look at such a system. Rather than take as a basis some sort of "equilibrium", one could think of an economy as made up of heterogeneous individuals interacting with each other. Many such systems have been analysed in other fields, and indeed, the first analogies that come to mind, are with the brain, the computer or with social insects. The actions and reactions of the components whether they are simple particles with a "spin", or neurons, or insects following simple and local rules can nonetheless generate complicated aggregate behaviour. The system may go through sudden large changes or "phase transitions" without being affected by some external force. For those not familiar with the way in which aggregate patterns can emerge from the interaction of simple components, John Conway's "Game of Life" is an excellent illustration.

Models are, by necessity, simplifications of reality, "the map is not the territory" to use Alfred Korzybski's famous phrase and taken up again more recently for economic models by John Kay (2011). How have economists achieved this simplification? They have done so by considering the individual actor and his (or her) preferences and the constraints that limit his choices. That individual has no control over his constraints which are determined by the actions of many other individuals. This individual is considered to make choices for every period in his future and typically lives for ever. Furthermore, this means that each individual can anticipate correctly what his constraints will be both now and at all later dates. If those constraints are determined by the prices of all goods in the future, then in a world without uncertainty, he will know all future prices of all goods. Given the number of different goods that exist, which with no time horizon must be infinite, this seems to be highly implausible. Of course, if we now recognise that future prices are only known with uncertainty, then we have to ask how individuals form their probability distribution over those prices and whether they will all have the same distributions. The way around this for economists has been to assume that all the agents do have identical distributions and more that these distributions are consistent with the evolution of the economy. They are said to have *rational expectations*. With this heroic assumption one can, it is argued, treat the economy as a whole, as behaving like an individual.

But if one were to start with statistical physics as one's model, then one would take a very different point of view, just as the contributions in this book do. Instead of reducing the model of the aggregate to one "representative individual", consider agents who may be very different but who are much simpler than the homo oeconomicus portrayed in standard macroeconomics. From the interaction between these individuals *emerges* aggregate behaviour which could not have been predicted

[1]See, e.g. Mirowski's (1989) account of the close relationship of economics with nineteenth-century physics.

by looking at individuals in isolation, any more than one could anticipate the structure and organisation of an ants' nest by examining individual ants. Simplifying our models of agents in this way, permits one to build models, such as agent-based models which permit one to simulate the system and to vary the parameters and rules to see how robust the conclusions about the aggregate are. Economists have been reticent to accept such an approach since it rarely allows one to prove "theorems" as to which causes will produce which effects. However, recently a number of policy makers have suggested that such an approach might be very useful. This was the opinion of Jean-Claude Trichet the former governor of the European Central Bank. He said:

> First, we have to think about how to characterise the homo economicus at the heart of any model. The atomistic, optimising agents underlying existing models do not capture behaviour during a crisis period. We need to deal better with heterogeneity across agents and the interaction among those heterogeneous agents. We need to entertain alternative motivations for economic choices. Behavioural economics draws on psychology to explain decisions made in crisis circumstances. Agent-based modelling dispenses with the optimisation assumption and allows for more complex interactions between agents. Such approaches are worthy of our attention. (Jean-Claude Trichet 2010)

But before turning to an examination of the contribution that complex system theory can make, it is worth reflecting for a moment on the evolution of the discipline of economics itself. For a considerable period, there was an increasing divergence between "pure theory" into which category many of the topics already mentioned would fall and institutional and behavioural economics. In Chap. 4 of this book, Rosser and Rosser trace the paths that have been followed in the latter areas, since Veblen and Simon. Veblen insisted firmly on the fact that individuals are embedded in a network which is in part determined by the institutions of the economy and which affects not only individual choices but also individual preferences. Simon was preoccupied by people's limited capacity to process information and developed the notion of "bounded rationality". Taking the emergence of institutions and their development into account and modelling individuals as less sophisticated than those portrayed in standard economic models are approaches which are at the heart of complex thinking in economics. An account of the complexity theory approach can be found in Kirman (2010), and here I shall illustrate how it can prove useful in analysing a number of aspects of economics and relate these issues to some of the chapters in this book. Let me first look at some aspects of what might call standard economic theory in the general equilibrium tradition.

## 1.1 Stability

Although often discussed as the problem of the stability of equilibria, what is really at issue here is what is the mechanism that leads an economy to an equilibrium when it is out of equilibrium? As Arrow (1972) noted in his Nobel lecture, there has been a sort of pervasive conviction that economies will be led by an invisible hand to an equilibrium. As he says:

> From the time of Adam Smith's *Wealth of Nations* in 1776, one recurrent theme of economic analysis has been the remarkable degree of coherence among the vast numbers of individual and seemingly separate decisions about the buying and selling of commodities. In everyday, normal experience, there is something of a balance between the amounts of goods and services that some individuals want to supply and the amounts that other, different individuals want to sell. Would-be buyers ordinarily count correctly on being able to carry out their intentions, and would-be sellers do not ordinarily find themselves producing great amounts of goods that they cannot sell. This experience of balance is indeed so widespread that it raises no intellectual disquiet among laymen; they take it so much for granted that they are not disposed to understand the mechanism by which it occurs. The paradoxical result is that they have no idea of the system's strength and are unwilling to trust it in any considerable departure from normal conditions. (K Arrow 1972, Nobel Prize Lecture)

What Arrow is arguing is that the empirical facts have, in general, been so convincing that there is no need for most people to worry about where the economy was before it came to its current state, since it seems to them that the economy is more or less constantly in equilibrium. But notice that Arrow explicitly suggests that when we do have a "considerable departure from normal conditions", people are immediately concerned about the economy's capacity to return to equilibrium. However, notice also that Arrow, himself, suggests that the system does have the strength to do this. Thus, economic theory seems to have moved from early doubts to what is now considered as being self-evident to the participants in the economy.

However, the theoretical difficulties that were then encountered in the 1970s almost immediately after Arrow wrote those words revealed that the general equilibrium model, as it had developed, did not allow us to show that the economy could achieve equilibrium. Until the results of Sonnenschein (1972), Mantel (1974) and Debreu (1974), there was a persistent hope that, with the standard assumptions on individuals and by specifying a mechanism by which prices adjusted, one could show that an economy starting from a disequilibrium state would tend to an equilibrium, reflecting a more precise statement of Smith's idea and a century later expressed in more formal terms, by Walras. Those who expressed scepticism about this were regarded as not having the analytical tools to show that equilibria were stable under reasonable assumptions on individuals. However, this accusation could hardly be levelled at the authors of the results just mentioned. They were proved by some of the most sophisticated mathematical economists of their time, and what they showed was, that, even under the stringent and unrealistic assumptions made on individuals in the standard theory, one could not show that equilibria were either unique or stable. This led Morishima (1984), also a distinguished economic theorist, to remark:

> If economists successfully devise a correct general equilibrium model, even if it can be proved to possess an equilibrium solution, should it lack the institutional backing to realize an equilibrium solution, then the equilibrium solution will amount to no more than a utopian state of affairs which bears no relation whatsoever to the real economy. (Morishima 1984, pp. 68–69)

These results left open the idea that economies out of equilibrium might self-organise into nonequilibrium states and not converge to any particular state at all. But this would have involved analysing the evolution of economies in nonequilibrium states. This would have meant sacrificing the basic theorems of welfare economics and would have had profound consequences. The first fundamental theorem of welfare economics contains the essence of the theoretical argument which justifies the interest in laissez faire. What it says is that an economy in a competitive equilibrium is in a Pareto optimal state, by which is meant a state in which making any participant in the economy better off would make some other individual worse off. This is a very weak criterion since a state in which one individual had all the goods would satisfy it, but even if we accept this rather weak result, it says nothing about how the economy would get to such a state. Even excellent economists have sometimes suggested that the theorem just mentioned justifies the "invisible hand" narrative. For example, Rodrik says:

> The First Fundamental Theorem is a big deal because it actually *proves* the Invisible Hand Hypothesis. (Dani Rodrik 2015)

Unfortunately, this is simply not true. All that the first theorem says is that if one is in a competitive equilibrium, the allocation of goods in the economy will have the Pareto optimal property. It says absolutely nothing about how the economy got there, and that is where the full weight of the Sonnenschein, Mantel and Debreu results is revealed.

Where then does the problem lie? A first reaction turned out to be to suggest that the adjustment process for prices might be modified so that we could then show that the invisible hand idea was, in fact, justified. Again, the sentiment was that it was only mathematical inadequacy that was preventing us in obtaining a solution to this problem. One idea was that the adjustment mechanism specified was inadequate. Therefore, a number of people have pursued alternative approaches (see, e.g. Herings 1997; Flaschel 1991). There are two possibilities, either one considers that, at each point in time, all the participants know all the prices of all the goods, and that they continue to do so as prices adjust, or more radically prices are seen to emerge from the interactions between economic agents who negotiate the terms on which they trade. This was the approach proposed by Edgeworth (1889) who had a running battle with Walras on the subject. It was later taken up by Fisher (1989). But, interestingly the struggle to overcome this "stability" problem revealed that a basic problem was that of the amount of information required to make the adjustments necessary.

## 1.2 Information

This brings us to the heart of the problem of the analysis of the self-organisation of complex systems which is the way in which information is processed and distributed and how much information is involved. This is the question that is also now being raised as a consequence of the technological revolution that has taken place, and it is curious that, when the discussion of the stability of economic equilibrium was discussed, little attention was paid to it. Yet, when the idea of modifying the way in which we model adjustment was mooted, it was Stephen Smale, a distinguished mathematician, who took up the challenge, and the question of information was already raised. Indeed, what became immediately clear, after the innovative work that he then undertook (Smale 1976), was that stability could only be achieved at the price of a significant increase in the amount of information needed. Smale's global Newton method did not solve the problem completely since it could not guarantee that the economy would move to an equilibrium, from any arbitrary starting prices and as already mentioned, it uses a great deal of information.

As a result, the informational efficiency of the competitive allocation mechanism, long vaunted as one of its most important merits, would no longer have held. Recall that, all that the market mechanism has to do is to transmit the equilibrium price vector corresponding to the aggregate excess demands submitted by the individual economic agents. The information required to make this system function at equilibrium is extremely limited. In fact, a well-known result of Jordan (1982) shows that the market mechanism is not only parsimonious in terms of the information that it uses, but, moreover, it is also the only mechanism to use so little information to achieve an efficient outcome in the sense of Pareto. This remarkable result depends, unfortunately, on one key assumption, which is that the economy is functioning *at equilibrium.*

However, as soon as one considers how the economy might function out of equilibrium, the informational efficiency property is lost. What is more, if one considers how an economy might adjust from any arbitrary starting point to equilibrium, looking at informational efficiency provides a key to the basic problem with equilibrium theory. Indeed, Saari and Simon (1978) put the final nail in the coffin by showing that an adjustment mechanism which would take an economy from any initial nonequilibrium prices to an equilibrium would necessarily use an infinite amount of information. But all of this is in the context of the standard general equilibrium economic model.

Now consider a system in which individuals receive their own limited and maybe local information and react to it by taking some action. Given the actions of the other individuals, they may react again. The one area in economics where this sort of thing has been taken seriously is game theory. But as Aruka points out in his chapter in this book, applying game theory to even a simple well-defined game with just two players quickly runs into informational problems. Thus, to think that this is the appropriate way to model the behaviour of the whole economy is, to be euphemistic, optimistic. Efforts to reduce the whole interactive system with all its feedbacks to a simple mechanistic model which can then be solved are doomed to failure, as Bookstaber (2017) says in his forthcoming book, because of the *computational irreducibility* of the overall problem. By this he means that "the complexity of our interactions cannot be unravelled with the deductive mathematics that forms the base—even the raison d'être—for the dominant model in current economics". What one can do is to simulate a large agent-based model in which the agents use simple rules to react to the evolution of the economy and to the behaviour of the other agents. One can then examine the outcomes when the economy evolves from some initial conditions and vary the parameters of the model to check for the robustness of the outcomes.

Of course, this means abandoning the idea of "proving that a causes b", but in economics such proofs are given for cases which involve such specific assumptions that the results have little general interest or applicability. Thus, the choice is whether to continue developing models which are abstract and proving results within their restrictive framework, or to start simulating systems which can capture some of the features of economies which are absent from current macroeconomic theory.

In particular, by adopting the latter strategy, one can keep an essential feature of complex systems which is that they are made up of components which are very different from one another. This heterogeneity cannot be systematically eliminated by some appeal to the law of large numbers since the components are far from independent of each other. Indeed, the transmission of information between different individuals can cause waves or "epidemics" of actions or opinions.

## 1.3 The Structure of Interaction: Networks

However, to take account of these more realistic features of the economy, one needs to specify more about the structure of the economy and how the different components interact with each other. This means, for example, starting with the network structure which links the components, whether consumers, firms or banks, for example, and analysing how the result of their interaction is channelled into changes in the overall system. Several of the papers in this book deal with this sort of problem. Chen and Venkatachalam in Chap. 5 trace the evolution of the relationship between agent-based models (ABM) and network analysis in economic models. They show that in the earliest versions of ABM, the networks aspect was limited to very simple structures such as the Moore neighbourhoods on a chessboard. Conway's Game of Life was the canonical example. In the second phase, the spatial aspect of the relationships between individuals largely disappeared, and there was anonymous interaction between any of the individuals. In the most recent literature, the network structure of the interaction and the form that it takes has become much more important. However, there remains a difference between those who use an ABM approach and those who use fully game theoretical reasoning to analyse interaction and its results. Some of the leading specialists in network theory such as Goyal (2007) and Jackson (2008), for example, have tended to remain with rather sophisticated agents, while others in finance theory (see, e.g. Bouchaud 2012; Battiston et al. 2012) have accepted the idea of rather simple agents using specific rules.

Interaction is not confined to individuals. Indeed, once the structure of the economy is decomposed, we can use network techniques to analyse the interdependence of the components. Consider, for example, the interdependence of the different sectors in the economy as measured, for example, by input–output tables. This has attracted some attention in the mainstream literature. There, one gets a glimpse of how understanding the network of linkages between asymmetric firms and the interactions between sectors may generate large systemic shocks in the work of

Acemoglu et al. (2011) and Gabaix (2011). But this has not penetrated modern macroeconomics other than as justification for a more fat-tailed distribution of aggregate shocks. The importance of the direct interaction between firms, which is an essential ingredient of the evolution of the economy, is not recognised. Macroeconomic models remain in the tradition of a set of actors taking essentially independent decisions with the mutual influence of these actions relegated to the role of inconvenient externalities.

Indeed, the main effort in economics, so far, has been directed at showing how relatively minor shocks to one sector can be transmitted to others, thus producing a major overall change. In fact, this approach could be pushed further to analyse the dynamics of the system as a whole and to show how small perturbations to one part of the economy can be translated into larger movements of the economy as a whole without the system ever reaching an "equilibrium" in the standard sense. Looking at the contribution of Sharma et al. in Chap. 11 of this book, we see how looking at the economy as an interdependent network through the lens of network analysis allows us to get a better hand on the evolution of both micro and macro level phenomena.

They use Indian Stock Exchange data to construct networks based on correlation matrices of individual stocks and analyse the dynamics of market indices. They use multidimensional scaling methods to visualise the sectoral structure of the stock market and analyse the co-movements among the sectoral stocks. They also examine the intermediate level between micro and macro by constructing a mesoscopic network based on sectoral indices. They use a specific tool, the minimum spanning tree technique in order to group technologically related sectors, and the mapping they obtain gives a good fit to empirical production relationships.

## 1.4 Sectoral Variations

The necessity to study the evolution of the different sectors in the economy and their interrelatedness is emphasised by the paper by Kubo et al. in Chap. 8 of this book which analyses the Survey on Employment Trends to propose methods to determine how regular are movements in unemployment rates across sectors. The relationship between sectors can be one of complementarity or substitutability, and this will have an important effect on the stability of economic systems. This was revealed long ago in the papers on the stability of the general equilibrium model and by subsequent re-examination of the input–output tables for different economies. In that work, a condition of "gross substitutability" or dominant diagonal of the matrix of derivatives of excess demand was imposed to rule out too much complementarity. The work here looks at how a measure of imbalance, unemployment, is related in different sectors and seeks to determine how regular variations in employment are across different industrial sectors. The problem is that there is a strong correlation across sectors as a result of macroeconomic movements and that this tends to dominate any idiosyncratic movements. They examined the different industrial sectors in Japan using the Survey on Employment Trends and found that most sectors were too similar for correlation analysis to provide satisfactory discrimination between sectors. They then used machine learning techniques to analyse the survey of economic trends and found significant co-movement in certain sectors. What is interesting here is the use of methods developed for non-numerical data to analyse numerical data, whereas, in general, the problem is the reverse. The results are interesting for they underline the correlated reaction of sectors which could be regarded as "essential" to the 2008 crisis. Such inter-sectoral developments are lost in modern macroeconomic models.

## 1.5 The Efficient Markets Hypothesis

Strongly related to the informational problem which is at the heart of complex systems analysis is the efficient markets hypothesis which claims that all the information relevant to the value of an asset is contained in the price of that asset. This hypothesis originated in the work of Bachelier (1900) and argued that individuals who receive private information about the fundamentals of some asset will then act upon that information by buying or selling that asset and, in so doing, will modify the price of that asset in such a way that the new information becomes visible to all. This is, of course, a very different vision than that of Walras, who posited a set of prices visible to, and known to, all the actors in the market but which were modified by some central authority or auctioneer and not by the trades effected by the market participants. This vision might be thought of as more consistent with the complex systems approach, but its defects were quickly pointed out by Poincaré (1908). As he explained, people do not look at their own information in isolation before acting, but rather have an inbuilt tendency to behave like sheep. This means that the observation of some piece of information by one market participant can rapidly be translated into a cascade of buying or selling with large-scale consequences and a substantial literature (see, e.g. Chamley (2004) for a good survey) has developed on this theme. Yet the efficient markets hypothesis still holds sway in some quarters, and paradoxically its leading defender Fama and one of its leading critics Shiller were awarded the Nobel Prize in economics in the same year. Why is this relevant here? It is because when there are important feedbacks from different actions, some of the latter, even if very small, can, in the complex systems framework, translate into large movements of the system as a whole without any convergence to fundamentals.

In Chap. 12 of this book, Myano and Kaizuji study the relationship between share prices and fundamentals for 8000 companies in America, Europe, Asia and the rest of the world. Their method consisted in regressing companies' share values against a number of factors some of which were idiosyncratic to firms and others which were constant across firms but which varied over time. The fundamentals were then taken to correspond to the idiosyncratic effects for the companies. Their results show that there were significant deviations from fundamentals and that these varied across regions. Nevertheless, there were significant correlations across regions. Before 2008 in all regions, prices were above the fundamentals, while in all regions in 2008, they were significantly below them. Of course, one could argue that the shift was essentially in people's expectations and perhaps they should have been more specifically taken into account in calculating the fundamentals in the first place. However, the study clearly shows how there are systematic deviations from fundamentals, and since fundamentals at any point in time depend on expected future returns, if there are significant differences in expectations across regions, this would appear to show deviations from any notion of "true" fundamentals. Once again, the interrelatedness of the system shows up clearly in the analysis.

Another informational question which arises in financial markets is the origin of information. To what extent is it due to external or exogenous news, and to what extent is it inferred from the movements of asset prices? Izumi et al., in Chap. 10 of this book, examine the relations between index futures and idiosyncratic shocks when some important news impacted the financial market. The methodology involved using the transfer entropy method and their data was order-book data, from the Tokyo Stock Exchange. They found that the information flows between assets were enhanced during the impact of major external shocks such as the Great East Japan Earthquake. Such a relationship reinforces a standard argument that asset prices are more correlated when there is important exogenous information. Second, order information became the source of information flow which enhanced the high-frequency relationship during the external shocks. This is, of course, related to Poincaré's argument that market participants are constantly monitoring other participants' activity, and what this study shows is that this is particularly true in periods of high volatility. Finally, index futures tended to have a significant influence on other stocks' price changes during the external shocks. This shows the importance of changing expectations. Izumi et al. propose a new analytical method that extracts the high-frequency relationship between assets using transfer entropy. To test the proposed method, they used the high-frequency data, order-book data in Tokyo Stock Exchange (time-series data of transactions and quotes of each stock).

As mentioned, believers in the efficient markets hypothesis argue that all relevant information concerning the "fundamental" value of an asset is contained in its price. One of the reasons, which is reinforced by Izumi et al.'s work, why fundamentals do not give a good explanation of asset prices is that agents are more concerned with the future values of fundamentals rather than their current values (see Engel et al. 2008). Since expectations vary considerably in time and across market participants, volatility is self-reinforcing, in contradiction with the idea of some sort of convergence to a set of long-term "equilibrium values" or "true" fundamental values.

## 1.6 Inequality

A topic which has long been neglected in theoretical macroeconomics is socio-economic inequality. However, this has come to the fore in public debate in recent years and is now frequently argued to be responsible not only for manifestations of frustration and hostility but has also led to the rise of political parties whose platforms would, until recently, have been considered "extreme". Many authors have analysed the increasing inequality in income and wealth in our societies (see, e.g. Atkinson 2015), and others have looked at the more general problems of equality of opportunity and access to certain occupations (see, e.g. Sen (1992). Thomas Piketty's (2014) recent book has reached a very large audience in many countries. There has been a long-standing concern with inequality in economics, but it has not penetrated modern macroeconomic models. The issue itself has been the subject of a substantial literature in economics, and the most widely cited early work is that of Pareto (1896, 1965) who defined a parametric form for income distributions in general and claimed that it was found in numerous empirical examples. His "Pareto Law" can be defined as

$$F(x) = Pr\,(X > x) = \begin{cases} \left(\frac{x_m}{x}\right)^{\alpha} & x \geq x_m \\ 1 \text{ x} < x_m \end{cases}$$

Research showed that closer examination of the proposed distribution suggests that the Pareto distribution fits best in the right tail of the distribution. This law has often been summarised as the 80/20 law, suggesting that the top 20% of a population have 80% of the total. In fact, Pareto found that in the case of the UK, the top 30% of the population had about 70% of income. It has since been observed that the Pareto distribution emerges in many settings, in economics, the size distribution of firms, sizes of cities as well as income and wealth distribution. In other social sciences, similar phenomena are revealed, the number of people speaking different language, family names, audiences for films, certain social network patterns and crimes per convicted individual. This sort of law emerges in various physical contexts, the sizes of large earthquakes, sizes of sand particles, sizes of meteorites, numbers of species per genus, [1]areas burnt in forest fires, etc. Of course, in each case the α parameter will be different.

The main indicator of inequality which has been widely adopted is the Lorenz curve named after an American economist at the beginning of the twentieth century. In the case of income, if one ranks people by their income, it shows the relation between the percentages of the population possessing a certain proportion of total income, for example, the 20% with the highest incomes might earn 80% of all income. In the case where income is perfectly equally distributed, the Lorenz curve is the diagonal illustrated in Fig. 1.1. More inequality moves the curve away from the diagonal.

It is of course clear that it is not necessarily possible using this measure to decide if one society or phenomenon is more unequal than another since the Lorenz curves of the two may cross.

As is often the case, in order to simplify things, economists have preferred to reduce the problem to one of finding an index of inequality, and the most common is the Gini coefficient. This coefficient is obtained by dividing the area between the Lorenz curve and the diagonal by the area under the diagonal as illustrated in Fig. 1.1.

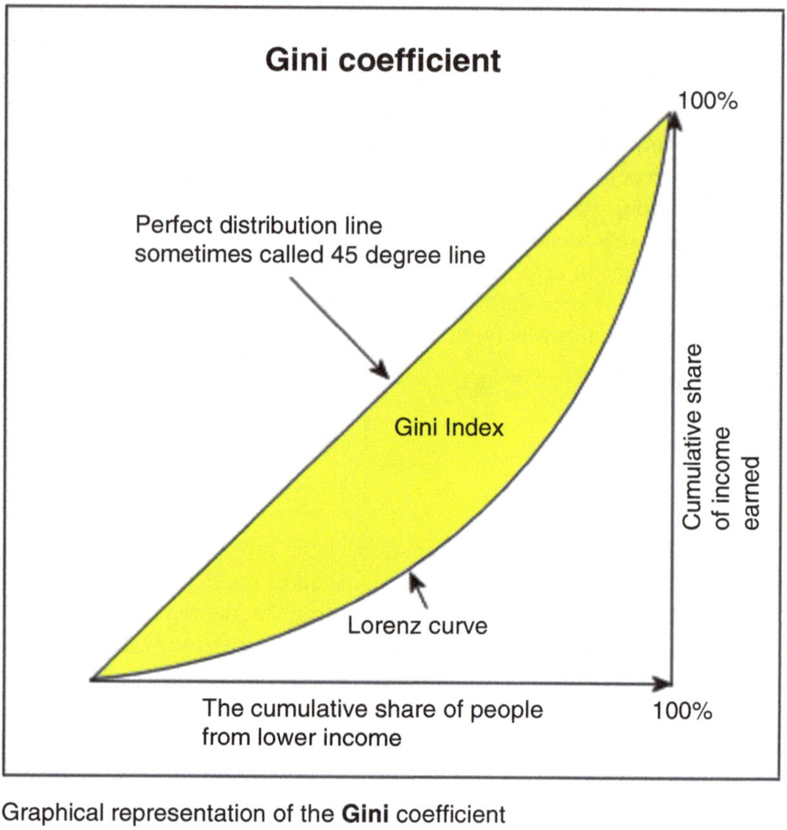


**Fig. 1.1** The Lorenz curve and the Gini coefficient

It is of some interest to note that the Gini coefficient for the Pareto distribution is given by

$$g = \frac{1}{2\alpha - 1}$$

As always, the recourse to a single index removes the discriminatory power of that index. There are simple examples of two situations where the Gini coefficient is the same but where there is a clear distinction between the two in terms of inequality.

Later, many other inequality measures have been discussed, and among them is the recently introduced Kolkata ($k$) index which gives a measure of the $1 - k$ fraction of population who possess the top $k$ fraction of wealth in the society recalling the 80/20 characterisation of the Pareto Law. Chapter 3 of this book by Chaterjee et al. reviews the character of such inequality measures, as seen from a variety of data sources, and discusses relationship between the Gini coefficient and the Kolkata index. They also investigate socio-economic inequalities in the context of man-made social conflicts or wars, as well as in natural disasters.

Yet it is not enough to find better and better parametric distributions in terms of their fit with the empirical data. What we need are explanations as to how these distributions arise. The simplest example is that of preferential attachment which gives rise to a Pareto-like distribution. Here, a simple example is that of city sizes. Suppose that the probability of a newcomer going to a city is proportional to the size of that city. This simple stochastic process will give rise to a power law. Another example is given by Chaterjee et al. in their chapter, who use a kinetic exchange model to obtain a distribution which gives a reasonable fit of their data.

In Chap. 7 of this book, Duering et al. explain how, in the last 20 years, physicists and mathematicians have developed models to derive the wealth distribution using both discrete and continuous stochastic processes (random exchange models) as well as related Boltzmann-type kinetic equations. In this literature, the usual concept of equilibrium in economics, as a solution of a static system of equations, is either replaced or completed by the notion of a statistical equilibrium (see, e.g. Foley 1994).

These authors present an exchange model to derive the distribution of wealth. They first discuss a fully discrete version (a stylised random Markov chain with finite state space).

They then study its discrete-time continuous-state-space version and prove the existence of the equilibrium distribution. One could, of course, argue that in an evolving system, there will be no convergence to a limit distribution and that what we need are dynamic models of the evolution of wealth distributions but which do not converge. This remains an ambitious target.

Another approach to the inequality problem is to take the physical analogy of a heat pump, for example. This is what Minkes does in Chap. 6 of this book, and he argues that just as a heat pump extracts heat from a cold environment and can then heat a house, so the economic activities of production, trade and banking may extract capital from a poor population and make a rich population richer. The rich and the poor in question can be within the same country or in two different countries linked by trade. Through foreign trade as well as the economic activity in internal markets, capitalistic economies like the USA, China and others get richer, and the efficiency of the machine, the difference between rich and poor, grows with time.

What Minkes argues is that the cooling effect of lower wages for the less well-off makes the system run efficiently. He suggests that Japan is no longer functioning efficiently since Japanese wages have now risen to the level of their US counterparts. In this case, he suggests the "economic motor" will run too hot and stall. The way back to productivity and competitivity in this view is for wages of the lower income classes in Japan to "cool down".

Those who argue for treating the economy as a complex system will be tempted to suggest that some of the feedbacks observed in modern economies and the world economy as a whole are ignored in this analogy, but it does provide a clear and simple physical analogy reminiscent of the hydraulic Philips machine and the much earlier Fisher machine inspired by the work of Gibbs (see Dimand and Betancourt 2012).

## 1.7 Language

Another interesting aspect of the information problem is taken up by Liu et al. in Chap. 9 of this book. They observe that the language in which information is conveyed may have a significant effect on the interpretation that is made of that information. Thus, the idea developed, in particular, by Hayek (1945) that individuals have local information which is transmitted by their actions to other individuals may suffer from the fact that even the language in which the information is available may lead to differences in transmission. To examine this problem, Liu et al. try to establish clusters of financial terms in Japanese and in English and to analyse the relation between them. As one could expect, the quality of a bigraph established in this way depends importantly on the "distance" of a term from its original source. But, once again, what might seem to be banal considerations, such as in which language a message is communicated, can have important consequences for societal reactions. This will not be a novel idea for political scientists, sociologists or anthropologists but has not been seriously been taken into account in macroeconomics.

## 1.8 Conclusion

Considering society in general, and the economy in particular, as a complex adaptive system leads to very different perspectives from those normally envisaged in modern macroeconomics, and the articles in this book underline that fact. Perhaps most interesting is the reappearance of notions that have been widely discussed in earlier periods. An example of this is the biological analogy studied here, in detail, in Chap. 2 by Aruka. Although many economists, notably Marshall (1890), argued that biology was a more relevant science with which to compare economics, physics and then mathematics tended to have a dominant influence on our discipline. Yet, as Frank Hahn (1991) observed:

> I am pretty certain that the following prediction will prove to be correct: theorising of the 'pure' sort will become both less enjoyable and less and less possible ... rather radical changes in questions and methods are required ... the signs are that the subject will return to its Marshallian affinities to biology.

Indeed, with developments in molecular biology, the study of immune systems, for example, is of particular interest to those interested in how systems react to different developments in their components. This is important for understanding the vulnerability of the system to such changes. In particular, it is essential to observe that the interaction of the components is stochastic but cannot always be analysed through recourse to standard statistical methods. As we move the focus from the reaction of independent "rational" individuals to the development of the system within which they function, but over which they have little control, to one in which individuals have more limited calculating capacity and more local information than their counterparts in standard economic models, we obtain new insights into the functioning of the system as a whole. Recognising explicitly that macro phenomena emerge from the interaction between the individuals in the system but cannot be understood by focusing attention on the characteristics of the individuals themselves is one of the keys to being less surprised at the evolution of markets and economies.

## References

D. Acemoglu, V.M. Carvalho, A. Ozdaglar, A. Tahbaz-Salehi, The network origins of aggregate fluctuations. Econometrica **80**(5), 1977–2016 (2012)

K.J. Arrow, General Economic Equilibrium: Purpose, Analytic Techniques, Collective Choice. Nobel Memorial Lecture, December 12, 1972

A.B. Atkinson, *Inequality What Can be Done?* (Harvard University Press, Cambridge, MA, 2015)

L. Bachelier. Théorie de la spéculation. Annales de l'École Normale Supérieure, 3e série, **17**, 21–86 (1900)

S. Battiston, D. Delli Gatti, M. Gallegati, B. Greenwald, J.E. Stiglitz, Liaisons dangereuses: increasing connectivity, risk sharing, and systemic risk. J. Econ. Dyn. Control. **36**(8), 1121–1141 (2012)

R. Bookstaber, *The end of Theory* (Princeton, Princeton University Press, 2017)

J.-P. Bouchaud, Crises and Collective socio-economic phenomena: simple models and challenges. J. Stat. Phys. **151**(3–4), 567–606 (2012)

C. Chamley, *Rational Herds* (Cambridge University Press, Cambridge, 2004)

G. Debreu, Excess demand functions. J. Math. Econ. **1**(1), 15–21 (1974)

R. Dimand, R.G. Betancourt, Irving Fisher's appreciation and interest (1896) and the Fisher relation. J. Econ. Perspect. **26**(4), 185–196 (2012)

F.Y. Edgeworth, The mathematical theory of political economy. Elements d'Economie Politique Pure. By Leon Walras. Nature **40**, 434–436 (1889)

C. Engel, N.C. Mark, K.D. West, Exchange rate models are not as bad as you think, in *NBER Macroeconomics Annual 2007* (National Bureau of Economic Research, Cambridge, MA, 2008)

F.M. Fisher, *Disequilibrium foundations of quilibrium economics* (Cambridge University Press, Cambridge/New York, 1989)

P. Flaschel, Stability—independent of economic structure? A prototype analysis. Struct. Chang. Econ. Dyn. **2**(1), 9–35 (1991)

D. Foley, A statistical equilibrium theory of markets. J. Econ. Theory **62**(2), 321–345 (1994)

X. Gabaix, The granular origins of aggregate fluctuations. Econometrica **79**, 733–772 (2011)

S. Goyal, *Connections: An Introduction to the Economics of Networks* (Princeton University Press, Princeton/Oxford, 2007)

F. Hahn, The next hundred years. Econ. J. **101**, 47–50 (1991)

F.A. Hayek, The use of knowledge in society. Am. Econ. Rev. **35**(4), 519–530 (1945)

J.-J. Herings, A globally and universally stable price adjustment process. J. Math. Econ. **27**, 163–193 (1997)
M.O. Jackson, *Social and Economic Networks* (Princeton University Press, Princeton/Oxford, 2008)
J.S. Jordan, The competitive allocation process is informationally efficient uniquely. J. Econ. Theory **28**, 1–18 (1982)
J. Kay, The Map is Not the Territory: An Essay on the State of Economics (2011). Posted at http://ineteconomics.org/blog/inet/john-kay-map-not-territory-essay-state-economics
A. Kirman, *Complex Economics: Individual and Collective Rationality* (Routledge, London, 2010)
R. Mantel, On the characterisation of aggregate excess demand. J. Econ. Theory **7**, 348–353 (1974)
A. Marshall, *Principles of Economics* (Macmillan, London, 1890)
P. Mirowski, *More Heat than Light: Economics as Social Physics, Physics as nature's Economics* (Cambridge University Press, Cambridge, 1989)
M. Morishima, The good and bad uses of mathematics, in *Economics in Disarray*, ed. by P. J. D. Wiles, G. Routh (Basil Blackwell, Oxford, 1984)
Pareto *Cours d'économie politique*, Lausanne, F. Rouge, 1896–97 (2 volumes) (1896)
V. Pareto, La Courbe de la repartition de la Richesse (Originally published in 1896), in *Oevres Completes de Vilfredo Pareto*, ed. by G. Busino (Librairie Droz, Geneva, 1965), pp. 1–5
T. Piketty, *Capital in the Twenty First Century* (Belknap Press, Cambridge, MA, 2014)
H. Poincaré, *Science et méthode* (Flammarion, Paris, 1908)
D. Rodrik, *Economics Rules* (W.W. Norton, New York, 2015)
D. Saari, C.P. Simon, Effective price mechanisms. Econometrica **46**, 1097–1125 (1978)
A.K. Sen, *Inequality Re-Examined* (Clarendon Press, Oxford, 1992)
S. Smale, Exchange processes with price adjustment. J. Math. Econ. **3**, 211–226 (1976)
H. Sonnenschein, Market excess demand functions. Econometrica **40**, 549–563 (1972)
J.-C. Trichet, Reflections on the nature of monetary policy non-standard measures and finance theory, Speech by President of the ECB, Opening address at the ECB Central Banking Conference Frankfurt (18 November) (2010)

# Part I
# Theoretical Foundations

# Chapter 2
# Systemic Risks in the Evolution of Complex Social Systems

**Yuji Aruka**

**Abstract** In this chapter, we have explored some philosophical designs for developing economics around systemic risks viewed in terms of their internal probability. This concept of probability is based on observations of living systems. An important point to note is that cognizing subjects are not limited to humans but also extend to materials. The material observer behaves as an "internal observer." Further, material interactions between the observer and the objects being observed must generate a complex process. Based on this conception, we aimed to develop an understanding of how systemic risks could be recognized in this kind of interaction. In line with this objective, we examined this in relation to immunology and immunoediting that occurs on the surface level of the immune system. Our exploration revealed the importance of checkpoints such as nodes within a networked structure that is formed based on the above complex interactions, enabling us to better understand systemic risks. Last, we focused on the network node or link of the input-output system of the economic system, and introduced network-related concepts such as alpha centrality, successfully identifying a measure of risk in Japan's current economic system.

## 2.1 Toward the Reconceptualization of a New Analytical Device

While the previous century witnessed consumption of physical goods and services on a massive scale, information utilization on a similar scale, in various contexts, is a feature of the current century. Intensive utilization of information as a result of pervasive information and communications technology (ICT) is a key characteristic of contemporary society. For example, high-frequency trading (HFT) markets and "smart grids" for the allocation of electricity are based on the idea of smart information management at astronomical scales that extend well beyond the scope

Y. Aruka (✉)
Faculty of Commerce, Chuo University, Higashinakano Hachioji-shi, Tokyo, Japan
e-mail: aruka@tamacc.chuo-u.ac.jp



Y. Aruka, A. Kirman (eds.), *Economic Foundations for Social Complexity Science*, Evolutionary Economics and Social Complexity Science 9,
DOI 10.1007/978-981-10-5705-2_2

of our imagination. However, given the scarcity of resources, the achievement of optimal performances is solely contingent on the deployment of human rationality.

The concept of myopic optimization in relation to scarce resources assumes the least use of available information. Evidently, myopic optimization was a winning strategy during a previous developmental stage when access to big data was denied. However, this strategy no longer seems tenable. The reality has shifted now that quantum computing is no longer a distant possibility. With the implementation of HFT, markets can even operate under essentially traditional rules, while producing a qualitatively different series of fluctuations from those of the past. A political revolution is not necessary; rather, a systemic revamping as suggested, for example, by "Industry 4," automatically entails a "re-domaining" of the socioeconomic system.

AlphaGo's victory over its human opponent in an Igo game competition in 2015 was a symbolic marker of the overwhelming gains of nonhuman intelligence over human intelligence. This was apparent given that Igo is the most difficult game in which the strategy tree comprises a total of nodes. With the advent of a new era coexisting with that of artificial intelligence, it has become imperative to revise our methods of inference within economics and, more widely, within the social sciences. A more realistic consideration will evidently expose us to a domain at an astronomical scale. This can be illustrated using the example of an iterative two player-two strategy game. Even in this game, a classical inference that is confined to a narrow set of rational principles for detecting a solution would not hold for a broader framework that is designed to allow players access to past memories of their moves. Memories relating to the preceding two periods would enable the formulation of possible strategic combinations of the two players to a value of $2^{21}$ by $2^{21}$ (4.4 trillion). Here, we simply introduced interactive decisions made during the preceding two turns associated with the burgeoning information structure.

H. von Stackelberg was the first to recognize inference based on the observation of complex interactions among participants in a duopolistic market. However, the procedure for examining the complex deployment of strategic interactions at such a vast scale is beyond our grasp. As this example demonstrates, the time has come to change our analytical approach from mathematically pedantic inference to agent-based inference.

Econophysics has become an established field following its emergence at the end of the last century as a result of the extensive collection of large amounts of data. With the availability of substantial amounts of simulated data, network analysis of heterogeneous interactions is being promoted through agent-based modeling that also enables human sentiments to be analyzed based on data/text mining. This mining reveals just how effective machine-based learning, which is not just limited to humans, actually is. This kind of learning will prove beneficial in discovering potential connections. Text mining will provide new insights for prediction. This kind of inference, in particular, may prove to be robust in an area wherein the traditional method does not work well. In addition, sensor technology, in combination with machine learning, is now, for example, drastically transforming our method of driving a car. Consequently, industries as well as industrial policies are also shifting toward intensive utilization of information.

To sum up, we can observe how the method of inference would be transformed as a result of adapting to the societal change that is currently underway. Our traditional method of inference should be revised to become information sensitive, marking a shift from myopic purification which does not entail any heavy commitment to a massive influx of information. The time is ripe for a change of theories as well as the formulation of analytical tools that are oriented toward this shift. Equipped with new insights resulting from considerations that are information sensitive, we aim to develop a new approach for handling the two-sided transformation of ontology and epistemology. This requires us to first attempt to construct the theoretical and ontological foundations of the new approach. Subsequently, empirical and epistemological studies can be conducted to examine sentiments and other movements within financial markets.

## 2.2 Some Philosophical Designs for Economics

I have previously discussed an alternative pathway for economics at some length (see Aruka 2015). Elsewhere, we have witnessed a number of debates centering on the polarity between Keynesian and Hayekian economic theories and on a "command and control" perspective versus an "unregulated market" perspective. A contemporary biologist, David Sloan Wilson, has proposed "e'*v*'onomics" (Wilson 2016) that highlights the significance of the multicellular organism:

> A multi-cellular organism is composed of trillions of cells that must be healthy at the cellular and organ levels for the multi-cellular organism to be healthy. In the same way, for a large-scale human society to work well, it must be organized as smaller-scale units that work well and are properly coordinated for the benefit of the whole. This is a daunting task that has been accomplished largely by unintended cultural group selection in the past, as Hayek was wise to note. But it needs to take place through intentional cultural evolution now more than ever before, realizing that this doesn't mean command and control. I think that Vincent and Elinor Ostrom were reaching in the same direction with their concept of polycentric governance.

Clearly, the biological notion of the multicellular organism entailing multilevel evolution in terms of computational complexity is helpful for understanding society. This perspective enables the identification of the institutional factors and integrative character of modern evolving systems. The computational complexity, which is based on algorithms, engenders the possibility of the complex emergence of higher orders of institutions based on cooperation. In the case of the economy, the notion of "Markomata" propounded by Mirowski (2007) plays a decisive role. Accordingly, "markets come to bits." As noted by Rosser (2011): "[t]he existence and competition between hierarchical economic institutions also implies problems of computational complexity, again with no definite direction or outcome [as] a

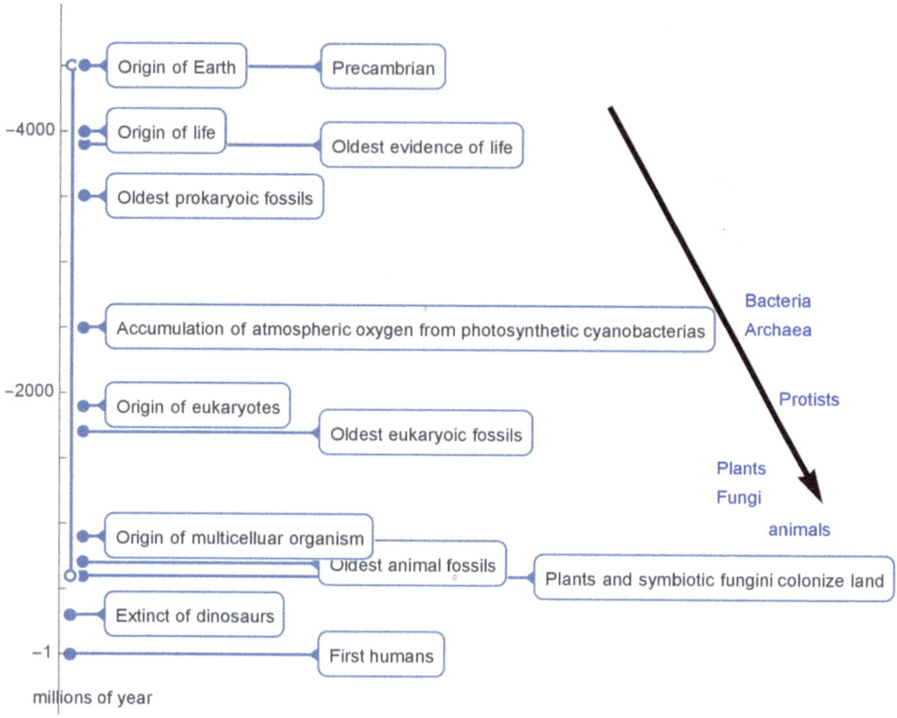


**Fig. 2.1** A phylogenetic and symbiogenetic tree of living organisms showing the origins of eukaryotes and prokaryotes

likely result. Evolution itself is a profoundly complex process, and so it is when it happens to economic institutions." Economic institutions must serve as the coordinating device. Conversely, this device will function as a sensor mechanism for sustaining the underlying system.

Before we proceed further, it would be helpful to broadly illustrate the earth's history by depicting a phylogenetic and symbiogenetic tree of living organisms that shows the origins of eukaryotes and prokaryotes. Figure 2.1 shows the evolutionary process from unicellular to multicellular systems.

### 2.2.1 *The Characteristic Features of Living Systems*

We shall now examine the inner workings of multicellular organisms. However, before we do so, we will introduce the application of probability within biology, that is, in relation to living systems. According to the law of large numbers, probability defines the limits of relative frequency. However, as established by Sober (2000), the definition is *circular* as a result of employing another probability that is either objective or hypothetical. Based on an examination of Karl Popper's propensity theory, Nakajima (2013) has proposed a new application of probability within biology. Briefly, Nakajima (2013) has used "cognizer system" modeling to demonstrate that the characteristic abilities of living systems, namely, **discriminability and selectivity**, enable them to engage with uncertainty. Moreover, these abilities could affect the relative frequencies of the occurrences of events. This influence occurs during different stages extending from the level of a molecule to that of a human being. Nakajima (2013) has, in fact, provided an apt instance to illustrate this:

> The molecular structure and chemical properties of proteins affect the probability that they will have encounters with other molecules, which in turn determines the rate of a certain molecular reaction in which proteins act as an enzyme.

Observations do not depend on any specific attributes of influencing factors such as a particular ligand. In the example of dice throwing, the physical conditions accompanying the throw, namely, the direction of gravity, the angle of a table surface, and others, will affect the frequencies. Consequently, we require the concept of internal probability in relation to the cognizer.

We must first assume that there is a **cognizer**. The nature of the cognizer may be either an objective material such as a molecule or a living cell or a person. Let us suppose that the state transition is cognized by the cognizer. As I subsequently demonstrate, the competence of the cognizer may depend on the quality of the analytical device. **The state transition** of comprising a set of the elements $\{a_1, a_2, \ldots, a_n\}$ can be described by the following configuration of states:

$$\{a_1, a_2, a_3, a_4, \ldots, a_1, a_2, a_5, a_6\}$$

"Cognition" is defined by the function:

$$f : (a_i, \cdot) \rightarrow a_k$$

Here $\cdot$ indicates an unspecified environment. This configuration of states includes a double and double, that is, several multiplicities of the same state(s). Thus, a "focal cognizer" is subsequently subjected to **the uncertainty** of events. We, therefore, need to focus on how successor $a_i$ will change or converge. This is a *cognitive process*. In this context, cognition is represented by **the motion function**. This motion function is constituted by several participating agents, whether material or human.

In the above world constituted by $A$, we will let a cognizer have the ability to *discriminate between* $e_2$ and $e_5$. Consequently, multiple (one-to-many) correspondences hold between a given cognition $(a_1, \cdot) \rightarrow a_2$ and subsequent cognitions, $(a_2, e_2) \rightarrow a_3$ and $(a_2, e_5) \rightarrow a_4$ for internal cognizers.

## 2.2.2 Internal Probability

Given the existence of uncertainties, interactions between cognizers and their environments must be significant. Each cognizer interacts with others, and the state changes to a subsequent state as the motion to cognize its environment. Internal probability subsequently emerges within the cognizer system. This is because necessary interactions affect "not only the degree of uncertainty, but [they] also affect the relative frequency of the occurrence of events after repeated trials" (Nakajima 2013:72).

It appears to be self-evident that a state $A$ depends on $E$ of variations of $\{e_1, \ldots, e_n\}$. However, a human or other subject cannot be an omniscient observer with the ability to discriminate between them. There appears to be a distinction between a state and an event resulting from a limited ability to discriminate. According to Nakajima (2013:75), in relation to internal cognizers,"[e]vents can occur indeterministically even within a deterministic world."

In sum, a cognizer cognizes or *discriminates* a state $a_i$ over time in the world which all existing cognizers have the ability to constitute. Now, let a state $a \in A$ occur in the environment $e \in E$ where $E$ is composed by a set of variations $\{e_1, e_2, \ldots, e_n\}$. Thus, a cognizer's cognition of $A$ is defined. We will subsequently examine how an extremely penetrative monitor could perceive the exact correspondence between $A$ and $E$.

#### 2.2.2.1 Discriminability

The ability to discriminate simply means distinguishing a specified environment $e_i$ or $e_k$ such as or from among different environmental configurations (Nakajima 2013:72).

#### 2.2.2.2 Selectivity

Selectivity is simply defined as the choice of one action from among many possibilities that affect a particular outcome. Moreover, the selecting subjects are not limited to humans (Nakajima 2013:72).

Thus, we can posit that both properties of selectivity and discriminability influence the degree of uncertainty as well as the relative frequency of the occurrence of events in an iterative cognitive process. Here, we should note that relative frequency exists as an objective fact for an external observer.[1]

[1]Nakajima (2013:72) has noted: "Unlike the demon, it is useful to define the external observer as an entity that includes both omniscient and non-omniscient observers. In this case, if the external observer is omniscient, measurements obtained by observation can correspond to system states in a one-to-one manner, irrespective of whether the observed system is deterministic."

A final consideration is that a more developed discriminative ability results in a reduced level of uncertainty for a particular cognizer. In other words, the limited discriminative ability of the cognizer generates uncertainties relating to successive occurrences.

## 2.3 Evolution and Heterogeneous Interactions in Light of Molecular Biology

The probability of events occurring is often calculated as a measure of their variance. In particular, when we observe the appearance of states and events in the process of self-organization, their variance can be considered a surrogate variable for measuring the degree of their order in the system's configuration. A higher level of cohesion will be formed when the system reflects a lower level of variance. An entity that entails a higher degree of cohesion will include some structural components. In the latter environment, the application of probability is restricted by its structure within which a particular pattern is to be sustained as the result of a nonlinear or heterogeneous interaction.

Taking an example from biology, the cell structure will be formed either through chemical reactions or through mutual interactions among constituent molecules or cells. Consequently, the probability of events occurring is affected by these attributes of living systems. This insight led Nakajima to formulate his internal probability theory in relation to biological or living systems. **Predator-prey modeling** is in fact cognizer modeling. While a predator endeavors to catch its prey, the prey attempts to flee from the predator. How they move is not a random motion; it is restricted by a structural relationship.

More importantly, the cognizing subjects in this process are not only limited to humans but also include materials. The material observer behaves as an "internal observer." It is a matter of common knowledge that observations are part of the material interactions between the observer and the objects being observed. Moreover, this kind of interaction appears within biological interactions.

Recent advances within molecular biology aimed at understanding **evolution** are remarkable. As evidenced by the history of economic thought, during the inception of the discipline, the first step taken to understand the economy had its origins in the conceptualization of blood circulation. One attempt to conceptualize the economy was made by the anatomist, William Harvey (May 1, 1578–June 3, 1657) in his *Theory of the circulation of the blood* published in 1628. A second attempt was made by another anatomist, William Petty (May 26, 1623–December 16, 1687). Based at Brasenose College, Oxford, Petty developed *political arithmetic* around 1676 and published this work in 1690. In continental Europe, *Tableau économique*, written by François Quesnay (June 4, 1694–December 16, 1774), a French economist of the Physiocratic school and a physician to the king, who rose to the rank of the first consulting physician, was published in 1758.[2]

### 2.3.1 The Evolutionary Paradox of Coagulation

Coagulation is indispensable for both lower and higher organisms. However, when coagulation occurs in higher organisms like the brain, it can be fatal (Mainzer 2007). This raises the question of whether this kind of immunoediting paradox is a Trojan horse. The coagulation system in fact overlaps with the immune system. Coagulation can physically trap invading microbes within blood clots. Moreover, some products from the coagulation system can contribute to the innate immune system through their ability to increase vascular permeability and act as chemotactic agents for phagocytic cells. The process of coagulation and fragmentation can be applied to social modeling and mathematically analyzed within a coagulation-fragmentation model. For example, Düring et al. (2017) have developed a stylized model for wealth distribution.

### 2.3.2 Immunoediting

The process of immunoediting refers to the modification of the immunogenicity of a cancer following immunosurveillance, often giving rise to immunodeficiency. An important point to note is that immunocompetency is lost in the immunocompetent host. The immune cell is usually capable of distinguishing between normal and abnormal cells. Taking cancer as an example, an antigen could be decomposed into peptides of amino acids. According to the Department of Immunotherapeutics at The University of Tokyo Hospital Department of Immunotherapeutics (2012), the following specific antigens have been identified:

1. Cancer, testicular antigens (cancer-testis antigens): MAGE, XAGE, and NY-ESO-1 proteins that only manifest a cancer cell and do not develop into a normal cell (other than in the testes).

[2]This part is suggested by Prof. Hiroshi Yoshikawa's final lecture at Faculty of Economics, University of Tokyo, March 5, 2016. In addition, another lineage from thermodynamics to apply a blood circuit, as Mimkes (2017) suggested:

| $C + O_2$ | $\Longrightarrow$ | $CO_2$ |
|---|---|---|
| Energy | | No energy |

2. Differentiation antigens: gp100, Melan/mart-1, tyrosinase, PSA, and PAP proteins which only develop into an organization produced by a cancer.
3. Proteins (mutated proteins): p53, K-ras, N-ras, H-ras, and BCR-ABL according to gene variation.
4. Overexpression proteins (overexpressed antigens): HER2, MUC-1, PSMA, survivin, and WT-1 proteins that are rare in normal cells and are frequently found in cancer cells.
5. Tumor fetus antigens (oncofetal antigens): CEA protein, which is the AFP that develops organizationally into a cancer cell of the fetus in the womb.
6. Proteins (viral proteins) that are derived from a cancer-causing virus such as EBV, HBV, HCV, HPV, and HERV.

The T-cell receptor (TCR) on an immune cell, which is specific to an antigen, subsequently reacts to the cancer cell. The antigen counters the antibody in an antigen-antibody reaction. The capacity of the immune system to provide immunity, which is known as immunoediting, functions bilaterally to suppress as well as shape abnormal cells.

According to Dunn et al. (2002), cancer immunoediting is characterized by three phases ("the three Es") which may occur either sequentially or independently. These are elimination, equilibrium, and escape.

#### 2.3.2.1 The Three Es of Cancer Immunoediting

(a) Elimination corresponds to cancer immunosurveillance, wherein immunity functions as an extrinsic tumor suppressor in naive hosts;
(b) Equilibrium represents the process whereby the immune system iteratively selects and/or checks the generation of tumor cell variants with increasing capacities to survive an immune system attack;
(c) Escape refers to the process wherein an immunologically sculpted tumor expands in an uncontrolled manner within an immunocompetent host. Here, tumor cell variants with dampened immunogenicity grow into clinically apparent cancers.

### *2.3.3 White Blood Cell Evolution*

As previously discussed, at the end of the seventeenth century, the theory of blood circulation was applied in the development of economics. A body of literature has now accumulated focusing on studies of blood at the molecular level. Here we consider the example of white blood. The following description is derived from

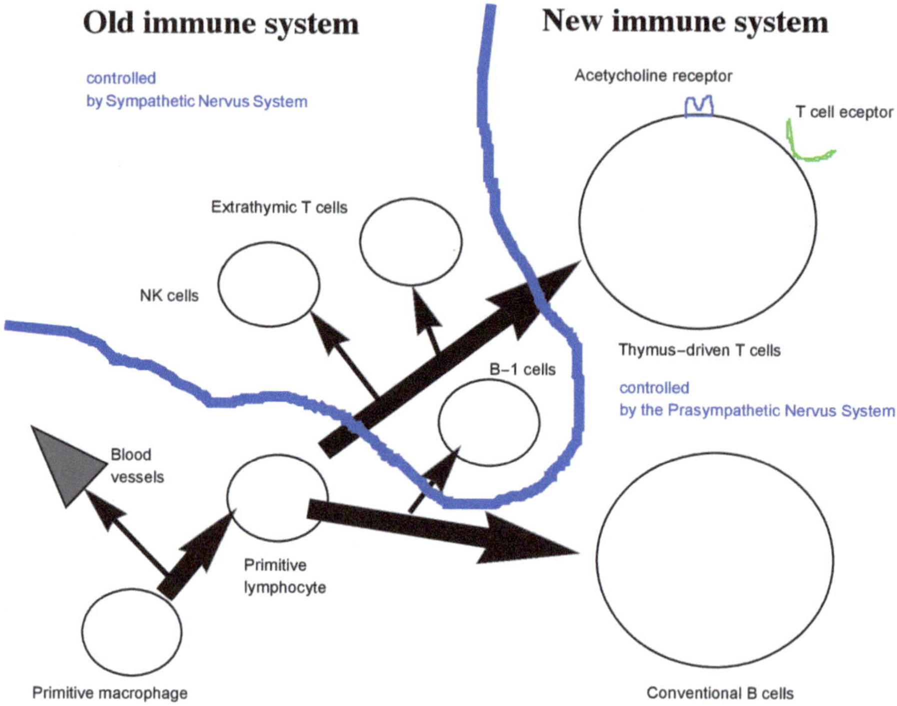


**Fig. 2.2** Evolution of a white blood cell (Derived from Abo 2010)

Abo (2010) and is also depicted in Fig. 2.2.[3] At an early stage in the evolution of white blood cells (WBCs), primitive lymphocytes are generated by primitive macrophages. Within the previously organized immune system, natural killer cells first branch off from primitive macrophages, followed by extrathymic T cells. Thus, the immune system is newly organized with thymus-derived T cells.

### *2.3.4 Risks to the Defense System*

In this chapter, we will demonstrate how a material cognizer initiates a complex action/reaction. It seemingly looks like an action by the use of a systemic device. Accordingly, the cognizer decides on its course of action by applying some sort of structural device, whether intentionally or unintentionally. Even at a biological

[3]See Abo and Balch (1981) and Abo et al. (1982) for the details.

cellular level, the structure of such a device must itself reflect an advanced level of complexity. Here we examine an instance of a complex interaction generated by the immune system.

#### 2.3.4.1 Co-stimulators

A complicated mechanism exists to activate T cells, which are usually antigen specific. These cells are activated as a result of being stimulated by a particular antigen that is bound to the TCR. The antigen stimulation constitutes the first signal that is received by the TCR. However, the first signal is not sufficient to activate the T cell, which requires a second signal for its activation. This second signal, which confirms activation, is known as the co-stimulatory factor. If notification by the second signal failed to occur, the TCR would enter an anergy state, indicating a state of inactivation. Immune tolerance would consequently occur. In sum, the integration of a signal for antigen stimulation in association with a co-stimulatory molecule around a particular antigen can activate an antigen-specific T cell.

#### 2.3.4.2 Co-inhibitors

A co-inhibitory molecule exists in parallel with a co-stimulatory molecule. This molecule, which is located in the same extracellular region as the co-stimulatory molecule, transmits a signal of restraint to a T cell. The inhibitory signal only operates when it is stimulated by an antigen. This membrane-type receptor is known as an inhibitory molecule.

Neither co-stimulatory nor inhibitory factors can ever influence the TCRs in the absence of antigen-specific T cells. Because of this characteristic, T cells have a considerable immunological advantage in being able to restrict their effects within a specific locale, regardless of whether their action relates to stimulation or inhibition.

The concept of **checkpoint** may be helpful for handling systemic risks.[4] The immune system defends biological cells. Here, we focus on the surface-level immune system of human cells. The surface-level cancer immune evasion mechanism that mediates between T cells and antigen-presenting cells (APCs) illustrates this. The basic mechanism comprises stimulation as well as inhibition occurring between T cells and APCs. Stimulation is associated with particular tags. A manifested tag will be targeted. A widely recognized tag is the major histocompatibility complex (MHC). As defined by Wikipedia (2016), this is "the

[4]The immune system has two kinds of checkpoints. One variety of checkpoints is found on the surface level of the immune system (see Pardoll 2016). The other type is associated with the cell division cycle which coordinates the reproduction of cells. For details on the latter, see Bartek and Lukas (2007) and Syljuaasen (2007).

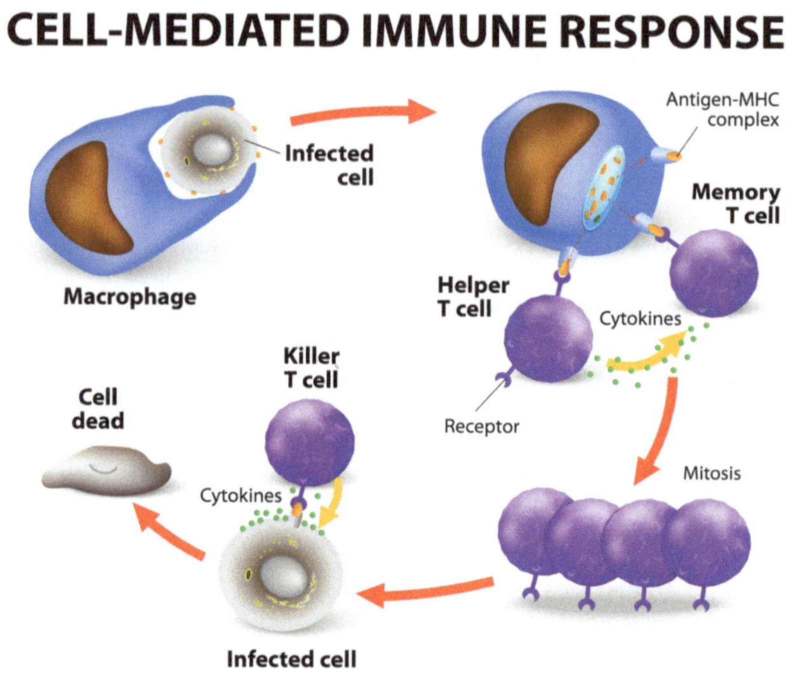


**Fig. 2.3** A cell-mediated immune response (Reproduced from an iStock vector image: 42359420)

tissue-antigen that allows the immune system (more specifically T cells) to bind to, recognize, and tolerate itself (autorecognition)."

If the tags do not manifest, T cells will lose their target. The relationship associated with inhibitors has been well studied. In this case, a cell's programmed death 1(**PD1**) is blocked by a programmed death-ligand 1(**PD-L1**), which is an inhibitor of the immune system. A number of asymmetrical matchings in which tags between stimulators and inhibitors are absent have been detected. Many blockades (not limited to one) against activation or inactivation are generated simultaneously within these matchings.

In the Figs. 2.3 and 2.4, inhibitors are indicated by male-plug shape $\triangleright$ while stimulators are indicated by female-plug shape $\supset$. The pairing of $\triangleright$ $\supset$ implies that the stimulator is successful in fetching the inhibitor. If the converse holds, the inhibitor will block the stimulator.

Distinct from the motion that occurs between T cells and APCs, another motion occurs in which activation of T cells generates cytokine irrespective of the existence of a particular target. The latter will generally contribute to inflammatory attacks on certain parts. In sum, the immune system is multi-dimensional in terms of its reactions, entailing sophisticated interactive motions at the bottom. The recent surge of works on agent-based modeling (**ABM**) or simulation in **medicine** is notable. These studies include Piras et al. (2011), Fullstone et al. (2015), and Altrock et al. (2015).

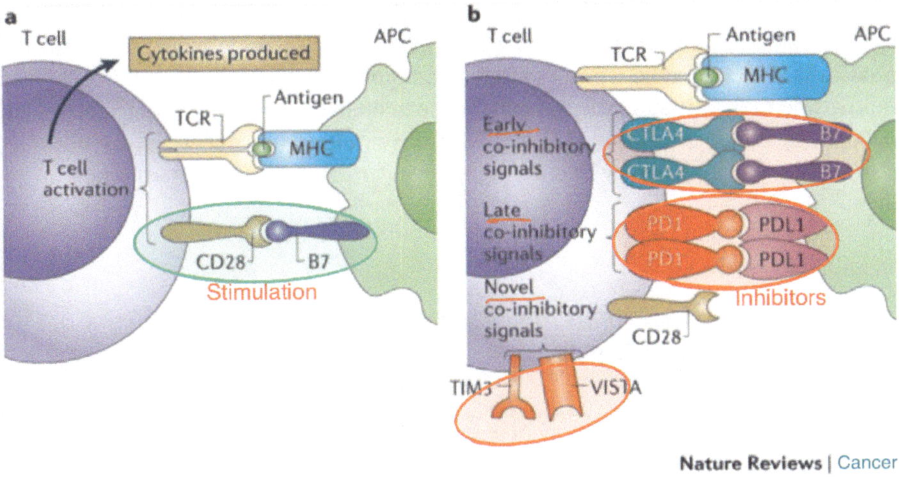


**Fig. 2.4** Novel cancer immunotherapy agents with survival benefits: recent successes and next steps (Source: Sharma et al. 2011)

### 2.3.5 *A Systemic Risk Generated in the Immune System*

**Immune cells**, both B and T cells, evolved from macrophages. The immune system is the result of biological evolution that has led to the development of the human body as a complex system entailing the immune system among other systems. More importantly, a particular cancer tumor, which represents a systemic risk for the human body, is composed of a collection of heterogeneous cells.

When a reproductive cell mechanism is partially disrupted, but is still capable of reproducing chromosomes within the kernel or nucleus of the cell, an excessive number of chromosomes will proliferate. These chromosomes will then produce newly degraded cells, resulting in further degradation. Consequently, the concerned tumor may be composed of a collection of heterogeneous variants. The interaction of such factors within a complex system may induce **immunodeficiency** that can be verified.

A single primitive cell may evolve into a more complex one. The antibody-antigen reaction does not necessarily function efficiently. Immunodeficiency may be generated as a subset entailing inadequate functioning of either antioxidant or immune stimulators to suppress cancer cells.

(a) Apoptosis
(b) Necrosis
(c) Autophagy
(d) Ferroptosis[5]

[5]See Xie et al. (2016).

Professor Yoshinori Ohsumi has been awarded a Nobel Prize in physiology or medicine for his discovery of autophagy. This process is defined as: “a normal process in which a cell destroys proteins and other substances in its cytoplasm (the fluid inside the cell membrane but outside the nucleus), which may lead to cell death. Autophagy may prevent normal cells from developing into cancer cells, but it may also protect cancer cells by destroying anticancer drugs or substances taken up by them.”[6]

## 2.4 A Measure of Systemic Risk Derived from the Input-Output Table

We now turn to a consideration of economic matters. In a previous section, we touched on the pioneering contributions of classical political scientists like William Harvey, William Petty, and François Quesnay, all of whom were medical doctors. These individuals conceptualized the economic system in a way that was similar to blood circulation as a kind of node-based network constituted by the circular processing of materials, resources, and services. Quesnay’s **economic table** was the first monumental achievement in this context. Over the course of the twentieth century, the forms of this table have changed, as evidenced in Leontief’s input-output table and von Neumann-Sraffa’s production system. The input-output table has been selected here for empirical reasons.

### *2.4.1 The Linear Economic Model*

A linear input-output relationship is commonly formulated as follows. Let $m \neq n$ in the number of process $i = 1, \cdots, m$, and in the number of good $j = 1, \cdots, n$. It then generally follows the process of production:

$$a_i = (a_{i1}, \cdots, a_{im}) \Rightarrow b_i = (b_{i1}, \cdots, b_{im}), i = 1, \cdots, n$$

in matrix forms, namely, $A \Rightarrow B$.

Let the number of productive processes $i$ be $m$. $a_{ij}$ representing the requirement of good $j$ per unit of activity to produce good $j$ while $b_{ij}$ is the output produced by a unit of activity $x_i$:

$$a_i x_i = a_{i1} x_i + \cdots + a_{in}, i = 1, \cdots, m$$

$$b_i x_i = b_{i1} x_i + \cdots + b_{in}, i = 1, \cdots, m$$

[6]National Cancer Institute (2016).

The configuration of the whole system's activities $x$ of a production system is represented by:

$$x = (x_1, \cdots, x_m)$$

Consequently, the net producibility in a matrix form is:

$$[B - A]x \geq 0$$

The input-output table holds when matrices $B$ and $A$ are square or a special case $m = n$. If it occurs, a nonnegative eigenvector $x^*$ is associated with a maximal eigenvalue $\lambda$. Meanwhile, even in a special case, we can find any complex vector, apart from the unique real eigenvector, in the remaining eigenvectors.

### 2.4.2 *The Location of Eigenvalues in the Complex Plane of the Production System*

Drawing on the findings of Brody (1970), Mariollis and Tsoulfidis (2016) have suggested that the eigenvalues of complex numbers could be considered as an index for the degree of capital heterogeneity. Here, we find the existence of a conjugate pair of two different industries except for its sign. Such a pair of industries can mutually influence each other. Aruka (1991) has shown that a price dynamics based on a linear production system may oscillate as a result of its own complex modes, although the dominant mode is influential. It would thus be feasible to make some allowances for the oscillation of values, in spite of the largest eigenvector, if deviations from some golden path occur.

In our classification of 13 sectors,[7] the largest real value greatly exceeded those of other real parts of the complex eigenmodes on the real eigenvalue axis, as shown in Fig. 2.5. In the largest real eigenmode, the most influential factor, which was the maximal coefficient, was derived from the service industry (no. 12), followed by that from the transport industry (no. 9). The influential factors in the largest complex modes appear to be the following industries: construction (no. 4), commerce (no. 6), and service (no. 11). These could cause *some nonlinear effects*, for example, network effects, to the production system.

As shown in Fig. 2.6, in our classification of 34 sectors, the largest of the real parts of the complex mode exceeded the largest real value. In the complex mode, the most influential dominant factor appeared to be the transport industry (no. 25). This could also result in some nonlinear effects being exerted on the production system. Meanwhile, in the actual mode of real values, metal products (no. 11), the construction industry (no. 19), and business services (no. 31) were evidently influential.

---

[7] Data provided by Director-General for Policy Planning (Statistical Standards 2016) were applied.

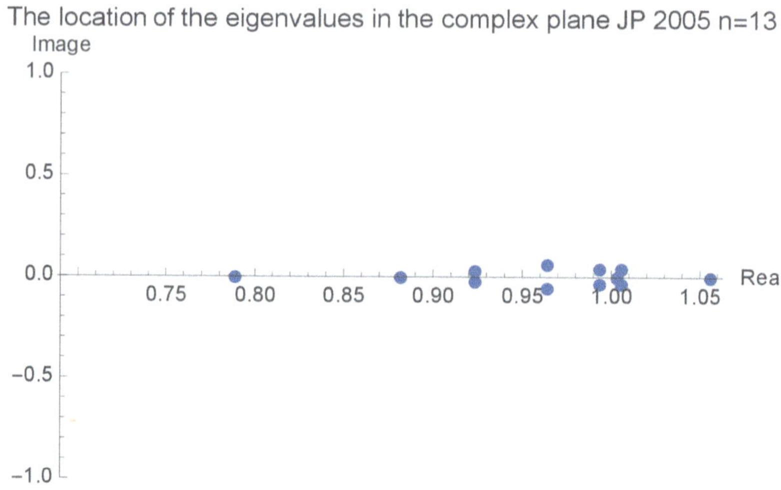


**Fig. 2.5** Eigenmodes in 13 sectors in 2005

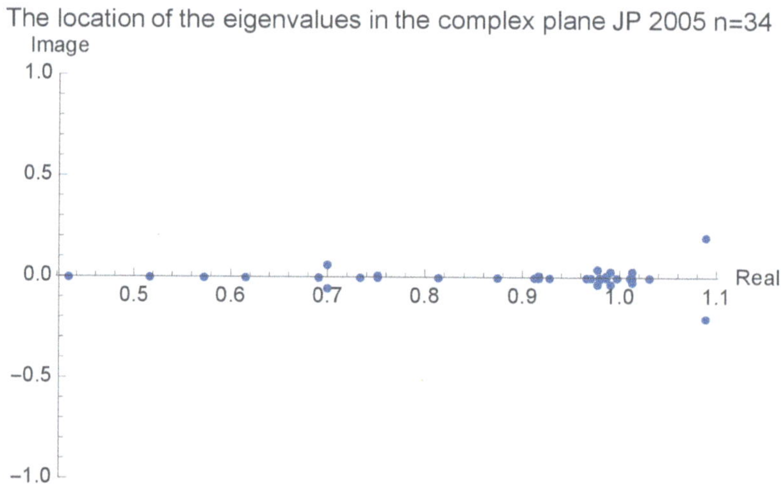


**Fig. 2.6** Eigenmodes in 34 sectors in 2005

### *2.4.3 A New Randomization of the Input-Output Table Based on Covariance Analysis*

Temporarily leaving aside the complex plane, another characterization of the input-output table can be considered in view of *correlated sectors* of the input-output system. The input coefficients with a transformation of subset $t^i$ can be reinterpreted as follows:

$$t_1^i, \ldots, t_T^i \to n^i$$

where $t$ denotes the kinds of inputs and $n$ denotes the kinds of outputs. Applying covariance analysis, we can analyze the correlations between and when $t = n$ in the input-output table.

#### 2.4.3.1 Extraction of Nonrandom Modes and the Largest Eigenmode

The subsequent procedure, known as **Iyetomi's method**,[8] entails the following steps:

(a) Construct a correlation matrix
(b) Check the randomness of this matrix
(c) Identify any strong correlations, namely, an influential hub group, among the $a_{ij}$s, evidently of nonrandom origin.

The largest eigenmode is considered a composite measure of system heterogeneity. The results showed that all of the components of the largest mode followed the same direction. These results for the input-output correlations resemble the market mode in relation to stock price fluctuations. Thus, we could employ the largest dominant mode of input-output correlations to estimate the main evolutionary pathway of the core system.

#### 2.4.3.2 Remarks on the Extraction of Nonrandom Modes

It is widely acknowledged that the **Hawkins-Simon Condition** (1949) in relation to the input-output table is required to fulfill the net-producible production system.[9] The first point to be noted is that the actual input-output production system must be normally identified with a system producing a net surplus. Hence the largest eigenmode supporting the net-producible system must exist. However, the Hawkins-Simon Condition may need to be imposed each time the matrix is randomized. Each random matrix derived from the input-output table should fulfill the Hawkins-Simon Condition. Accordingly, the area of random distributions will be restricted by the Hawkins-Simon Condition. However, while some random modes may fail to be identified, nonrandom modes, and especially the largest eigenvector, may be accurately detected. Thus, the results obtained usually cannot be drastically modified.

An alternative perspective is that strict application of the Hawkins-Simon Condition is not required. Our current focus is on the evolutionary change of the production system. In this process, several permutations may occur within a

[8]See Iyetomi et al. (2011a,b).

[9]See Nikaido (1968:90–92) for details.

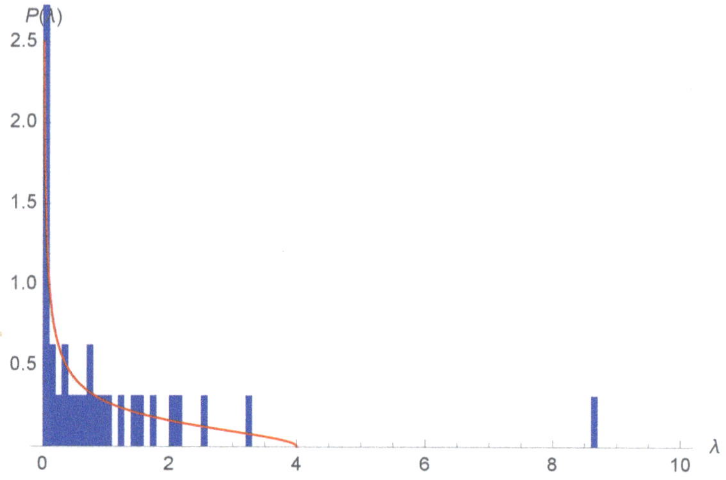


**Fig. 2.7** A comparison of the actual modes in the 108 sectors with a random matrix distribution

given system without fulfilling the Hawkins-Simon's Condition. While Hildenbrand (1981) elaborated a novel alternative approach, here our focus is on a random network of the production system (Fig. 2.7).

### *2.4.4 An Alternative Way of Measuring the Difference Between a Weighted Random Graph Model and Real Distributions*

We have identified the real system with a subsystem containing nonrandom modes. It may be useful to elucidate the difference between random and real distribution. There are several ways of representing the characteristics of random distribution. We begin by demonstrating how to generate a weighted random graph model. This model is generated based on the probability $q(w)$ that a weight $w$ (i.e., a number of links) occurs between any pair of vertices as follows: $q(w)=p^w\ (1-p^w)$.

An **adjacency matrix** is a binary matrix where $\alpha_{\text{ij}}$ means that vertices are connected, while a weighted adjacency matrix is a matrix where $w_{\text{ij}} = n$ means that there are $n$ links between vertices. It then follows that there are three different distributions of the weighted adjacency matrix:

- $P_>(w)$ is the cumulative **weight** distribution.
- $P_>(k)$ is the cumulative **degree** distribution where $k_i = \sum_{j=1}^{N} \alpha_{\text{ij}}$.
- $P_>(s)$ is the cumulative **strength** distribution where $s_i = \sum_{j=1}^{N} w_{\text{ij}}$.

Here, $w_{ij}$ was interpreted as a covariance matrix coefficient of the input-output table $a_{ij}$. Consequently, three different distributions of the randomly generated matrix of 108 nodes were compared with **the real distribution of 108 sectors** of the covariance matrix of the input-output table in 2011. Based on a comparison of

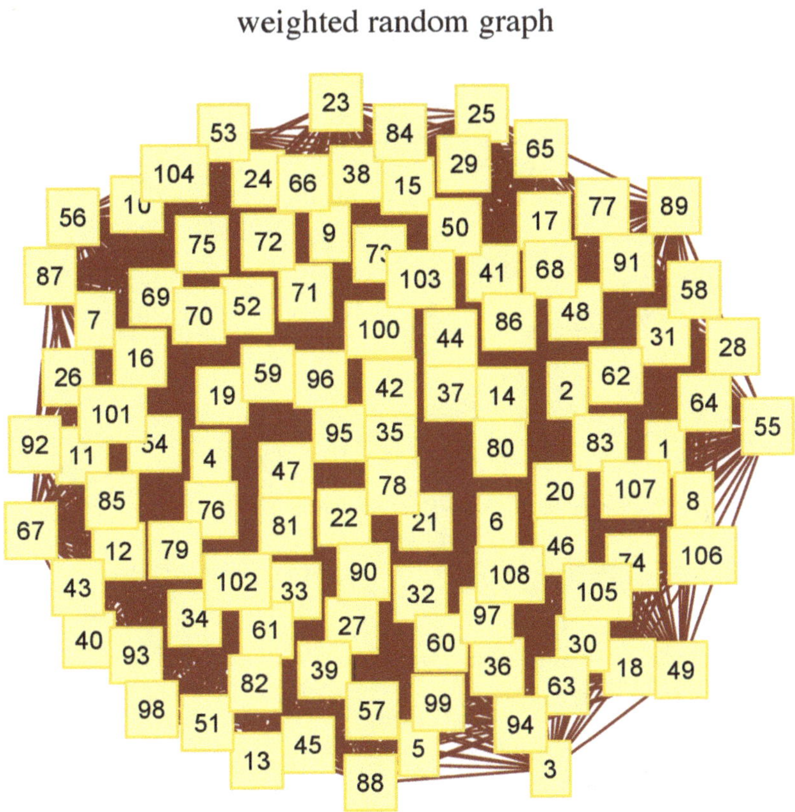


**Fig. 2.8** A weighted random graph of 108 nodes with Adapted from Tiziano and Garlaschelli (2016)

the random and real distributions, and using Iyetomi's method,[10] differences in the degree to which real distribution was influenced by some of the nonrandom modes discussed above were easily verified. The results appeared to be natural, because the real system had been shaped not only by **histocompatible adaptation** but also by some **intelligent institutional designs**. Each $w_{ij}$ was interpreted as **an input coefficient** $a_{ij}$. Hence, the sum with respect to each $i$ indicated the demand sum of each industry, $i$. Figures 2.8, 2.9, and 2.10 were produced based on the application of an extended version of the "**Weighted Random Graph**" depicted in the **Wolfram demonstration project** (Tiziano and Garlaschelli 2016). Strength distribution was subsequently selected. Figure 2.8 shows the formation of random distribution with the probability of the vertices being conjugated set at 55%. The second figure shows the real distribution.

[10]These data have been kindly provided by Prof. Hiroshi Iyetomi.

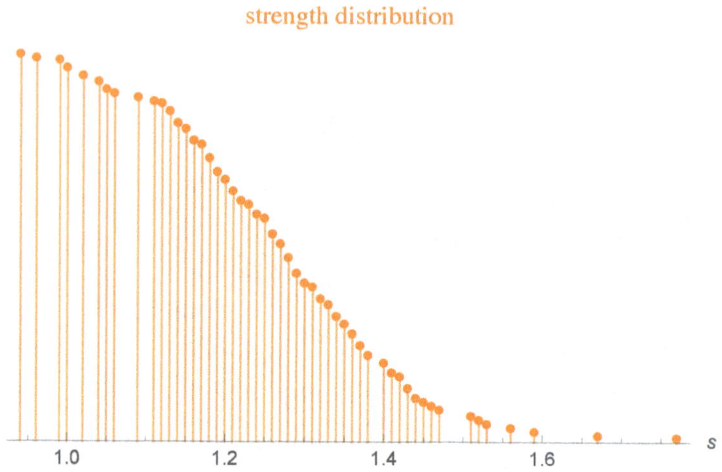


**Fig. 2.9** The strength distribution of the weighted random graph

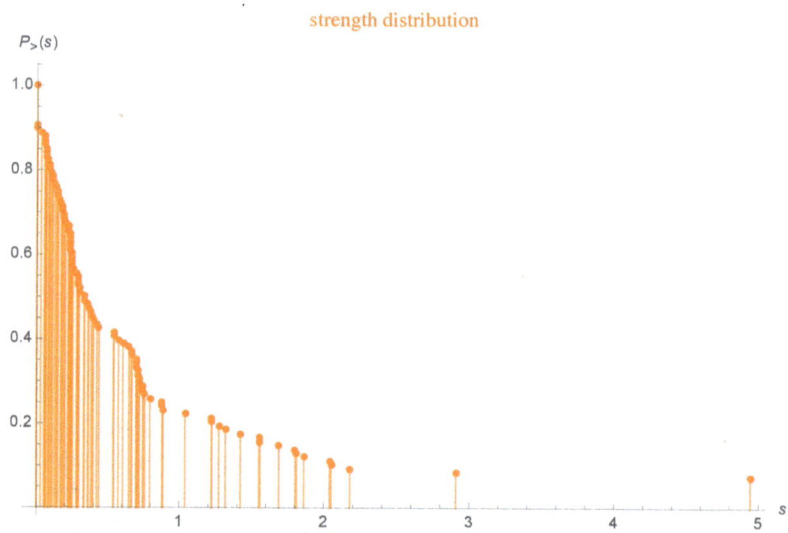


**Fig. 2.10** The strength distribution of the real system of 108 sectors in 2011

## *2.4.5 Alpha Centrality Characterizing the Growth Pathway*

We have so far applied Iyetomi's method to examine the nonrandomness of the characteristics based on a covariance matrix of the input-output system. We now return to the real input-output matrix. Given the problem of complex numbers, these were classified into industries generating nonlinear effects on the entire system as discussed above.

The definition of the $\alpha$ **centrality** of the system[11] follows from the input-output system. It is accepted that $\alpha$ **centrality** depends not only on the adjacency matrix (network connections) $A = [a_{ij}]$ but also on important exogenous influences on node $j$: $e_j$. In other words, $e_j$ was incorporated into the overall influence during each round. Applying the input-output matrix, positive surplus or net output could be generated, or the Hawkins-Simon Condition could be met. Given that a nonnegative square matrix is nonnegatively invertible, it follows that the process will converge during the next state as shown below:

$$x_j = \alpha \sum a_{ij} x_i + e_j$$

The solution for this system is:

$$x = (I - \alpha A^T)^{-1} e$$

If $\alpha$ =0, there is no adjacency effect in the system, which is only dominated by external influences. Meanwhile, if $\alpha$ is infinitely large, the system will be an eigenvector-centric one.

Alpha centrality will be employed as a measure of spreading risk within the system in the following manner. Based on the adjacency matrix (network connections), each incoming spread received from *inward* neighbors will be distributed among *outward* neighbors. The process will then be repeated until convergence is achieved as follows:

$$x_i = (1/\lambda) \sum a_{ij} x_j$$

where $x_i$ is the extent of influence that node $i$ carries. $A^T$ is the adjacency matrix, and $\lambda$ is the principal eigenvalue. The outward spreads are balanced by inward spreads. This balance is equivalent in meaning to Sraffa's **standard commodity** (Sraffa 1960). This suggests that input-output analysis and network analysis are the same. The dual forms of quantity and price systems are verifiable.[12] While this kind of equilibrium formation has been discussed elsewhere,[13] our focus, in this chapter, is on the alpha centralities of the world input/output table.

Quantity system $x_j = \alpha \sum a_{ij} x_i + e_j$
Price system $\rho_i = \alpha \sum a_{ij} p_j + l_i$

---

[11]See Bonacich and Paulette (2001) for a discussion on $\alpha$ **centrality**. **Eigenvector centrality** may be used to characterize the survival pathway based on the rate of profit/growth. The latter seems to be related to the idea of a **cell death cycle (CDC)**.

[12]Ultimately, the valuation by producers' prices underlies **the international equilibrium price system** allowing multiple intermediate products to be traded. This kind of equilibrium remained unproven over a period of two centuries until Shiozawa (2015, 2007) presented an elegant proof using **sub-tropical geometry**.

[13]The empirical foundation of Shiozawa's theory, formulated by Fujimoto (2007) and Fujimoto and Shiozawa (2012) has been widely explored.

Viewed in relation to these dual forms, **alpha centrality** indicates the growth factor $\alpha = 1/(1+g)$ where $g$ is defined as the rate of growth. Thus, it is clearly relevant to the growth or degeneration of the system. Although the extent of centrality reflects some degree of proportionality to the system, it is only considered as a parameter within network analysis and, therefore, requires redefinition. The value of $\alpha$ is usually set at 1.

In the field of social networks, the concept of **alpha centrality** is viewed as an **amplification factor** and/or **attenuation parameter**. It can be applied as an asymmetric weighted adjacency matrix to the input-output table. Here $a_{ij}$ denotes the amount of good used for the production of product $j$. $\omega_{ij}$ denotes the share of product in relation to the total amount of intermediate inputs employed by industry $j$. Hence measures of the inward/outward spread of external disturbances are defined as follows:

Outward spreads: AMPLIFICATION INDEX (AI) $x = (I - \alpha A^T)^{-1}e = (I + \alpha A^T + \alpha^2 A^{T^2} + \cdots + \alpha^t A^{T^t} + \cdots)e$
Inward spreads: VULNERABILITY INDEX (VI) $x = (I - \alpha A)^{-1}e = (I + \alpha A + \alpha^2 A^2 + \cdots + \alpha^t A^t \cdots)e$

#### 2.4.5.1 The Three Types of Input-Output Table at Producers' Prices

Before performing an empirically based calculation of $\alpha$ centrality, it is necessary to select one of the three types of input-output tables that have been officially published by Statistics Bureau (2016).

- The $[I - A]^{-1}$ input-output table
  The construction of input-output table $A$ is based on the assumption of competitive imports. In this table, domestic and imported products are viewed neutrally, and no distinction is drawn between them.
- The $[I - [I - M]A]^{-1}$ input-output table
  Construction of the import coefficient matrix $M$ entails the removal of imported products from the above-described competitive table $A$. This is based on the common ratio of the imported product to the total input among all of the products. All of the export outputs are assumed to be domestic products.
- The $[I - A_D]^{-1}$ input-output table
  Based on the assumption of noncompetitive imports, the actually detected ratio of the imported products to the total input of each industry can be calculated. Subsequently, a noncompetitive table $A_D$ was constructed according to the extent of our knowledge.

#### 2.4.5.2 Industries Exposed to Systemic Risk in View of *VI*

By definition, the vulnerability index *VI* in relation to industry represents inward directed ripple effects such that industries from the upper value stream are forced to reduce their downward activities. Based on an explicit consideration of **the import factor,** the amplitude index $AI$ for the last two cases could be demonstrated.[14] $A$ was replaced with $A_D$ in the case of industries entailing competitive imports and with $[I - M]A$ in the case of industries entailing noncompetitive imports, respectively.

- $VI_M$ was calculated for industries with competitive imports according to the $[I - [I - M]A]^{-1}$ input-output table.
- $VI_D$ was calculated for industries with noncompetitive imports according to the $[I - A_D]^{-1}$ input-output table.

According to the concept of alpha centrality, the industry with the largest index value may be regarded as the industry that is most sensitive to inward directed disturbances, that is, the **vulnerability index** (VI). In line with a standard sociological convention, we initially assumed that $\alpha = 1$ and then listed VI values for industries with both competitive and noncompetitive imports.

Top ten industries in relation to vulnerability indices

| Ranking | | |
|---|---|---|
| 1 | 73 Commerce | 73 Commerce |
| 2 | 74 Finance and insurance | 74 Finance and insurance |
| 3 | 101 Other business services | 101 Other business services |
| 4 | 28 Petroleum refinery | 59 Motor vehicle parts/accessories |
| 5 | 59 Motor vehicle parts/accessories | 38 Steel products |
| 6 | 38 Steel products | 69 Electricity |
| 7 | 69 Electricity | 37 Pig iron and crude steel |
| 8 | 37 Pig iron and crude steel | 28 Petroleum refinery |
| 9 | 93 Research | 93 Research |
| 10 | 100 Repair of motor vehicles and machines | 100 Repair of motor vehicles and machines |

Thus, the list identifies industries susceptible to **the upper stream** of industries around the top ten industries. Notably, the index for noncompetitive imports exceeded that of competitive imports in the case of the petroleum refinery industry.

[14]Tran et al. (2016) have attempted to apply $\alpha$ centrality to the classical matrix without explicitly considering the import factor. However, this article focuses on an international comparison of the indices driven by the centrality among the advanced countries using global **input-output data**.

#### 2.4.5.3 Identification of Industries Generating Nonlinear Effects on the Entire System Based on the Vulnerability Index

The existence of industries generating nonlinear effects on the entire system has already been covered in our discussion of the alpha centrality of the weighted adjacency matrix, that is, the real input-output system. As long as the Hawkins-Simon Condition holds, the real nonnegative eigenvalue can always be guaranteed. However, other eigenvalues may contain complex numbers. These can be identified in relation to industries generating nonlinear effects on the entire system. We have, therefore, focused on these industries in the ranking results calculated for alpha centrality. Two industries were identified in relation to *VI* according to whether imports were competitive or noncompetitive. Nonlinear effects generated by these industries were as follows:

| | |
|---|---|
| 7 Nonmetallic ores | 7 Nonmetallic ores |
| 20 Chemical fertilizers | 20 Chemical fertilizers |

#### 2.4.5.4 Industries Exposed to Systemic Risk in View of AI

By definition, the *VI* in relation to industry represents outward directed ripple effects such that industries from the lower stream are forced to reduce their downward activities. As in the *VI* case, by explicitly considering **the import factor**, the amplitude index (*AI*) was demonstrated for cases of industries entailing competitive and noncompetitive imports, respectively (Figs. 2.11 and 2.12).

#### 2.4.5.5 Industries Exposed to Systemic Risk in View of *AI*

By definition, the *AI* in relation to industry represents outward directed ripple effects such that industries from the lower value stream are forced to reduce their downward activities according to the calculation shown below:

$$x = (I - \alpha A^T)^{-1} e = (I + \alpha A^T + \alpha^2 A^{T^2} + \cdots + \alpha^t A^{T^t} + \cdots)e$$

$A$ was replaced with $A^T$ in the case of competitive imports and with $[I - M]A^T$ in the case of noncompetitive imports, respectively.

As in the case of *VI*, the *AI* was demonstrated for industries entailing competitive and noncompetitive imports, respectively. The industry with the highest index value *AI* was considered the industry most sensitive to outward directed disturbances, that

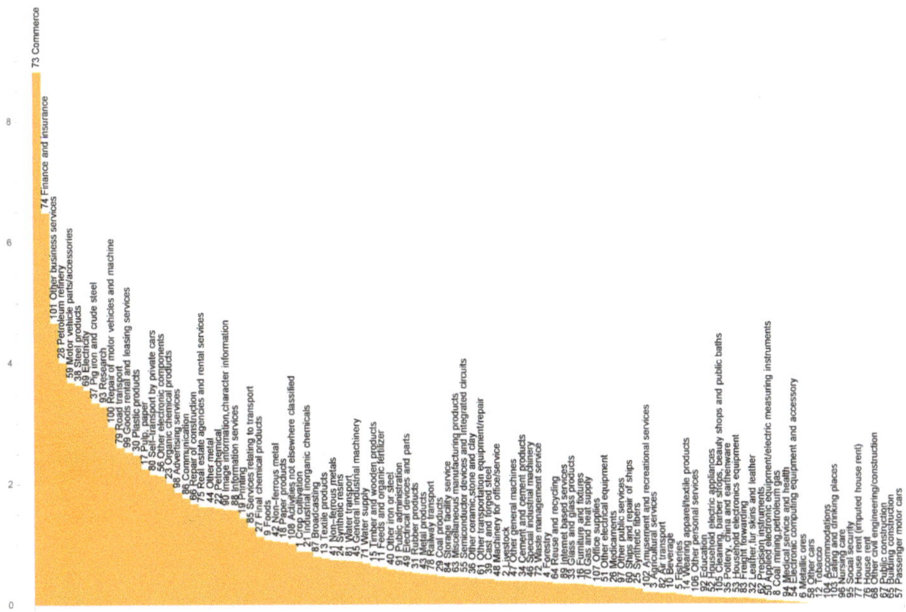


**Fig. 2.11** The ranking of indices for cases of industries entailing competitive imports

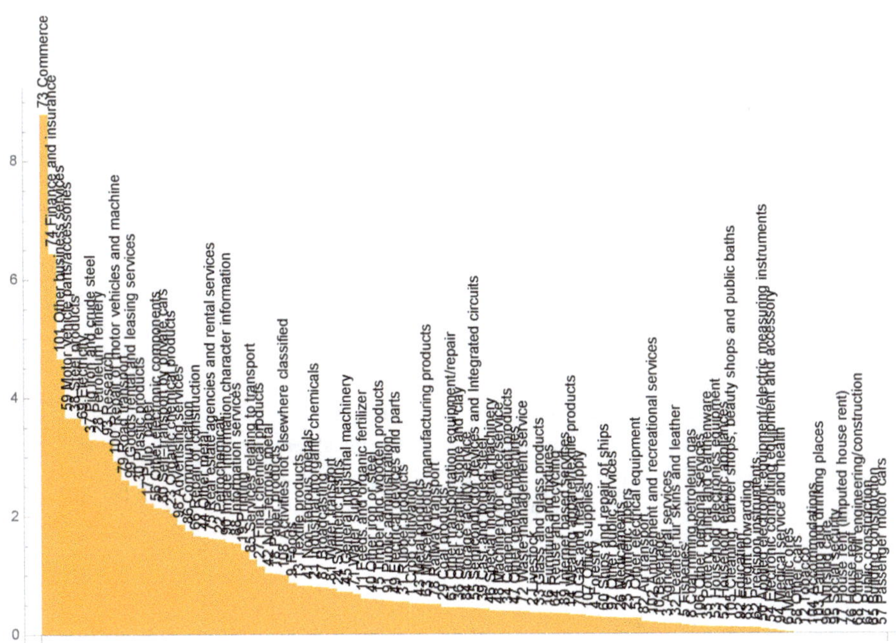


**Fig. 2.12** The ranking of indices in the case of industries entailing noncompetitive imports

is, **amplitude index**, *AI*. In line with the sociological convention, we first assumed that $\alpha = 1$ and subsequently listed *AI* values for industries with both competitive and noncompetitive imports.

Top ten industries in relation to amplitude indices

| Ranking | | |
|---|---|---|
| 1 | 57 Passenger motor cars | 57 Passenger motor cars |
| 2 | 58 Other cars | 58 Other cars |
| 3 | 107 Office supplies | 108 Activities not classified elsewhere |
| 4 | 108 Activities not classified elsewhere | 107 Office supplies |
| 5 | 59 Motor vehicle parts/accessories | 59 Motor vehicle parts/accessories |
| 6 | 23 Organic chemical products | 48 Machinery for offices/services |
| 7 | 60 Ships and repair of ships | 24 Synthetic resins |
| 8 | 24 Synthetic resins | 23 Organic chemical products |
| 9 | 48 Machinery for offices/services | 2 Livestock |
| 10 | 2 Livestock | 30 Plastic products |

Thus, we distinguished industries that were influential to **the lower stream** of industries around the top ten industries in the list. It is noteworthy that entries in both cases were almost common but several cases.

#### 2.4.5.6 Identification of Industries Generating Nonlinear Effects on the Entire System Based on the Amplitude Index

Relative to *VI*, there are many industries that can possibly cause some *nonlinear* effects to the entire system. These could be identified in the ranking results calculated for alpha centrality. In view of *AI*, there were 13 industries in the category of competitive imports and 17 industries in the category of noncompetitive imports. These industries could generate nonlinear effects (Figs. 2.13 and 2.14).

| | |
|---|---|
| 22 Petrochemicals | 5 Fisheries |
| 28 Petroleum refinery | 20 Chemical fertilizers |
| 29 Coal products | 22 Petrochemicals |
| 37 Pig iron and crude steel | 28 Petroleum refinery |
| 38 Steel products | 29 Coal products |

(continued)

| | |
|---|---|
| 39 Cast and forged steel | 37 Pig iron and crude steel |
| 40 Other types of iron or steel | 38 Steel products |
| 43 Metal products | 39 Cast and forged steel |
| 44 Other metals | 40 Other types of iron or steel |
| 47 Other general machines | 43 Metal products |
| 79 Road transport | 44 Other metals |
| 80 Self-transport by private cars | 47 Other general machines |
| 81 Water transport | 60 Ships and repair of ships |
| | 64 Reuse and recycling |
| | 79 Road transport |
| | 80 Self-transport by private cars |
| | 81 Water transport |

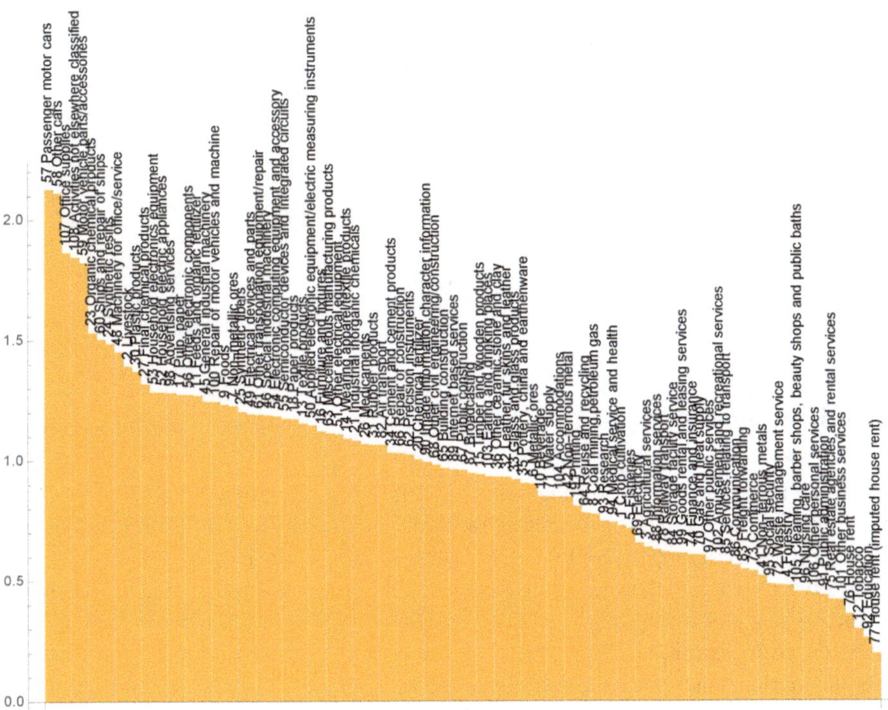


**Fig. 2.13** The ranking of indices for industries characterized by competitive imports

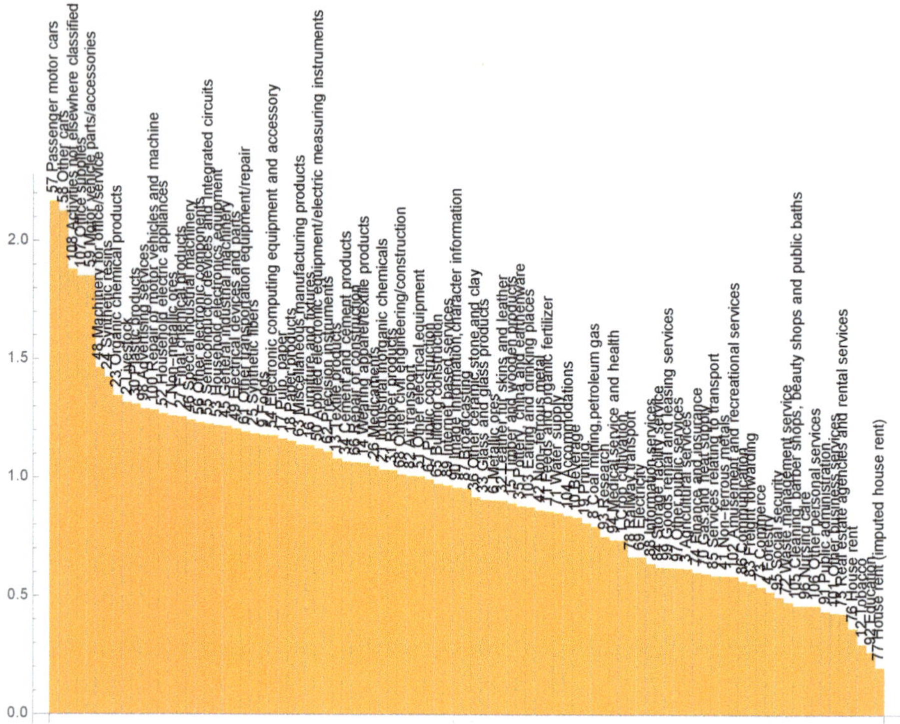


**Fig. 2.14** Ranking of indices for industries characterized by competitive imports

## 2.5 Conclusion

As shown in the above, the biological notion of the multicellular organism entailing multilevel evolution in terms of computational complexity is helpful for understanding society. This perspective enables the identification of the institutional factors and integrative character of modern evolving systems. The computational complexity engenders the possibility of the complex emergence of higher orders of institutions based on cooperation. Here we demonstrated how a material cognizer initiated a complex action/reaction. It seemingly looks like an action by the use of a systemic device. Accordingly, the cognizer decides on its course of action by applying some sort of structural device, whether intentionally or unintentionally. Here we examined an instance of a complex interaction generated by the immune system in relation to immunology and immunoediting that occurs on the surface level of the immune system. Our exploration then revealed the importance of **checkpoints** such as nodes within a networked structure that is formed based on the above complex interactions, enabling us to better understand systemic risks.

Finally we focused on the network node or link of the input-output system of the economic system and introduced network-related concepts such as **alpha**

**centrality**, successfully identifying a measure of risk in Japan's current economic system. According to the concept of alpha centrality, the industry with the largest index value may be regarded as the industry that is most sensitive to inward directed disturbances, that is, the **vulnerability index (VI)**. The existence of industries generating nonlinear effects on the entire system has already been covered in our discussion of the alpha centrality of the weighted adjacency matrix, that is, the real input-output system. In general, eigenvalues may contain complex numbers. These can be identified in relation to industries generating nonlinear effects on the entire system. Two industries were commonly identified in relation to VI according to whether imports were competitive or noncompetitive. Nonlinear effects generated by these industries were nonmetallic ores and chemical fertilizers

As in the VI case, by explicitly considering the import factor, the amplitude index (AI) was demonstrated for cases of industries entailing competitive and noncompetitive imports, respectively. In view of AI, there were 13 industries in the category of competitive imports and 17 industries in the category of noncompetitive imports. These industries could generate nonlinear effects.

## References

T. Abo, *Dr. Abo's Immunology lecture*. Sanwa-shobo (2010). http://www.sanwa-co.com/94/

T. Abo, C.M. Balch, A differentiation antigen of human NK and K cells identified by a monoclonal antibody (HNK-1). J. Immunol. **127**(3), 1024–1029 (1981). http://www.jimmunol.org

T. Abo, M.D. Cooper, C.M. Balch, Characterization of HNK-1+ (Leu-7) human lymphocytes. I. Two distinct phenotypes of human NK cells with different cytotoxic capability. J. Immunol. **129**(4), 1752–1757 (1982). http://www.jimmunol.org

P.M. Altrock, L.L. Liu, F. Michor, Affiliations, the mathematics of cancer: integrating quantitative models. Nat. Rev. Cancer 730–745 (2015). doi:10.1038/nrc4029

Y. Aruka, Generalized goodwin theorems on general coordinates, in *Structural Change and Economic Dynamics*, vol. 2 (Oxford University Press, Oxford, 1991), pp. 69–91

Y. Aruka, Network analysis of production and its renewal, chapter 3, in *Evolutionary Foundations of Economic Science* (Springer, 2015), p. 219. ISBN:978-4-431-54844-7

J. Bartek, J. Lukas, DNA damage checkpoints: from initiation to recovery or adaptation. Curr. Opin. Cell Biol. **19**, 238–245 (2007). doi:10.1038/sj.onc.1210402

P. Bonacich, L. Paulette, Eigenvector-like measures of centrality for asymmetric relations. Soc. Netw. **23**, 191–201 (2001)

A. Brody, *Proportions, Prices and Planning. A Mathematical Restatement of the Labor Theory of Value* (Akademiai Kiado, Budapest/North-Holland, London, 1970)

Director-General for Policy Planning (Statistical Standards), in *2011 Input-Output Tables for Japan*. Ministry of Internal Affairs and Communications Japan (2016). http://www.soumu.go.jp/english/dgpp_ss/data/io/index.htm

G.P. Dunn, A.T. Bruce, H. Ikeda, L.J. Old, R.D. Schreiber, Cancer immunoediting: from immunosurveillance to tumor escape. Nat. Immunol. **3**(4), 991–998 (2002). doi:10.1038/ni1102-991

B. Düring, N. Georgiou, E. Scalas, A stylized model for wealth distribution, in *Economic Foundations for Social Complexity Science* (Springer, Tokyo, 2017)

T. Fujimoto, Architecture-based comparative advantage – a design information view of manufacturing. Evol. Inst. Econ. Rev. **4**(1), 55–112 (2007). doi:10.14441/eier.4.55

T. Fujimoto, Y. Shiozawa, Inter and intra company competition in the age of global competition: a micro and macro interpretation of Ricardian trade theory. Evol. Inst. Econ. Rev. **8**(2), 193–231 (2012). doi:10.14441/eier.8.193

G. Fullstone, J. Wood, M. Holcombe, G. Battaglia, Modelling the transport of nanoparticles under blood flow using an agent-based approach. Sci. Rep. **5**(10649) (2015). doi:10.1038/srep10649.

D. Hawkins, H. Simon, Some conditions of macroeconomic stability. Econometrica **17**(3/4), 245–248 (1949). doi:10.2307/1905526, http://www.jstor.org/stable/1912746.

W. Hildenbrand, Short-run production functions based on microdata. Econometrica **49**(5), 1095–1125 (1981). doi:10.2307/1912746, http://www.jstor.org/stable/1912746

H. Iyetomi et al., What causes business cycles? Analysis of the Japanese industrial production data. J. Jpn. Int. Econ. **25**, 246–272 (2011a)

H. Iyetomi et al., Fluctuation-dissipation theory of input-output interindustrial correlations. Phys. Rev. E **83**(12), 016103 (2011b)

K. Mainzer, Kapitel 7: Zufall in Kultur, Wirtschaft, und Gesselschaft. In *Der kreative Zufall: Wie das Neue in die Welt kommt?* CH Beck (2007)

T. Mariolis, L. Tsoulfidis, Capital theory 'paradoxes' and paradoxical results: resolved or continued? Evol. Inst. Econ. Rev. **13**(2), 297–322 (2016)

J. Mimkes, Calculus based econophysics: applications to the Japanese economy, in Chap. 2 *Economic Foundations for Social Complexity Science* (Springer, Tokyo, 2017)

P. Mirowski, Markets come to bits: evolution computation and markomata in economic science. J. Econ. Behav. Organ. **63**(2), 209–242 (2007). doi:10.1016/j.jebo.2005.03.015, http://dx.doi.org/10.1016/j.jebo.2005.03.015

T. Nakajima, Probability in biology: overview of a comprehensive theory of probability in living systems. Prog. Biophys. Mol. Biol. **113**(1), 67–79 (2013). doi:10.1016/j.pbiomolbio.2013.03.007, http://dx.doi.org/10.1016/j.pbiomolbio.2013.03.007

National Cancer Institute, Autophagy. Homepage (2016). http://www.cancer.gov/Common/PopUps/popDefinition.aspx?id=CDR0000561319&version=Patient&language=English

H. Nikaido, *Convex Structures and Economic Theory* (Academic Press, New York, 1968)

D.M. Pardoll, The blockade of immune checkpoints in cancer immunotherapy. Nat. Rev. Cancer **12**, 252–264 (2016). doi:0.1038/nrc3239

V. Piras, K. Hayashi, M. Tomita, K. Selvarajoo, Enhancing apoptosis in TRAIL-resistant cancer cells using fundamental response rules. Sci. Rep. **1**(144) (2011). doi:10.1038/srep00144, http://www.nature.com/articles/srep00144

B. Rosser, Computational and dynamic complexity in economics, in *Handbook on the Economic Complexity of Technological Change* (Edward Elgar, London, 2011)

P. Sharma, K. Wagner, J.D. Wolchok, J.P. Allison, Novel cancer immunotherapy agents with survival benefit: recent successes and next steps. Nat. Rev. Cancer **11**, 805–812 (2011). doi:10.1038/nrc3153

Y. Shiozawa, A new construction of Ricardian trade theory – a many-country, many-commodity case with intermediate goods and choice of production techniques. Evol. Inst. Econ. Rev. **3**(2), 141–187 (2007)

Y. Shiozawa, International trade theory and exotic algebras. Evol. Inst. Econ. Rev. **12**(1), 177–212 (2015). doi:10.1007/s40844-015-0012-3

E. Sober, *Philosophy of Biology*, 2nd edn. (Westview Press Boulder, Colorado, 2000)

P. Sraffa, *Production of Commodities by Means of Commodities* (Cambridge University Press, Cambridge/New York, 1960)

Statistics Bureau, in *2011 Input-Output Tables for Japan*, e-Stat: Portal site of official statics of Japan. Ministry of Internal Affairs and Communications Japan (2016). http://www.e-stat.go.jp/SG1/estat/List.do?bid=000001060671&cycode=0

R.G. Syljuaasen, Checkpoint adaptation in human cells. Oncogene 5833–5839 (2007). doi:10.1038/sj.onc.1210402

The University of Tokyo Hospital Department of Immunotherapeutics, Basic Knowledge on tumor immunology. Homepage (2012). http://immunoth.umin.jp/tumor_immunity/index_03.html

S. Tiziano, D. Garlaschelli, The Weighted Random Graph Model. Wolfram Demonstrations Project, (2016). http://demonstrations.wolfram.com/WeightedRandomGraph/. Diego Garlaschelli, arXiv:0902.0897v1

K. Tran, H. Sato, A. Namatame, Y. Aruka, *Network Approach to Find Key Industrial Sectors in World Economy*. Mimeo, Yokosuka (2016)

Wikipedia, Major histocompatibility complex (2016). https://en.wikipedia.org/wiki/Major_histocompatibility_complex

D.S. Wilson, Stop crying about the size of government. Start caring about who controls it. The role of the state in the economy. Personal Blog (2016). http://evonomics.com/stop-crying-size-of-government/

Y. Xie, W. Hou, X. Song, Y. Yu, J. Huang, X. Sun, R. Kang, D. Tang, Ferroptosis: process and function. Cell Death Differ. **23**, 369–379 (2016). doi:10.1038/cdd.2015.158

# Chapter 3
# Socioeconomic Inequality and Prospects of Institutional Econophysics

**Arnab Chatterjee, Asim Ghosh, and Bikas K. Chakrabarti**

**Abstract** Socioeconomic inequality is measured using various indices. The Gini ($g$) index, giving the overall inequality, is the most commonly used, while the recently introduced Kolkata ($k$) index gives a measure of $1-k$ fraction of population who possess top $k$ fraction of wealth in the society. This article reviews the character of such inequalities, as seen from a variety of data sources, the apparent relationship between the two indices, and what toy models tell us. These socioeconomic inequalities are also investigated in the context of man-made social conflicts or wars, as well as in natural disasters. Finally, we forward a proposal for an international institution with sufficient fund for visitors, where natural and social scientists from various institutions of the world can come to discuss, debate, and formulate further developments.

## 3.1 Introduction: Socioeconomic Inequality

The complex dynamics of human social interactions lead to interesting phenomena, and inequalities at various levels often show up in course. The recent availability of huge amount of data (empirical data from databases, electronic footprints, and

A. Chatterjee (✉)
Condensed Matter Physics Division, Saha Institute of Nuclear Physics, 1/AF Bidhannagar, 700064, Kolkata, India

TCS Innovation Lab, Delhi, India
e-mail: arnabchat@gmail.com

A. Ghosh
Department of Computer Science, Aalto University School of Science, P.O. Box 15400, FI-00076, Aalto, Finland
e-mail: asim.ghosh@aalto.fi

B.K. Chakrabarti
Condensed Matter Physics Division, Saha Institute of Nuclear Physics, 1/AF Bidhannagar, 700064, Kolkata, India

Economic Research Unit, Indian Statistical Institute, 203 B.T. Road, 700108, Kolkata, India
e-mail: bikask.chakrabarti@saha.ac.in



Y. Aruka, A. Kirman (eds.), *Economic Foundations for Social Complexity Science*, Evolutionary Economics and Social Complexity Science 9,
DOI 10.1007/978-981-10-5705-2_3

sometimes large surveys) for various forms of human social interactions has made it easier to uncover certain patterns present, to analyze them and investigate the reasons behind various socioeconomic inequalities manifested in them. The Age of *Big Data* has opened up new avenues and challenges, and scientists are in the quest to understand *why* certain things look like as they do and *how* do they happen. Researchers are pooling in knowledge and techniques from various disciplines (Lazer et al. 2009), e.g., statistics, applied mathematics, information theory, and computer science, while tools of statistical physics have proved to be quite successful in better understanding of the precise (spatiotemporal) nature and origin of socioeconomic inequalities prevalent in our society. The more the data that is acquired and analyzed, the more we become confident in addressing the *why*s and *how*s.

Statistical physics tells us that systems of a large number of interacting dynamical units collectively exhibit a behavior which is solely determined by only a few basic dynamical properties of the individual constituent units and of the embedding dimension but is independent of all other details. This feature, which is specific to "critical phenomena," as in continuous phase transitions, is known as *universality* (Stanley 1971). There is no shortage of empirical evidence that several social phenomena are characterized by simple emergent behavior out of the interactions of many individual constituent units. In recent times, a growing community of researchers have been analyzing large-scale social dynamics to uncover universal patterns and proposing simple microscopic models to describe them, very similar to the minimalistic models used in statistical physics to understand physical phenomena. These studies have revealed quite a few interesting patterns and behaviors in social systems, as in elections (Fortunato and Castellano 2007; Chatterjee et al. 2013; Mantovani et al. 2011), population growth (Rozenfeld et al. 2008) and economic growth (Stanley et al. 1996), income and wealth distributions (Chakrabarti et al. 2013), financial markets (Mantegna and Stanley 2000), languages (Petersen et al. 2012), etc. (see Castellano et al. 2009; Sen and Chakrabarti 2013 for a review).

Socioeconomic inequality (Arrow et al. 2000; Stiglitz 2012; Neckerman 2004; Goldthorpe 2010; Chatterjee 2015) usually concerns the existence of unequal "wealth" and "fortunes" accumulated due to complex dynamics and interactions within the society. Usually containing structured and recurrent patterns of unequal distributions of goods, wealth, opportunities, and even rewards and punishments, this is classically measured in terms of *inequality of conditions* and *inequality of opportunities*. The former refers to the unequal distribution of income, wealth, assets, and material goods, while the latter refers to the unequal distribution of "life chances." This is reflected in levels of education, health status, treatment done by the criminal justice system, etc. Socioeconomic inequalities are mostly responsible for conflicts, wars, crises, oppressions, criminal activities, instability in political scenario, and sociopolitical unrest, which in turn affects economic growth (Hurst 1995). Historically as well as traditionally, economic inequalities have been extensively studied in the context of income and wealth (Yakovenko and Barkley Rosser 2009; Chakrabarti et al. 2013; Aoyama et al. 2010), although it is also measured for many quantities like energy consumption (Lawrence et al.

2013). The studies of inequality in society (Piketty and Saez 2014; Cho 2014; Chin and Culotta 2014; Xie 2014) have been always very important and are also a topic of contemporary focus and immediate global interest, drawing attention of researchers across disciplines – economics, sociology, mathematics, statistics, demography, geography, graph theory, computer science, and, not surprisingly, theoretical physics.

Quantifying socioeconomic inequalities is a challenge but is done in numerous ways. The probability distributions of various quantities, of course, provide the most detailed measures. It is very common to find that most quantities display broad distributions – most common are lognormals, power laws, or their combinations. For example, the distribution of income is usually found to be exponential for the bulk followed by a power law (Drăgulescu and Yakovenko 2001; Chakrabarti et al. 2013) for the top income range. However, such distributions can widely differ in their forms and details, and as such they are rather difficult to handle. This leads to the introduction of various *indices* like the Gini (1921), Theil (1967), Pietra (2010), and other socio-geometric indices (Eliazar 2015a,b), which try to characterize various geometric features of these distributions using a single number. Of course, each of these indices comes with certain merits, and certain indices are more useful than others, depending on the context they are used in. In this article we will focus on the most common one, the Gini index and a recently proposed $k$ index ($k$ = Kolkata) which has a nice, useful socio-geometric interpretation.

The most commonly used measure to quantify socioeconomic inequality is the Gini index. To compute this, one has to consider the "Lorenz curve" (Lorenz 1905), which shows the cumulative proportion $X$ of (poor to rich) ordered individuals (entries) in terms of the cumulative share $Y$ of their wealth. $Y$ can of course represent income or assets of individuals but it can as well represent citation of articles, votes in favor of candidates, population of cities, etc. It is first computed from a given statistical distribution or a dataset. The Gini index ($g$) is defined as the ratio of the area, enclosed between the Lorenz curve and the equality line, to that below the equality line, taking values 0 for absolute equality and 1 for absolute inequality. Let the area between (i) the Lorenz curve and the equality line be represented as $\mathscr{A}$ and (ii) that below the Lorenz curve be $\mathscr{B}$ (See Fig. 3.1). Then the Gini index is $g = \mathscr{A}/(\mathscr{A} + \mathscr{B}) = 2\mathscr{A}$. The recently introduced Kolkata index (symbolizing the extreme nature of social inequalities in Kolkata) or "$k$-index" (Ghosh et al. 2014b) is defined as the fraction $k$ such that $(1 - k)$ fraction of people (or papers) possess $k$ fraction of highest incomes (or citations) (Inoue et al. 2015; Chatterjee et al. 2016a; Ghosh et al. 2016).

The empirical data on Gini index from World Bank data (World Bank) for incomes over several years are given in Fig. 3.2. The values seem to be mostly between 0.2 and 0.6. In the later part of our article, we will argue that the simple kinetic exchange models can even reproduce this feature.

We also discuss here the specific case of the citation distributions. It was shown earlier (Radicchi et al. 2008) that the distribution of citations $c$ to papers within a discipline has a broad distribution, which is universal across broad scientific disciplines, by defining a relative indicator $c_f = c/\langle c \rangle$, where $\langle c \rangle$ is the average

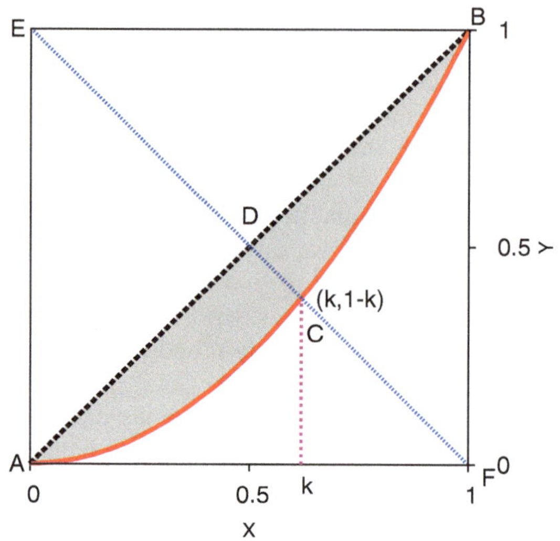


**Fig. 3.1** Lorenz curve is shown in *solid red line* for a typical probability distribution function and the equality line in *dotted black diagonal*. The Lorenz curve shows the cumulative fraction of "wealth" possessed by the corresponding fraction of poorer population. The $g$-index is given by area of the *shaded region*, while the $k$-index is computed from the coordinate of the point of intersection C $(k, 1-k)$ of the Lorenz curve and the diagonal perpendicular to the equality line. Thus, while the $g$-index measures the overall inequality in the system, the $k$-index gives the fraction $k$ of wealth possessed by the $1-k$ fraction of richer population

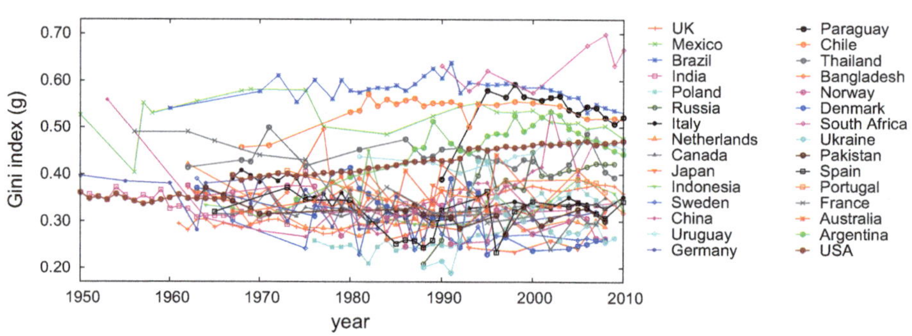


**Fig. 3.2** Gini index from World Bank data (World Bank) for income for several countries over years

citation within a discipline. Our study (Chatterjee et al. 2016a) confirmed this case for academic institutions as well as journals across disciplines.

Studies on the statistics of human deaths from wars, conflicts, and natural disasters show that the form of the probability distribution for the number of people killed exhibits power-law decay for the largest sizes, the exponent values being quite similar. We argue if a common mechanism is responsible for similarity that is manifested.

## 3.2 Introduction: Institutional Econophysics and Sociophysics

In view of the truly interdisciplinary nature of econophysics and sociophysics, it can be argued that some interdisciplinary visiting facilities for social and natural scientists are absolutely necessary today. These will provide scientists from different disciplines to interact over some long period, discuss, and debate and develop in their own discipline. In the concluding part of this article, we argue about the need to establish a research institute dedicated to socioeconomic problems with an interdisciplinary character, with some specific model in mind.

## 3.3 Inequality in Citations for Academic Institutions and Journals

In a recent study (Chatterjee et al. 2016a), we were able to conclude that the citation distributions for articles published in different journals (Fig. 3.3b), as well as from different academic institutions (Fig. 3.3a), followed the same functional form, irrespective of time (the year they are published) and space (institution). One has to carefully scale the probability distributions by their average, and the rescaled curves show excellent scaling collapse. The most of the resulting scaling curve fits to a lognormal function

$$F(x) = \frac{1}{x\sigma\sqrt{2\pi}} \exp\left[-\frac{(\log x - \mu)^2}{2\sigma^2}\right], \tag{3.1}$$

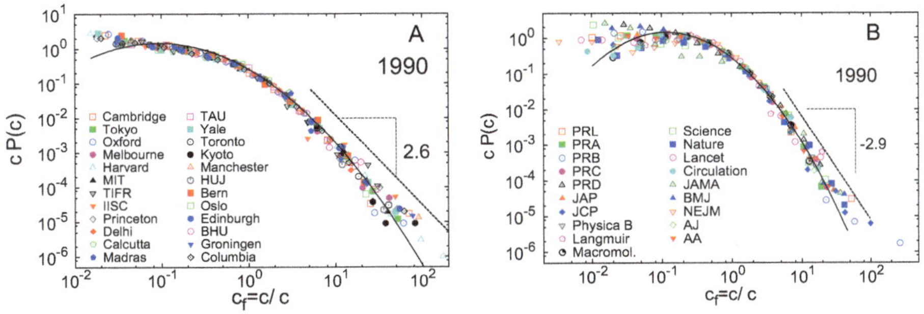


**Fig. 3.3** (**a**) Probability distribution $P(c)$ of citations $c$ rescaled by average number of citations $\langle c \rangle$ to publications from 1990 for several academic institutions. The scaled distribution fits very well to a lognormal for most of its range, with $\mu = -0.73 \pm 0.02$, $\sigma = 1.29 \pm 0.02$. The largest citations do not follow the lognormal behavior and seem to follow a power law: $c^{-\alpha}$, with $\alpha = 2.8 \pm 0.2$. (**b**) Probability distribution $P(c)$ of citations $c$ rescaled by average number of citations $\langle c \rangle$ to publications from 1990 for several academic journals. The scaled distribution function fits well to a lognormal function with $\mu = -0.75 \pm 0.02$, $\sigma = 1.18 \pm 0.02$, while $\langle c \rangle P(c) \to$ const. as $c/\langle c \rangle \to 0$ for the lower range of $c$. The largest citations fit well to a power law: $c^{-\alpha}$, with $\alpha = 2.9 \pm 0.3$ (Data is taken from Chatterjee et al. 2016a)

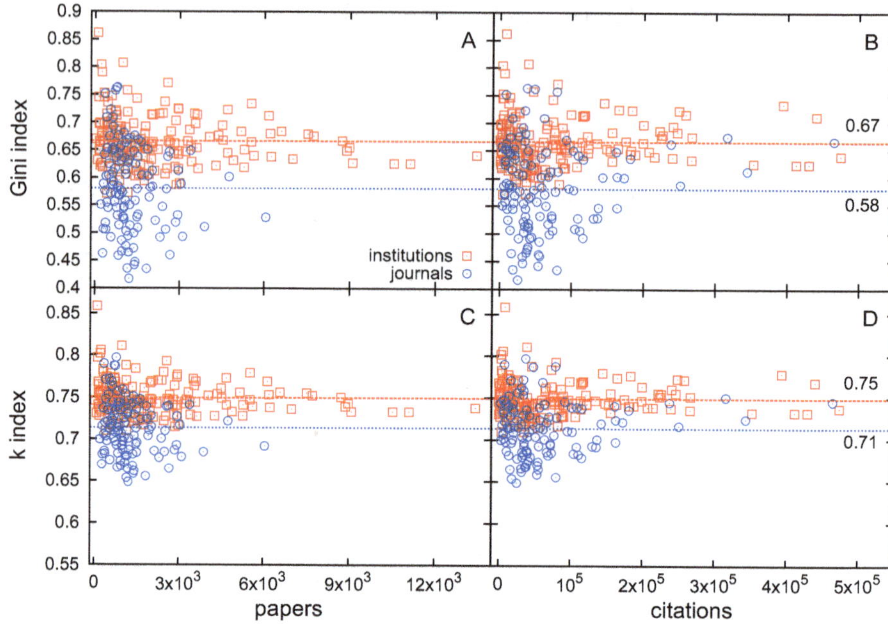


**Fig. 3.4** Variation of Gini and $k$ indices with number of papers and citations for academic institutions and journals. (**a**) Gini index on institutions/journals. (**b**) Gini index on citations. (**c**) $k$ index on papers. (**d**) $k$ index on citations. For larger number of papers or citations, the values seem to fluctuate less or converge around the mean values $\bar{g}$ and $\bar{k}$ respectively. For academic institutions, the values are $\bar{g} \approx 0.67$ for Gini and $\bar{k} \approx 0.75$, while for the journals, the values are $\bar{g} \approx 0.58$ and $\bar{k} \approx 0.71$. Note that squares exhibit institutions, and circles exhibit journals (Figure adapted from Chatterjee et al. 2016a)

while the extreme right tail deviates from this and seem to fit more to a power law with a decay exponent around $2.6 - 2.8$. We additionally observed that for the academic institutions, Gini index was $g = 0.67 \pm 0.10$ and $k = 0.75 \pm 0.04$, which means around 75% citations come from the top 25% papers. For academic journals, $g = 0.58 \pm 0.15$, $k = 0.71 \pm 0.08$ which means about 71% citations come from the top 29% papers.

We further noted that Gini and $k$ indices fluctuate less around respective mean values $\bar{g}$ and $\bar{k}$ as the number of articles or the number of citations became large (Fig. 3.4). For academic institutions, the values were $\bar{g} \approx 0.66$ for Gini and $\bar{k} \approx 0.75$. For journals, the values are $\bar{g} \approx 0.58$ and $\bar{k} \approx 0.71$.

## 3.4 Empirical Findings on $g-k$ Relationship

The huge variety of socioeconomic data suggest that there might be a simple relation between the two seemingly different inequality measures (Chatterjee et al. 2017). Analysis of the following was carried out: (i) citations of papers published from academic institutions and journals (data from ISI Web of Science (2014) and reported in Chatterjee et al. 2016a), (ii) consumption expenditure data of India (NSSO), Brazil (IBGE 2015a,b), and Italy (Banca d'Italia 2015) income data from the USA (Internal Revenue Service 2014), (iii) voting data from open list proportional elections (Chatterjee et al. 2013) of Italy, the Netherlands, and Sweden and *first past the post*-election data for Indian Parliamentary elections and Legislative Assembly elections (Election Commission of India), the UK (UK General Elections), Canada (Election Canada), Bangladesh (Bangladesh Election Commission), and Tanzania (National Election Commission Tanzania), and (iv) city population data from Ghosh et al. (2014a).

The $g-k$ relation seems to be perfectly linear for smaller values while it becomes nonlinear at the limit of extreme inequality, i.e., as $g$ or $k$ approaches unity (Fig. 3.5). The most striking feature is that the data from a variety of these sources hardly depart from a seemingly smooth curve.

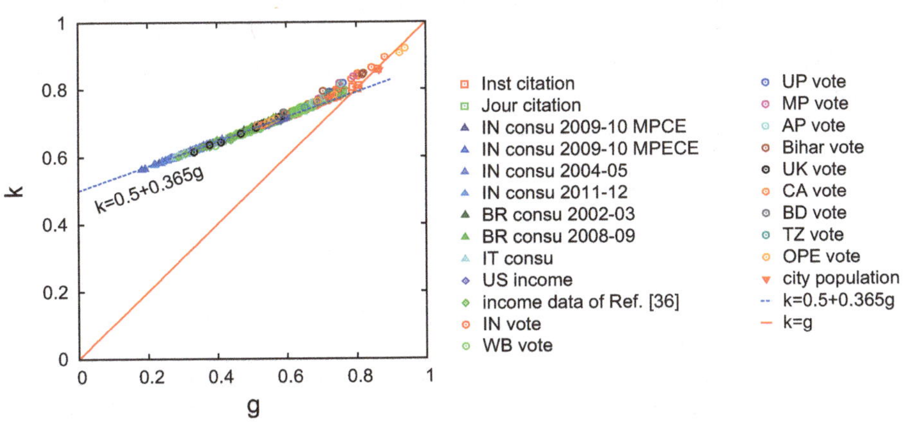


**Fig. 3.5** Estimated values of $k$-index and $g$-index from the various datasets: citations (retrieved from ISI Web of Science (ISI 2014), analyzed in Chatterjee et al. (2016a); *Inst* = institutions, *Jour* = journals) expenditure (Data taken from Chatterjee et al. (2016b) and Chakrabarti et al. (2016); *IN* = India, *BR* = Brazil, *IT* = Italy), income (Data taken from Inoue et al. 2015), voting data from proportional elections (Data taken from Chatterjee et al. 2013; OPE), voting data from first-past-the-post elections (*IN* = India, *WB* = West Bengal, *UP* = Uttar Pradesh, *MP* = Madhya Pradesh, *AP* = Andhra Pradesh, *UK* = United Kingdom, *CA* = Canada, *BD* = Bangladesh, *TZ* = Tanzania), and city population (Data taken from Ghosh et al. 2014a). Data details are given in Chatterjee et al. (2017). The dotted straight line represents $k = 0.5 + 0.365\,g$

The $k$-index and $g$-index obey a linear relationship

$$k = \frac{1}{2} + \gamma.g, \quad \text{for } 0 \leq g \lesssim 0.70, \tag{3.2}$$

with $\gamma = 0.365 \pm 0.005$ (Chatterjee et al. 2017).

There has been an attempt to explain the slope of the $g - k$ curve for small values. by approximating the Lorenz curve as an arc of a circle (Chatterjee et al. 2017). This linear relationship (with the value of the slope $\gamma \approx 0.363$) can be argued to be more generally valid. If the Lorenz curve $L(x)$ in Fig. 3.1 is taken as a parabola ($L(x) = x^2$, as in case of normalized uniform distribution $P(m)$ of income/wealth $m$; $L(x) = \int_0^x 2mP(m)dm$), one gets $g = 2\int_0^1 (x - L(x))dx = \frac{1}{3} \approx 0.33$ and $1 - k = L(k) = k^2$, giving $k = \frac{1}{2}(\sqrt{5} - 1) \approx 0.62$, the values of $g$ and $k$ satisfy the above relationship very well.

## 3.5 Estimates of $g - k$ Relation from Kinetic Exchange Models

The market models developed by physicists, specifically the kinetic exchange models (Chatterjee and Chakrabarti 2007; Chakrabarti et al. 2013), can provide an estimate of the relation between the inequality indices. In the CC model (Chatterjee and Chakrabarti 2007), where an agent retains a (same for all) fraction $\lambda$ of their income or wealth before going for any (stochastic) exchange (call it trade or scattering) with another agent, the dynamics is defined by

$$m_i(t+1) = \lambda m_i(t) + r(1-\lambda)\left[m_i(t) + m_j(t)\right]$$

$$m_j(t+1) = \lambda m_j(t) + (1-r)(1-\lambda)\left[m_i(t) + m_j(t)\right],$$

where $r$ is a random fraction in [0, 1], drawn at each time step (trade or exchange). $m_i(t)$ and $m_i(t+1)$ are the wealth of the $i$th agent at trading times $t$ and $(t+1)$, respectively. The steady-state distribution of wealth is argued to be Gamma distribution (Patriarca et al. 2004; Chatterjee and Chakrabarti 2007) with the peak position shifting to higher income or wealth as $\lambda$ increases ($\lambda = 0$ corresponds to Gibbs or exponential distribution and $\lambda \to 1$ approaches $\delta$-function). The $g - k$ relationship for such distributions is found to be linear (Fig. 3.6a), obeying $k = \frac{1}{2} + \gamma.g$ with $\gamma \approx 0.365 \pm 0.005$.

In the CCM model (Chakrabarti et al. 2013; Chatterjee and Chakrabarti 2007), each agent $i$ has a saving fraction $\lambda$ drawn from a (quenched) distribution $\Pi(\lambda) = (1+\delta)(1-\lambda)^\delta$. Following similar stochastic dynamics as in CC model,

$$m_i(t+1) = \lambda_i m_i(t) + r\left[(1-\lambda_i)m_i(t) + (1-\lambda_j)m_j(t)\right]$$

$$m_j(t+1) = \lambda_j m_j(t) + (1-r)\left[(1-\lambda_i)(m_i(t) + (1-\lambda_j)m_j(t)\right],$$

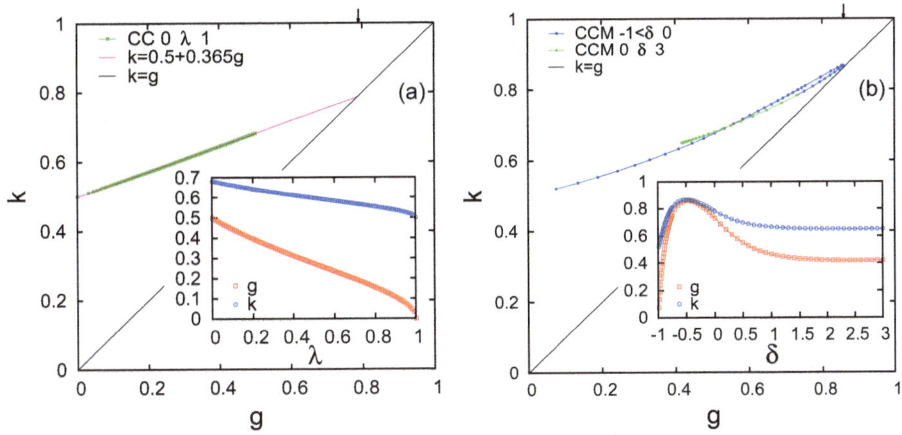


**Fig. 3.6** Monte Carlo simulation results for $g$ vs. $k$ in CC and CCM models (for 1000 agents). (**a**) For CC model, varying parameter $\lambda$. The inset shows the plots of $g$ and $k$ in the range of $0 \leq \lambda \leq 1$. The points fit to $k = \frac{1}{2} + \gamma.g$ with $\gamma \approx 0.365 \pm 0.005$. (**b**) For CCM model, varying parameter $\delta$. The inset shows the variation of $g$ and $k$ in the range of $-1 < \delta \leq 3$ (Figure adapted from Chatterjee et al. 2017)

one gets a steady-state distribution of income or wealth with power-law tails $P(m) \sim m^{-(2+\delta)}$ for large $m$ (Chatterjee and Chakrabarti 2007). $g$ and $k$ computed for such distributions (Ghosh et al. 2016) are given in inset of Fig. 3.6b for varying range of $\delta$. The $g - k$ relationship here is found to be nonlinear (see Fig. 3.6b) but very much around a similar linear relationship.

## 3.6 Universality in the Statistics of Deaths in Conflicts and Disasters

The history of human civilization has been frequently shaped by events of wars, conflicts, and disasters. In recent times, the scale of disaster events has increased remarkably. Growing population around the world has been seen as one of the reasons for the increase in counts of people affected by disaster events. A study on the statistics of human deaths from wars, conflicts, as well as natural disasters shows that the probability distribution of number of people killed in natural disasters as well as man-made situations exhibit similar universality in statistics with power-law decay for the largest sizes, the exponent values being quite similar (Chatterjee and Chakrabarti 2016), in the range of 1.5−1.8. Comparing with natural disasters, where event sizes are measured in terms of physical quantities, like the energy released in earthquake, the volume of rainfall, the land area affected in forest fires, etc., also show striking similarities. These universal patterns in their statistics might suggest some subtle similarities in their mechanisms and dynamics (Fig. 3.7).

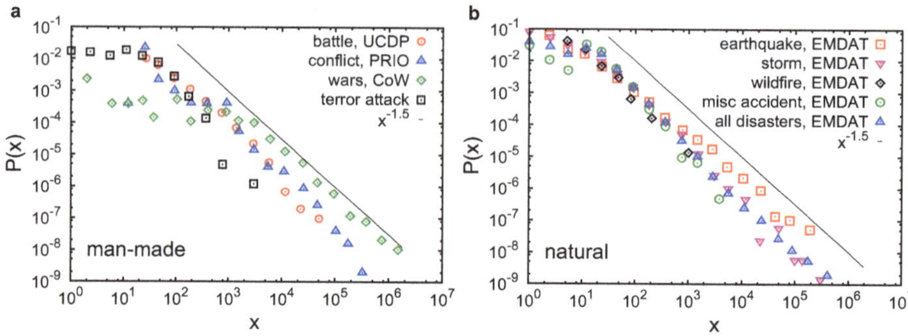


**Fig. 3.7** The probability distributions $P(x)$ of event size $x$, measured by the number of human deaths in the corresponding event. (**a**) *Man-made events*: human deaths from conflicts during 1946–2008, according to the PRIO database (Peace Research Institute Oslo 2016) (using lowest estimates), dead according to the Correlators of Wars (CoW) database (The Correlates of War Project 2016) during 1816–2007, dead in terror attack (Wikipedia) during 1910 till July 2016, and battle deaths according to UCDP database (Uppsala Conflict Data Program 2016) during 1989–2014. Except the terrorist attack data, all these distributions seem to have a power law tail with similar exponents. The *straight line* is a guide to the exponent value 1.5 for comparison. (**b**) *Natural disasters*: human death from earthquakes, storms, wildfires, miscellaneous accidents, as well as all natural disasters listed in the EMDAT database (EM-DAT 2014) during 1900–2013. The values of the exponents are in 1.5–1.8 (Details in Chatterjee and Chakrabarti 2016)

## 3.7 Discussions on Citations and Relationship Between Inequality Measures in General

The Gini index $g$ is the most popular among economists and sociologists, since it gives an overall measure of the inequality in a society. As evident from Fig. 3.1, it requires accurate data for the entire Lorenz curve to provide a measure of the shaded area enclosed by it and the equality line. The Kolkata index $k$ being given by the intersection of the Lorenz curve and the cross diagonal to the equality line. The $g - k$ linear relationship is extremely robust for not so high values of inequality and fits different forms of Lorenz curve, and hence, distributions of income, wealth, citations, etc. and this robustness are also observed empirically (Fig. 3.5). We could even compare these findings with simple kinetic exchange models of wealth distributions, where the scaling relation between $g$ and $k$ was found to be also true. The $g - k$ relationship would be extremely useful to translate from one inequality measure to the other; since $1 - k$ fraction of people possess precisely $k$ fraction of the total wealth, translation of social inequality measures into $k$-index language can be of major significance.

One of our recent focus had been the inequality in citations for academic institutions and journals. Although institutions and journals have their own ranking depending on the "quality" of research and publications that come out, get noticed, and cited, we find that the form of the distribution function for citations is invariant with respect to the average citations, holding across institutions and over time as well. In terms of absolute inequality measures, roughly 75% citations come from the top 25% papers in case of academic institutions, and 71% citations come from the top 29% papers for journals.

We also discussed how the inequality statistics of deaths in social conflicts or wars compare with those in natural disasters.

## 3.8 Concluding Remarks: Some Random Thoughts About Prospects of Institutional Econophysics

Twenty years have passed since the formal coining of the term and hence the launch of econophysics as a research topic (since 1995; see the entry by Barkley Rosser on Econophysics in The New Palgrave Dictionary of Economics Barkley Rosser 2008). Furthermore, econophysics has been assigned the Physics and Astronomy Classification Scheme (PACS) number 89.65 Gh by the American Institute of Physics. However, regular interactions and collaborations between the communities of natural scientists and social scientists are rare. Though interdisciplinary research papers on econophysics and sociophysics are regularly being published at a steady and healthy rate (more than 1000 documents containing the explicit term "econophysics" and more than 240 documents containing the explicit term "sociophysics" in the years 2014 and 2015 according to Google Scholar) and published mostly in physics journals, and a number of universities (including Universities of Leiden, Bern, Paris, and London) are offering the interdisciplinary courses on econophysics and sociophysics, not many clearly designated professor or other faculty positions for that matter are available yet (except for econophysics in Universities of Leiden and London). Neither there are any designated institutions on these interdisciplinary fields nor separate departments or centers of studies, for instance. We note, however, happily in passing, a recently published highly acclaimed ("landmark" and "masterful") economics book (Shubik and Smith 2016) by Martin Shubik (Seymour Knox Professor of Mathematical Institutional Economics, Emeritus, at Yale University) and Eric Smith (Santa Fe Institute) discusses extensively on econophysics approaches and in general on the potential of interdisciplinary researches inspired by the developments in natural sciences.

In view of these, it seems it is time to try for an international center for interdisciplinary studies on complexity in social and natural sciences, specifically on econophysics and sociophysics.[1] The model of the Abdus Salam International Centre for Theoretical Physics (ICTP), Trieste (funded by UNESCO and IAEA), could surely be helpful to guide us here. We are contemplating if an ICTP-type

[1]Although the presentation in the conference (by BKC) was mainly on the materials discussed in the earlier sections, extensive discussions with several participants, including the conference organizers, had been on this point.

interdisciplinary research institute could be initiated for researches on econophysics and sociophysics.

We note that Helbing (ETH, Zurich) and colleagues have been trying for an European Union-funded "Complex Techno-Socio-Economic Analysis Centre" or "Economic and Social Observatory" for the last 5 years (see Helbing and Balietti (2011) containing the white papers arguing for the proposed center). We are also aware that Indian Statistical Institute had taken a decision to initiate a similar center in India (see the "Concluding Remarks" in Ghosh 2013). Also there was an attempt for a similar Asian Centre in Singapore, initiated in Nanyang Technological University. In view of some recent enthusiasms at the Japan-India heads of states or prime minister level, and signing of various agreements (predominantly for business deals, infrastructure development, technical science, and also cultural exchanges) by them, possibility of an Indo-Japan center for studies on complex systems is also being explored. In such bilateral (Indo-Japan) initiatives, there are explicit memorandum of understandings already signed by the prime ministers. It did not have any economic or sociological study centers ever planned under such bilateral efforts.

These proposals are for regular research centers on such interdisciplinary fields, where regular researchers will investigate such systems. However, in view of the extreme interdisciplinary nature of econophysics and sociophysics, such efforts may be complemented by another visiting center model. Unlike the abovementioned kind of centers, therefore, this proposed center may be just a visiting center where natural and social scientists from different universities and institutions of the world can meet for extended periods to discuss and interact on various interdisciplinary issues and collaborate for such researches, following the original ICTP model.

Here, as in ICTP, apart from a few (say, about ten to start with) promising young researchers on econophysics and sociophysics as permanent faculty will continue active research, and active visiting scientist programs (in physics, economics, sociology), etc. can be pursued. The faculty members, in consultation with the advisers from different countries, can choose the invited visitors and workshops or courses, on economics and sociological complexity issues, can be organized on a regular basis (as for basic theoretical sciences in ICTP or in Newton Centre, Cambridge, etc.).

We think that it is an appropriate time for the healthy growth of these "new or evolving economic and sociological thinkings" including econophysics and sociophysics. We believe that Tokyo would be the ideal location for such an international center. In such new studies on social sciences, econophysics, and sociophysics in particular, Japan has already significant large, active, and established groups, and hence, Tokyo could be its natural location.

**Acknowledgements** We are extremely thankful to Yuji Aruka and Taisei Kaizoji for sharing with us many enthusiastic ideas on several collaboration projects. We are also grateful to Sudip Mukherjee for his help in preparing the manuscript.

## References

H. Aoyama, Y. Fujiwara, Y. Ikeda, *Econophysics and Companies: Statistical Life and Death in Complex Business Networks* (Cambridge University Press, Cambridge, 2010)

K.J. Arrow, S. Bowles, S.N. Durlauf, *Meritocracy and Economic Inequality* (Princeton University Press, Princeton, 2000)

Banca d'Italia, Distribuzione dei microdati (Distribution of microdata). http://www.bancaditalia.it/statistiche/tematiche/indagini-famiglie-imprese/bilanci-famiglie/distribuzione-microdati/index.html. Accessed Sept 2015

Bangladesh Election Commission. http://www.ec.org.bd. Accessed June 2013

J. Barkley Rosser Jr., Economics and economic complexity, in *The New Palgrave Dictionary of Economics*, ed. by S.N. Durlauf, L.E. Blume, vol. 2 (Palgrave Macmillan, Basingstoke/New York, 2008), pp. 729–732

C. Castellano, S. Fortunato, V. Loreto, Statistical physics of social dynamics. Rev. Mod. Phys. **81**, 591–646 (2009)

B.K. Chakrabarti, A. Chakraborti, S.R. Chakravarty, A. Chatterjee, *Econophysics of Income and Wealth Distributions* (Cambridge University Press, Cambridge, 2013)

A.S. Chakrabarti, A. Chatterjee, T.K. Nandi, A. Ghosh, A. Chakraborti, Quantifying invariant features of within-group inequality in consumption across groups (2016). arXiv:1601.02463

A. Chatterjee, Socio-economic inequalities: a statistical physics perspective, in *Econophysics and Data Driven Modelling of Market Dynamics*, ed. by F. Abergel, H. Aoyama, B.K. Chakrabarti, A. Chakraborti, A. Ghosh. New Economic Windows (Springer, Cham, 2015), pp. 287–324

A. Chatterjee, B.K. Chakrabarti, Kinetic exchange models for income and wealth distributions. Eur. Phys. J. B **60**(2), 135–149 (2007)

A. Chatterjee, B.K. Chakrabarti, Fat tailed distributions for deaths in conflicts and disasters. Rep. Adv. Phys. Sci. **1**, 1740007 (2017)

A. Chatterjee, M. Mitrović, S. Fortunato, Universality in voting behavior: an empirical analysis. Sci. Rep. **3**, 1049 (2013)

A. Chatterjee, A. Ghosh, B.K. Chakrabarti, Universality of citation distributions for academic institutions and journals. PloS one **11**(1), e0146762 (2016a)

A. Chatterjee, A.S. Chakrabarti, A. Ghosh, A. Chakraborti, T.K. Nandi, Invariant features of spatial inequality in consumption: the case of india. Physica A **442**, 169–181 (2016b)

A. Chatterjee, A. Ghosh, B.K. Chakrabarti, Socio-economic inequality: relationship between Gini and Kolkata indices. Physica A **466**, 583–595 (2017)

G. Chin, E. Culotta, What the numbers tell us. Science **344**(6186), 818–821 (2014)

A. Cho, Physicists say it's simple. Science **344**(6186), 828 (2014)

A.A. Drăgulescu, V.M. Yakovenko, Exponential and power-law probability distributions of wealth and income in the United Kingdom and the United States. Physica A **299**(1), 213–221 (2001)

Election Canada. http://www.elections.ca. Accessed June 2013

Election Commission of India, General election results and statistics. www.eci.nic.in. Accessed June 2013

I. Eliazar, The sociogeometry of inequality: part I. Physica A **426**, 93–115 (2015a)

I. Eliazar, The sociogeometry of inequality: part II. Physica A **426**, 116–137 (2015b)

I.I. Eliazar, I.M. Sokolov, Measuring statistical heterogeneity: the Pietra index. Physica A **389**(1), 117–125 (2010)

EM-DAT: the international disaster database. http://www.emdat.be/database. Retrieved July 2014

S. Fortunato, C. Castellano, Scaling and universality in proportional elections. Phys. Rev. Lett. **99**(13), 138701 (2007)

A. Ghosh, Econophysics research in India in the last two decades. IIM Kozhikode Soc. Manag. Rev. **2**(2), 135–146 (2013)

A. Ghosh, A. Chatterjee, A.S. Chakrabarti, B.K. Chakrabarti, Zipf's law in city size from a resource utilization model. Phys. Rev. E **90**(4), 042815 (2014a)

A. Ghosh, N. Chattopadhyay, B.K. Chakrabarti, Inequality in societies, academic institutions and science journals: Gini and k-indices. Physica A **410**(14), 30–34 (2014b)
A. Ghosh, A. Chatterjee, J.-I. Inoue, B.K. Chakrabarti, Inequality measures in kinetic exchange models of wealth distributions. Physica A **451**, 465–474 (2016)
C. Gini, Measurement of inequality of incomes. Econ. J. **31**(121), 124–126 (1921)
J.H. Goldthorpe, Analysing social inequality: a critique of two recent contributions from economics and epidemiology. Eur. Sociol. Rev. **26**(6), 731–744 (2010)
D. Helbing, S. Balietti, Complex techno-socio economics. Eur. Phys. J. Spl. Top. **195**, 1–186 (2011)
C.E. Hurst, *Social Inequality: Forms, Causes, and Consequences* (Allyn and Bacon, Boston, 1995)
IBGE, Instituto Brasileiro de Geografia e Estatística – Pesquisa de Orçamentos Familiares (Consumer Expenditure Survey) 2002–2003. http://www.ibge.gov.br/home/estatistica/populacao/condicaodevida/pof/2002aquisicao/microdados.shtm, 2002–2003. Accessed Sept 2015
IBGE, Instituto Brasileiro de Geografia e Estatística – Pesquisa de Orçamentos Familiares (Consumer Expenditure Survey) 2008–2009. http://www.ibge.gov.br/english/estatistica/populacao/condicaodevida/pof/2008-2009/perfil-despesas/microdados.shtm, 2008–2009. Accessed Sept 2015
J.-I. Inoue, A. Ghosh, A. Chatterjee, B.K. Chakrabarti, Measuring social inequality with quantitative methodology: analytical estimates and empirical data analysis by Gini and $k$ indices. Physica A **429**, 184–204 (2015)
Internal Revenue Service (IRS) – individual tax tables. http://www.irs.gov/uac/SOI-Tax-Stats-Individual-Statistical-Tables-by-Size-of-Adjusted-Gross-Income. Retreived June 2014
ISI Web of Science, Retreived Apr 2014. http://portal.isiknowledge.com
S. Lawrence, Q. Liu, V.M. Yakovenko, Global inequality in energy consumption from 1980 to 2010. Entropy **15**(12), 5565–5579 (2013)
D. Lazer, A. Pentland, L. Adamic, S. Aral, A.-L. Barabási, D. Brewer, N. Christakis, N. Contractor, J. Fowler, M. Gutmann, T. Jebara, G. King, M. Macy, D. Roy, M. Van Alstyne, Computational social science. Science **323**(5915), 721–723 (2009)
M.O. Lorenz, Methods for measuring the concentration of wealth. Am. Stat. Assoc. **9**, 209–219 (1905)
R.N. Mantegna, H.E. Stanley, *Introduction to Econophysics: Correlations and Complexity in Finance* (Cambridge University Press, Cambridge, 2000)
M.C. Mantovani, H.V. Ribeiro, M.V. Moro, S. Picoli Jr., R.S. Mendes, Scaling laws and universality in the choice of election candidates. Europhys. Lett. **96**(4), 48001 (2011)
National Election Commission Tanzania. http://www.nec.go.tz. Accessed June 2013
K. Neckerman, *Social Inequality* (Russell Sage Foundation, New York, 2004)
NSSO, Household consumer expenditure from the National Sample Survey Office (NSSO). http://mail.mospi.gov.in/index.php/catalog/CEXP. Accessed June 2015
M. Patriarca, A. Chakraborti, K. Kaski, Statistical model with a standard $\Gamma$ distribution. Phys. Rev. E **70**(1), 016104 (2004)
Peace Research Institute Oslo (PRIO): the battle deaths dataset version 3.0. https://www.prio.org/Data/Armed-Conflict/Battle-Deaths/The-Battle-Deaths-Dataset-version-30/. Retreived July 2016
A.M. Petersen, J. Tenenbaum, S. Havlin, H.E. Stanley, Statistical laws governing fluctuations in word use from word birth to word death. Sci. Rep. **2**, 313 (2012)
T. Piketty, E. Saez, Inequality in the long run. Science **344**(6186), 838–843 (2014)
F. Radicchi, S. Fortunato, C. Castellano, Universality of citation distributions: toward an objective measure of scientific impact. Proc. Natl. Acad. Sci. **105**(45), 17268–17272 (2008)
H.D. Rozenfeld, D. Rybski, J.S. Andrade, M. Batty, H.E. Stanley, H.A. Makse, Laws of population growth. Proc. Natl. Acad. Sci. **105**(48), 18702–18707 (2008)
P. Sen, B.K. Chakrabarti, *Sociophysics: An Introduction* (Oxford University Press, Oxford, 2013)
M. Shubik, E. Smith, *The Guidance of an Enterprise Economy* (MIT Press, Cambridge, 2016)

H.E. Stanley, *Introduction to Phase Transitions and Critical Phenomena* (Oxford University Press, Oxford/New York, 1971)

M.H.R. Stanley, L.A.N. Amaral, S.V. Buldyrev, S. Havlin, H. Leschhorn, P. Maass, M.A. Salinger, H.E. Stanley, Scaling behaviour in the growth of companies. Nature **379**(6568), 804–806 (1996)

J.E. Stiglitz, *The Price of Inequality: How Today's Divided Society Endangers Our Future* (WW Norton & Company, London, 2012)

The correlates of war project: COW war data, 1816–2007 (v4.0). http://www.correlatesofwar.org/data-sets/COW-war. Retreived July 2016

H. Theil, *Economics and Information Theory* (North-Holland, Amsterdam, 1967)

UK General Elections since 1832. http://www.politicsresources.net/area/uk/edates.htm. Accessed June 2013

Uppsala Conflict Data Program (UCDP), Battle-related deaths dataset v.5-2015, 1989–2014. http://www.pcr.uu.se/research/ucdp/datasets/ucdp_battle-related_deaths_dataset/. Retreived July 2016

Wikipedia, List of battles and other violent events by death toll, terror attacks. Retreived July 2016

World Bank, All the Ginis dataset. http://siteresources.worldbank.org/INTRES/Resources/469232-1107449512766/allginis_2013.xls. Retrieved June 2014

Y. Xie, Undemocracy: inequalities in science. Science **344**(6186), 809–810 (2014)

V.M. Yakovenko, J. Barkley Rosser Jr., Statistical mechanics of money, wealth, and income. Rev. Mod. Phys. **81**(4), 1703 (2009)

# Chapter 4
# The Evolution of Behavioural Institutional Complexity

J. Barkley Rosser and Marina V. Rosser

## 4.1 Introduction

This essay considers how behavioural economics and institutional economics have coevolved in their development, with understanding their links central to understanding how evolutionary processes within economies dynamically develop. Key to this is that a most important function of economic institutions is to aid humans in overcoming the limits imposed by their bounded rationality. To do this, we shall consider the ideas of the respective founders of institutional economics and behavioural economics, Thorstein Veblen and Herbert Simon, both of whom will also be shown to have complex evolutionary views of how the economy operates. Veblen's work was earlier (1898, 1899); however, he prefigured Simon's work in many ways, with Simon tying the concept of behavioural economics, a term he coined, with that of bounded rationality, a term he also coined (1947, 1955, 1957).

Veblen not only called for economics to be an *evolutionary science* (1898), but introduced certain ideas that have since proven to be important in understanding the nature of complexity in economics, particularly that of *cumulative causation*, often thought to have been introduced later by either Allyn Young (1928) or Gunnar Myrdal (1957), with the latter making the term widely known among economists, and Nicholas Kaldor (1972) drawing out its negative implications for equilibrium economics (Rosser and Rosser, 2016). Among the various forms of complexity that are relevant to economics, cumulative causation is most obviously tied to *dynamic complexity*, which leads to increasing returns, multiple equilibria and a

J.B. Rosser (✉) • M.V. Rosser
Department of Economics, James Madison University, Harrisonburg, VA, USA
e-mail: rosserjb@jmu.edu; rossermv@jmu.edu



Y. Aruka, A. Kirman (eds.), *Economic Foundations for Social Complexity Science*, Evolutionary Economics and Social Complexity Science 9,
DOI 10.1007/978-981-10-5705-2_4

variety of bifurcations in economic dynamical systems. However, it can be seen to be connected also to *computational complexity*, as well as *hierarchical complexity* due to Simon (1962).

Simon's formulation of the bounded rationality concept provided the foundation for his later views, and he developed in the context of considering problems of administrative behaviour in organisations, with *Administrative Behavior* (1947) being the title of his first book where he first presented the idea, if not the term. It was thinking about the nature of human bounded rationality that led Simon into studying computer science and artificial intelligence, with this then leading into his considering the problems of computational complexity (1969). All of this was evident in his idea of hierarchical complexity, in which evolutionary emergence is a central concept.

An important issue for the matter of how evolutionary theory relates to institutional economics in its early formulation involves Veblen's relations with John R. Commons and Joseph Schumpeter. Veblen developed ideas of Darwinian evolutionary economics in the early twentieth century in the USA, while Schumpeter is widely viewed as a strong supporter of an evolutionary approach to economic development, particularly regarding the evolution of technology, even as he criticised institutional economics and the application of biological ideas. Also not widely known, Commons (1924) also supported an evolutionary view, although he had more of a teleological perspective on that than did either Veblen or Schumpeter, both of whom saw no necessary direction to technological evolution and change (Papageorgiou et al. 2013). Dealing with a complexity issue, Schumpeter strongly advocated a discontinuous, or *saltationalist* view of evolution (Schumpeter 1912; Rosser 1992), which Veblen agreed with regarding technological change. Regarding institutional evolution, Veblen mostly saw it proceeding in a more continuous manner through cumulative causation, thus being somewhat closer to Commons on that matter, even as he argued that it was fundamentally unstable and would experience crises and breakdowns.

An important element of evolutionary processes is the emergence of higher-level structures out of lower-level and simpler ones. This closely links with Simon's (1962) view of hierarchical complexity, which also links to his views of bounded rationality. This fits with the issue of multilevel evolution, long controversial in evolutionary theory (Henrich 2004). Within human systems, this becomes tied to cooperation, with Ostrom (1990) developing how such cooperation can arise through particular institutions. This process of emergence is linked to deep concepts of complexity, with Simon (1962) a crucial developer of this line of thought.

The evolution of complex behavioural and institutional dynamics extends to a deeper matter of epistemic issues arising from hierarchical emergence that may show deep links between computational and dynamic complexity (Koppl and Rosser 2002). Memes involve information structure systems that are understood by computational complexity concepts, with this form of complexity exhibiting levels. Competition between such structures in markets can see the emergence of higher-order institutional forms in markets as analysed by Mirowski (2007) that show ever-expanding bounds on rationality as higher levels emerge. Thus, we see a deep unification between Veblen's cumulative causation and Simon's bounded rationality as explaining profound forms of evolutionary dynamics.

## 4.2 Forms of Complexity

A discussion regarding the relationship between "complexity" and something else clearly requires some discussion of what is meant by this term, or at least what this observer means by it. Indeed, this is arguably a weasel term, one that has no clearly agreed-on meaning more generally. The MIT engineer, Seth Lloyd, some time ago famously gathered a list of various different meanings, and this list was at least 45 before he stopped bothering with this effort or at least making it publicly known (Horgan 1997, p. 303). It may be useful therefore to refer to the broadest possible view of complexity that includes all of these and any others as being *meta-complexity*. The definition of this may simply amount to listing all possible meanings that any have ever claimed should be on the list.

If one seeks general definitions or concepts, something often appears in such general definitions is the idea that somehow something that is complex involves a whole that is "greater than the sum of its parts", as the old cliché puts it. Such an idea can be traced as far back as Aristotle, with many since contributing to it. We shall see below that not all the items on Seth Lloyd's list might agree with this, particularly the many that relate to *computational complexity*, arguably the subcategory of complexity with more variations than any other. That those concerned with this subcategory might not have such a view might explain why John von Neumann (1966) did not distinguish complexity from mere *complicatedness*. While some may not wish to make this distinction, many do, with Israel (2005) noting that the two words come from different roots in Latin, *complecti* and *complicare*, respectively, the former meaning "to enfold" and the latter "to entangle". Thus, while close and possibly from an identical deeper origin, the former implies some completing in a higher order, whereas the latter implies more simply "to confuse" due to the bringing together of many different elements.

In any case, perusing Lloyd's list allows one to lump many of his definitions into higher-order subcategories. Arguably the subcategory with the most items on it can be considered forms of *computational complexity*, with at least as many as 15 of them fitting in this category, possibly more. If there is a linking concept through this set of definitions, it involves ideas of size or length, how long a programme is or how many distinct units there are within the object such as bits of information. However, the many variations on this do not map onto each other readily. Nevertheless, many of these definitions have the virtue of being clearly measurable, even if there are many such definitions. Thus, if one gloms onto one of these, one can argue that it may have a stronger claim to being "scientific" due to this specific clarity than some other fuzzier alternatives. Interestingly, among those fuzzier alternatives listed by Lloyd is the *hierarchical complexity* concept introduced by Herbert Simon (1962), which is relevant to several disciplines.

Within economics and arguably several other disciplines, the strongest rival to the varieties of computational complexity can be called *dynamic complexity*, although no item called precisely this appears on Lloyd's list, with perhaps the closest being "self-organisation" and "complex adaptive systems". More precisely, Day (1994) defined (dynamic) complexity as arising in nonlinear dynamical systems that due to endogenous causes do not asymptotically approach a point, a non-oscillating growth or decline or two-period oscillation. Thus, such a system will exhibit some form of erratic dynamic behaviour arising endogenously from within itself, not due to an erratic exogenous driver. Rosser (1999) adopted this definition for his "broad-tent" complexity that is clearly dynamic.

Within this broad-tent form of dynamic complexity, one can observe four well-known subcategories that were identified as being "the four Cs" of *chaoplexity*, according to Horgan (1997, Chapter 11). These were cybernetics, catastrophe theory, chaos and "small-tent" or agent-based or Santa Fe complexity. Horgan argued that these have all constituted a succession of intellectual fads or bubbles, beginning in the 1950s with Norbert Wiener's cybernetics and moving on successively, with agent-based complexity simply the latest in this succession that was overhyped and then discarded after being shown to be overhyped. However, an alternative view is that these represent an accumulating development of knowledge regarding the nature of nonlinear dynamics and that students of this development should take Horgan's ridicule and turn it on its head, much as such art movements as Impressionism were originally named critically, only to have them become widely admired. Let the "four Cs" be the focus of a successful ongoing intellectual system.

Norbert Wiener (1948) introduced *cybernetics*, which strongly emphasises the role of positive and negative feedback mechanisms. Wiener emphasised issues of control, which made cybernetics popular in the Soviet Union and other socialist planned economies long after it had faded from attention in Western economies. While Wiener did not emphasise nonlinear dynamics so much, certain close relatives of cybernetics, *general systems theory* (von Bertalanffy 1950) and *systems dynamics* (Forrester 1961), did so more clearly, with Forrester particularly emphasising how nonlinearities in dynamical systems can lead to surprising and "counterintuitive" results. However, the discrediting of cybernetics and its relatives may have come most strongly from the failure of the limits to growth models based on systems dynamics when they forecast disasters that did not happen (Meadows et al. 1972). Much of the criticism of the cybernetics approaches, which emphasised computer simulations, focused on the excessive levels of aggregation in the models, something that more recent agent-based models are not guilty of, with these arguably representing a new improved revival of the older cybernetics tradition.

*Catastrophe theory* developed out of broader bifurcation theory, and to the extent that formal catastrophe theory may not be applicable in many situations due to the strong assumptions required for it to be applied, broader bifurcation theory can analyse the same fundamental phenomenon, that of smoothly changing underlying control variables having critical values where values of endogenous state variables may change discontinuously. Formal catastrophe theory, based on Thom (1972), provides generic forms for these bifurcation conditions on equilibrium manifolds according to the number of control and state variables, and Zeeman (1974) provided the first application in economics to the analysis of stock market crashes using the cusp catastrophe model that has two control variables and one state variable. Empirical analysis of such models requires the use of multimodal statistical methods (Guastello 2009). A backlash developed as critics argued that the theory was applied to situations that did not fulfil the strict assumptions necessary for the application, but Rosser (2007) has argued that this backlash was overdone, with many avoiding its use who should not do so.

While *chaos theory* can be traced back at least to Poincaré (1890), it became prominent after the identification of sensitive dependence on initial conditions, aka "the butterfly effect", by the climatologist, Edward Lorenz (1963), probably the most important idea associated with the phenomenon. Applications in economics followed after an important paper by May (1976) that initially suggested some of them. Debates over empirical measurements and problems associated with forecasting have reduced some of the earlier enthusiasm for chaos theory in economics, which probably peaked during the 1980s. However, the fundamental insights derived from it continue to influence economic thinking as well as that in other disciplines.

Coming on the heels of the popularity of chaos theory would be agent-based (or "small-tent") dynamic complexity, strongly associated with the Santa Fe Institute. However, its origin is generally traced to the urban segregation model of Schelling (1971), who used a go board rather than a computer to work out the dynamics of a city starting out racially integrated and then segregating with only the slightest of incentives through the nearest neighbour effects. Such systems are famous for exhibiting self-organisation and do not generally converge on any equilibrium, also showing cross-cutting hierarchical interactions and ongoing evolutionary change (Arthur et al. 1997). Substantial active research in economics using such models is ongoing.

We note that these are only a small subset of the full array of complex dynamics that nonlinear systems can exhibit. Others include *nonchaotic strange attractors* (Lorenz 1982), *fractal basin boundaries* (Abraham et al. 1997), *flare attractors* (Hartmann and Rössler 1998; Rosser et al. 2003a) and more.

A central point that should be clear is that the presence of such dynamic complexities in economic systems greatly complicates the problem for economic agents of forming rational expectations regarding the future path of such systems. In their presence, it becomes highly unlikely that agents can fulfil the conventional assumption of full information and complete rationality in their decision-making. Complexity is a foundational source of bounded rationality.

## 4.3 Veblen, Simon and Complexity

At the time when Thorstein Veblen was writing his most important works when the nineteenth century was turning into the twentieth, there was no clear or general awareness of what we now call *complexity*, even as many ideas we now associate with it had been floating around in various disciplines for many years, especially in mathematics and even somewhat in economics (Rosser 2009). We have no reason to believe that Veblen was particularly aware of these strands, although evolution itself is now viewed as a complexity process par excellence (Hodgson and Knutsen (2006), which Veblen would strongly advocate. In any case, central to Veblen's approach to economic evolution was his invocation of the idea of *cumulative causation*, which he was the first to introduce. What is important is to note that cumulative causation can lead to such dynamic complexities through increasing returns, which Veblen recognised as present in industrial technology.

Computational complexity has long been argued that there are levels of this, with linear programmes that can be solved in a short time often labelled as not complex and ones that involve polynomial or non-polynomial time being viewed as higher levels of complexity. Highest of all are problems or programmes that cannot be solved that suffer from a halting problem that leaves them running for an infinite time, the highest order of computational complexity. The greater precision involved in these computational complexity definitions has led some economists to advocate focusing on this sort of complexity in economics (Velupillai 2000; Axtell 2005), with Simon's work a great inspiration for this. Like Simon later, Veblen was concerned with information systems, and we shall argue that computational complexity may be involved in behavioural institutional evolution.

Finally among the broader categories of complexity that are relevant to economics, we have hierarchical complexity, initially formulated by Herbert Simon (1962). This is concerned with decomposability of systems and also ultimately of the emergence of new levels of hierarchy in systems whether through evolutionary or physical processes. We shall argue that indeed this is centrally tied to institutional evolution as emergence of higher levels of institutions (and also organisations) is a key part of the dynamics of institutional evolution.

## 4.4 Herbert Simon and Bounded Rationality

The late Herbert A. Simon is widely considered to be the father of *modern behavioural economics*, at least it was his work to which this phrase was first applied. He was also an early theorist of complexity economics, if not the father per se, and also was one of the founders of the study of artificial intelligence in computer science. Indeed, he was a polymath who published well over 900 academic papers in numerous disciplines, and while he won the Nobel Prize in economics in 1978 for his development of the concept of *bounded rationality*, his PhD was in political science, and he was never in a department of economics. We must use the term "modern" before "behavioural economics" because quite a few earlier economists can be seen as focusing on actual human behaviour while assuming that people do not behave fully in what we would now call an "economically rational" manner (Smith 1759; Veblen 1899).

We must at this point be clear that by "behavioural economics" we are not assuming a view similar to that of "behavioural psychology" of the sort advocated or practised by Pavlov or B.F. Skinner (1938). The latter does not view studying what is in peoples' minds or consciousness as of any use or interest. All that matters is how they behave, particularly how they respond to repeated stimuli in their behaviour. This is more akin to standard neoclassical economics, which also purports to study how people behave with little interest in what is going on inside their heads. The main difference between these two is that conventional economics makes a strong assumption about what is going on inside peoples' heads: that they are rationally maximising individual utility functions derived from their preferences using full information. In contrast, behavioural economics does not assume that people are fully rational and particularly does not assume that they are fully informed. What is going on inside their heads is important, and such subjects as *happiness economics* (Easterlin 1974) are legitimate topics for behavioural economics.

In any case, from the beginning of his research with his path-breaking PhD dissertation that came out as a book in 1947, *Administrative Behavior* and on through important articles and books in the 1950s (Simon 1955, 1957), Simon saw people as being limited in both their knowledge of facts as well as in their ability to compute and solve the difficult problems associated with calculating optimal solutions to problems. They face unavoidable limits to their ability to make fully rational decisions. Thus, people live in a world of *bounded rationality*, and it was this realisation that led him into the study of artificial intelligence in computer science as part of his study of how people think in such a world (Simon 1969).

This led Simon to the concept of *satisficing*. People set targets that they seek to achieve and then do not pursue further efforts to improve situations once these targets have been reached, if they are. Thus a firm will not maximise profits, but its managers will seek to achieve an acceptable level of profits that will keep owners sufficiently happy. This idea of satisficing became the central key to the behavioural study of the firm (Cyert and March 1963) and entered into the management literature, where it probably became more influential than it was in economics, for quite a long time.

Some economists, notably Stigler (1961), have taken Simon's position and argued that he is actually a supporter of full economic rationality but only adding another matter to be optimised, namely, minimising the costs of information. People are still optimising but take account of the costs of information. However, Stigler's argument faces an unavoidable and ineluctable problem: people do not and cannot know what the full costs of information are. In this regard, they face a potential problem of infinite regress (Conlisk 1996). In order to learn the costs of information, they must determine how much time they should spend in this process of learning; they must learn what the costs of information are. This then leads to the next higher-order problem of learning what the costs of information are, and there is no end to this regress in principle. In the end they must use the sorts of *heuristic* (or "rule of thumb") devices that Simon proposes that people facing bounded rationality must use in order to answer the question. Full rationality is impossible, and the ubiquity of complexity is a central reason why this is the case.

Simon (1976) distinguishes *substantive rationality* from *procedural rationality*. The former is the sort of rationality traditionally assumed by most economists in which people are able to achieve full optimisation in their decision-making. The latter involves them selecting procedures or methods by which they can "do their best" in a world in which such full optimisation is impossible, the heuristics by which they manage in a world of bounded rationality. In this regard, it is not the case that Simon views people as being outright irrational or crazy. They have interests, and they generally know what those are and they pursue them. However, they are unavoidably bounded in their ability to do so fully, so they must adopt various essentially ad hoc methods to achieve their satisficing goals.

Among these heuristics that Simon advocated for achieving procedural rationality were trial and error, imitation, following authority, unmotivated search and following hunches. Pingle and Day (1996) used experiments to study the relative effectiveness of each of these, none of which clearly can achieve fully optimal outcomes. Their conclusion was that each of these can be useful for improving decision-making; however, none of them are clearly superior to the others. It is advisable for agents to use a variety of these heuristics.

## 4.5 Imitation and the Instability of Markets

While this list of procedures that can support a boundedly rational pursuit of procedural rationality, a point not clearly made is that excessive focus on one of these rather than others can lead to problems. Clearly following authority can lead to problems when the authority is flawed, as many unfortunate examples in history have shown. Any of these can lead to problems if too intensively followed, but one that has particularly played an unfortunate role in markets is imitation, even though it is a widely used method by many people with a long history of being evolutionarily successful. The problem is particularly acute in asset markets, where imitation can lead to speculative bubbles that destabilise markets and can lead to much broader problems in the economy, as the crisis of 2008 manifestly shows.

A long literature (MacKay 1852; Baumol 1957; Zeeman 1974; Rosser 1997) has recognised that while agents focusing on long-term fundamental values of assets tend to stabilise markets by selling them when their prices exceed these fundamentals and buying when they are below those, agents who chase trends can destabilise markets by buying when prices are rising, thus causing them to rise more and vice versa. When a rising price trend appears, trend chasers will do better in returns than fundamentalists, and imitation of those doing well will lead agents who might have followed stabilising fundamentalist strategies to follow destabilising trend chasing strategies, which will tend to push the price further up. And when a bubble finally peaks out and starts to fall, trend chasers can then push the price down more rapidly as they follow each other in a selling panic.

That such a tendency to engage in trend chasing speculation is deeply rooted in the human psyche was initially established by Smith et al. (1988), with many subsequent studies supporting this observation. Even in situations with a finite time horizon and a clearly identified payment that establishes the fundamental value of the asset being traded, in experimental markets, it has been repeatedly shown that bubbles will appear even in these simplified and clear-cut cases. People have a strong tendency to speculate and to follow each other into such destabilising speculation through imitation. Procedures that can support procedural rationality in a world of bounded rationality can lead to bad outcomes if pursued too vigorously.

We note that such patterns regularly take three different patterns. One is for price to rise to a peak and then to fall sharply after hitting the peak. Another is for price to rise to a peak and then decline in a more gradual way in a reasonably symmetric manner. Finally, we see bubbles rising to a peak, then declining gradually for awhile, finally collapsing in a panic-driven crash. Kindleberger's classic *Manias, Panics, and Crashes* (2001) shows in its Appendix B that of 47 historical speculative bubbles, each of the first two has five examples, while the remainder, the vast majority, follow the final pattern, which requires heterogeneous agents who are not fully rational for it to occur (Rosser 1997). This shows that complexity is deeply involved in most speculative bubbles.

All of these three patterns described above happened during the run-up and beginning of the Great Recession (Rosser et al. 2012). The first is for oil, which peaked at $147 per barrel in July 2008, the highest nominal price ever observed, and then crashed hard to barely over $30 per barrel in the following November. It seems that commodities are more likely to follow this pattern than other assets (Ahmed et al. 2014).

The second pattern was followed by the housing bubble, which peaked in mid-2006. This sort of pattern historically is often seen with real estate market bubbles. The more gradual decline than in the other patterns, nearly symmetric with the increase, reflects certain behavioural phenomena. People identify very personally and intensely with their homes and as a result tend not to easily accept that their home has declined in value when they try to sell it during a downturn. As a result, they have a tendency to offer prices that are too high and then refuse to lower their prices readily when they fail to sell. The upshot is a more dramatic decline in volume of sales on the downswing compared to the other patterns as people hang on and refuse to lower prices.

The third case was followed by the US stock market as exhibited by the Dow Jones average, which peaked in October 2007, only then to crash in September 2008. Such patterns seem to be more common in markets for financial assets. Such patterns show heterogeneity of agents with different patterns of imitation, a smarter (or luckier) group that gets out earlier at the peak, followed by a less smart (or less lucky) group that hangs on hoping the price will return to rising, only to panic later en masse for whatever reason.

## 4.6 The Discontinuity Debate in Evolutionary Theory

It was Leibniz who initially coined the phrase *natura non facit saltum* or "nature does not take a leap". It would be picked up by Darwin himself who repeated it and applied it to his theory of natural selection, and Marshall would follow Darwin in applying to economics, repeating it in the Prefaces of all eight editions of his *Principles of Economics*. For Darwin (1859, pp. 166–167):

> Natura non facit saltum . . . Why should not Nature take a leap from structure to structure? On the theory of natural selection we can clearly understand why she should not: for natural selection can only act by taking advantage of slight successive variations; she can never take a leap, but must advance by the shortest and slowest steps.

This was a strong statement for Darwin to make given that he did not understand the underpinnings of how the process of mutation through changes in genes worked, but indeed many evolutionary theorists since Darwin have been impressed by the idea that only minor changes in genes can occur at a time for species to be viable and survive and reproduce, thus setting up at least most evolutionary processes to be slow and gradual as asserted by Darwin. However, until the understanding of genetics was fully integrated into Darwinian theory with the neo-Darwinian synthesis in the 1930s, there was more of an opening for more noticeable discontinuous change in the Lamarckian perspective that allowed for the inheritance of acquired characteristics and thus more rapid evolutionary change.

After the 1930s, the more dramatic reassertion of the possibility for rapid change in the form of *punctuated equilibrium* would come with Eldredge and Gould (1972), whose arguments remain controversial among evolutionary biologists. However, the groundwork for their arguments was laid in the development of the neo-Darwinian synthesis itself during the 1930s, even if it was not clearly recognised at the time. A central part of the neo-Darwinian synthesis, especially as formulated by Fisher (1930), involved focusing on the gene, with natural selection operating at the level of the gene, which contrasted with theories that saw natural selection operating at higher levels on wholes. Changes at the level of a gene must be fairly small to be viable, but a method of studying this through fitness landscapes as introduced by Sewall Wright (1932) opened the door for a broader perspective, one that can be carried over to the study of institutional evolution (Mueller 2014).

A piece of groundwork always there regarding Wright's fitness landscape framework that opened the door to such saltationalist discontinuities or punctuations was that Wright from the beginning allowed for multiple local optima or equilibria within those landscapes. While he himself did not see a dramatic discontinuities happening at the genetic level, he recognised that rapid environmental changes could shift the landscapes so that a former peak could fairly quickly become a valley with the nearest peak reachable by a gradient that might be some distance away, which would imply some rapid evolution, if not necessarily discontinuous in genotype and phenotype.

Regarding the application of these ideas to economic evolution and more specifically institutional evolution, it is generally accepted that while Marshall may have agreed with Leibniz and Darwin that *natura non facit saltum*, Veblen tended to accept the idea that institutional evolution could be discontinuous or at least that institutional equilibria were not stable and could change suddenly. Thus he declared (Veblen 1919, pp. 242–243):

> Not only is the individual's conduct hedged about and direct by his habitual relations to his fellows in the group, but these relations, being of an institutional character, vary as the institutional scene varies. The wants and desires, the end and the aim, the ways and the means, the amplitude and drift of the individual's conduct are functions of an institutional variable that is of a highly complex and unstable character.

Curiously while Schumpeter strongly supported the idea of discontinuous technological change and used the language of evolution in the context of economic development, he rejected the use of biological analogies in such discussions, declaring that (Schumpeter 1954, p. 789) "no appeal to biology would be of the slightest use". He dismissed selective mechanisms whether of a Darwinian or Lamarckian sort, using the word "evolution" in a simply developmental way (Hodgson 1993).

While Wright did not spell it out, a key to the existence of multiple local equilibria in his fitness landscapes is the presence of some sort of increasing returns. This brings in Arthur's (1994) emphasis on increasing returns and its link to the existence of multiple equilibria and dynamic complexity, which carries over to institutional evolution. Minniti (1995) used a variation of the Arthur et al. (1987) urn model to show how low- and high-crime equilibria can arise in a society, with social interactions providing positive feedbacks the key to such an outcome, with potential discontinuities arising as the amount of crime can shift very suddenly from one state to another. Rosser et al. (2003b) applied this model to informal economies in transition economies, with there also being multiple equilibria as seen by large differences in this variable among the transition economies of Eastern Europe, with the degree of inequality of income feeding into the determination of where an economy ends up.

## 4.7 Institutions, Organisations and the Locus of Economic Evolution

If economies are evolutionary systems, then the question of what is the locus of that evolution is important. Hodgson and Knudsen (2006) argue that there are three crucial characteristics involved in truly Darwinian evolution: variability, natural selection and inheritance. For something to qualify as a locus of evolution, it must exhibit all three of these. In biological evolution, the gene certainly fulfils all of these: mutation provides random variability, natural selection determines whether an organism containing a gene will survive or not, and genes pass from one organism to another through reproduction if the organism is able to survive and attract mates to effectuate this. Critics of evolutionary economics argue that there is no definitive unit or element in economies that fulfil all three of these, even if many fulfil some of them.

Given the long advocacy by institutionalist followers of Veblen for making economics an evolutionary science, these issues have been central to debates within this area. A focus on organisations has long attracted attention, with this arguably more important to Commons than to Veblen. For Commons, directed or artificial selection was more important than strictly random natural selection, and he noted that Darwin himself spent much time discussing both random natural selection and artificial breeding (Commons 1934, p. 657; Vanberg 1997). Commons saw organisations as being subject to direction and thus appropriate objects for this sort of directed evolution, which had a goal of general human improvement. In his argument for evolution as the fundamental force in microeconomics, Armen Alchian (1950) emphasised the competition of firms, with the survival of the fittest involving which firms can come closest to maximising profits, even if they do not know precisely how they are doing so, with firms clearly the locus of evolution.

A criticism of the idea of firms, or more generally organisations, serving as the key locus of evolution in economics is that while they are subject to random variability as they experience shocks from the system, and natural selection clearly operates in their competition with each other, with unprofitable firms failing to survive, the missing piece is that of inheritance. Firms and organisations do not essentially reproduce themselves. All they do is survive, although they may change while doing so. These changes may reflect these evolutionary forces of natural selection, but the inheritance element of their doing so must be operating at some lower level than that of the firm or organisation itself.

The leading alternative for serving as the evolutionary meme, an idea due initially to Dawkins (1976), is habits or practices within an organisation. While they were not driven to this argument by trying to fit new institutional economics into an evolutionary framework per se, this is how North (1990) and Williamson (2000) define institutions. They are habits or practices, not organisations. This is also what Nelson and Winter (1982) came to in their search for the key to evolutionary economics, although they labelled these memes to be "routines". But prior to any of these and prior to Commons and his emphasis on organisations, Veblen identified habits, including habits of thought, as the central locus of evolution in economic institutions, declaring (Veblen 1899, pp. 190–191):

> The situation of today shapes the institutions of tomorrow through a selective, coercive process, by acting upon men's habitual view of things, and so altering or fortifying a point of view or a mental attitude handed down from the past.

Given that as he put it the individual's conduct is "hedged about by his habitual relations with his fellows in the group", with these relations of an "institutional character", it is habits and habitual relations that are at the foundation of the evolution of institutions, even if he sees these institutions as being higher-order social structures. It is the habits that are at the foundations, and habits can change, leading to new habits that may be inherited by the individuals and organisations using them.

## 4.8 Emergence and Multilevel Evolution

Among the ideas most strongly associated with complexity are that of *emergence*, that a higher-order entity arises out of a lower-level one that is not simply the sum of the parts of the lower-level one and that the emergent entity is something qualitatively different. While the idea of a whole being greater than the sum of its parts has been around for a long time, a scientific formalisation of it is probably due to J.S. Mill (1843) in his discussions of logic in which he characterised situations where something qualitatively different from its parts appears as representing *heteropathic laws*. His original examples involved chemistry such has how salt appears when one combines sodium with chlorine, with salt not being at all like either of them separately. Lewes (1875) applied the term *emergence* to such phenomena. This led to the "British emergentist" school of thought that especially in the 1920s (Morgan 1923) would apply this concept to evolution, in particular to such problems as how multicellular organisms arose out of unicellular ones. It would be applied to how larger social groups would organise themselves to act together out of previously smaller separate groups, an idea clearly important in the evolution of institutions.

In biological evolutionary theory, this view fell out of favour in the 1930s with the rise of the neo-Darwinian synthesis, which put the focus on the gene as the locus of evolution, the meme, as Dawkins (1976) labelled it. The idea that natural selection occurred at levels above the gene, at the level of "wholes" or groups, was specifically rejected (Williams 1966). The obvious counter to this in biological evolution involves the social insects (Wilson 2012), in which individuals are subordinated to the good of the colony, with the colony becoming the vehicle of evolution. Most attribute the mathematical understanding of how this can arise to the work of Price (1970) and Hamilton (1972). However, in fact, the original formalisation of this understanding in terms of within-group versus between-group selection was due to Crow (1955).

Let $B$w be the within-group genic regression on the fitness value of the trait as defined by Wright (1951); $B$b be the between-group genic regression to the fitness value; $V$w be the variance among individuals within a group, and $V$b be the variance among means across groups. For an *altruistic gene*, one would expect $B$w to be negative (that the behaviour within the group damages the individual), while $B$b would be positive (the behaviour of the individual helps the group). From this a sufficient condition for the altruistic gene to increase in frequency is given by

$$B\mathrm{b}/\left(-B\mathrm{w}\right) > V\mathrm{w}/V\mathrm{b}. \qquad (4.1)$$

Within biology there it has been widely argued that this condition rarely holds. However, it has also been recognised that it appears to hold for the social insects, and as Wilson (2012) argues, this implies that even though only a minority of species show this characteristic, they end up constituting a huge portion of the animal biomass on the earth (especially if one includes human beings in that calculation).

Indeed, this formulation can be carried over to humans to resolve the problem of cooperation versus cheating within a prisoner's dilemma game theoretic context (Henrich 2004). The specific problem for humans becomes one of recognising who is a cooperator and who is not within social groups, with successfully doing so being the condition for cooperation and a higher-level coordination to come about. Considering in detail how such cooperation can arise in numerous contexts for dealing with common property resources was the central focus of the work of Ostrom (1990). This can more generally be viewed as a condition for the emergence of higher-level institutions out of lower-level ones, with these ideas further pursued by Sethi and Somanathan (1996) and Rosser and Rosser (2006).

Somewhat parallel to this is a formulation of emergence in biological evolution due to Eigen and Schuster (1979) known as the *hypercycle*, which involves information preservation and transmission, tying this more to computational forms of complexity. They define a "threshold of information content" which if exceeded for a system will lead to a degeneration of information due to an "error catastrophe". Let $V_m$ be the number of symbols, $\sigma_m > 1$ be the degree of selective advantage superiority of the "master copy" and $q_m$ be the quality of symbol copying. The threshold is then given by

$$V_m < \ln \sigma_m / (1 - q_m) . \tag{4.2}$$

This can be seen as linked to *self-organisation* as initially formulated by Turing (1952) in the form of *morphogenesis*. When such morphogenesis involves emergence at a higher level this become *hypercyclic morphogenesis* (Rosser 1991, Chap. 6), with Radzicki (1990) applying such arguments to the question of the formation of institutions out of underlying chaotic dynamics. Within evolution the emergence of higher hierarchical levels was also the central focus of Simon (1962).

Another strand of emergent evolutionary processes is associated with the neo-Schumpeterian view strongly associated with Nelson and Winter (1982) and their study of what are the key memes in evolutionary economics. They are known for their advocacy of the idea that routines are the key meme that is the locus of such evolutionary developments. Nelson and Winter themselves were less focused on this matter of emergent higher orders that become the locus of evolution, but some of their followers have pursued such ideas. In particular has been the development of the idea of *mesoeconomics* by Dopfer et al. (2004), originally due to Ng (1986). This is a level of economics that is intermediate in level between the microeconomics of the firm where the Nelson and Winter processes presumably mostly operate and the fully aggregated level of macroeconomics. The mesoeconomic level is more at the industry or sector level where a meme may have diffused across firms within a sector or even a set of related sectors. Such developments can lead to this being the most important part of the economy from the standpoint of growth and evolutionary development.

In terms of institutional evolution operating at higher levels of emergent structures, a possibly surprising supporter of this view is an Austrian economist, Friedrich Hayek. This would appear to be at least partly associated with his open embrace of complexity (Hayek 1967) and especially in connection with this the concept of emergence, harking openly back to the British emergentists of the 1920s. His opening to this strand of thought came from his early work in psychology that culminated in his *The Sensory Order* (Hayek 1952). In this work he specifically saw human consciousness as an emergent property arising from the nervous system and the brain (Lewis 2012). Crucial in his formulating this was the influence of systems theory as developed by Ludwig von Bertalanffy (1950), who in turn was influenced by the *cybernetics* of Norbert Wiener (1948), thought by many to be another early form of dynamic complexity. Lying more deeply behind cybernetics was the development of the "universal system of organisations" or *tektology* of A.A. Bogdanov (1925–1929), arguably a form of evolutionary institutional economics stressing emergence.

Indeed, Hayek (1988) in his final work, *The Fatal Conceit*, applied his view of emergent complexity involving evolution in a higher-order way, with such emergent institutional structures competing with each other and evolving as wholes competing with each other and surviving or not through a process of systemic natural selection. Some would argue that this embrace of natural selection operating at the level of higher-order societal wholes constituted a contradiction with the methodological individualism of the Austrian School, although in fact in this he harked back to evolutionary ideas of the founder of that school, Carl Menger (1923), that like Hayek he developed late in his career.

## 4.9 Hierarchical Complexity and the Question of Emergence

While we can see Herbert Simon's discovery of bounded rationality as an indirect claim to being a "father of complexity", his most direct claim, recognised by Seth Lloyd in his famous list, is his 1962 paper to the American Philosophical Society on *The Architecture of Complexity*. In this transdisciplinary essay, he deals with everything from organisational hierarchies through evolutionary ones to those involving "chemico-physical systems". He is much concerned with the problem of the decomposability of higher-order systems into lower-level ones, noting that productions ones, such as for watchmaking, as well as organisational ones, function better when such decomposability is present, which depends on the stability and functionality of the lower-level systems.

However, he recognises that many such systems involve *near decomposability*, perhaps a hierarchical complexity equivalent of bounded rationality. In most of them, there are interactions between the subsystems, with the broader evolution of the system depending on aggregated phenomena. Simon provides the example of a building with many rooms. Temperature in one room can change that in another, even though their temperatures may fail to converge. But the overall temperatures that are involved in these interactions are determined by the aggregate temperature of the entire building.

Simon also deals with what many consider to be the most fundamental issue involving complexity, namely, that of emergence. His most serious discussion of the emergence of higher levels of hierarchical structure out of lower levels involves biological evolution, where these issues have long been most intensively discussed. He argues that how these higher levels emerged has not reflected teleological processes but strictly random processes. He also argues that even in closed systems, there need be no change in entropy in the aggregate when subsystems emerge within that system. But he also recognises that organisms are energetically open systems, so that "there is no way to deduce the direction, much less the rate, of evolution from classical thermodynamic considerations" (Simon 1962, p. 8). However, it is the development of stable intermediate forms that is the key for the emergence of yet higher forms.

Simon does not cite this older literature, but this issue was central to the British "emergentist" literature that came out of the nineteenth century to become the dominant discourse in the 1920s regarding the broader story of biological evolution, all embedded within a broader vision fitting this within the emergence of physical and chemical systems from particles through molecules to such higher levels above biological evolution in terms of human consciousness, social systems, and yet higher systems. Simon dealt with this multiplicity of processes without drawing their interconnection as tightly as did these earlier figures. In the 1930s with the *neo-Darwinian synthesis* (Fisher 1930; Wright 1932; Haldane 1932), the emphasis returned to near-continuous Darwinian process of gradual changes arising the level of probabilistic changes arising from mutations at the gene level, with the gene the ultimate focus of natural selection (Dawkins 1976).

While Simon avoided dealing with this issue of emergence in biological evolution in 1962, when the reductionist neo-Darwinian synthesis was at the highest level of its influence, soon the emergence view would itself re-emerge, based on multilevel evolutionary process (Crow 1955; Price 1970; Hamilton 1972). This would further develop with the study of nonlinear dynamics and complexity in such systems, with such figures as Stuart Kauffman (1993) and James Crutchfield (1994, 2003), who draw on computational models for their depictions of *self-organisation* in biological evolutionary systems.

This view remains questioned by many evolutionists (Gould 2002). While the tradition going through catastrophe theory from D'Arcy Thompson (1917) has long argued for form arising deep structures in organic evolution, critics have argued that such self-organising processes are ultimately teleological ones that replicate old pre-evolutionary theological perspectives such as Paley's (1802) in which all things are in their place as they should be due to divine will. Others have criticised that such process lacks invariance principles (McCauley 2005). Others coming from a more computational from such processes (Moore 1990). There is no easy resolution of this debate, and even those advocating the importance of emergent self-organisation recognise the role of natural selection. Thus, Kauffman (1993, p. 644) has stated, "Evolution is not just 'chance caught on a wing.' It is not just a tinkering of the ad hoc, of bricolage, of contraption. It is emergent order honored and honed by selection".

While the mechanisms are not the same, the problems of emergent self-organisation apply as well to socio-economic systems. Simon's focus tended to be on organisations and their hierarchies. While he may well have sided with the more traditional neo-Darwinian synthesisers when it came to emergence of higher-order structures in biological evolution, the role of human consciousness within human socio-economic systems means that the rules are different there, and the formation of higher-order structures can become a matter of conscious will and planning, not mere randomness.

## 4.10 Epistemic Issues and Institutional Hierarchical Emergence

So far the form of complexity that has been the main focus of our discussion of institutional evolution has been that of dynamic complexity. However, especially as one considers the issue of the emergence of hierarchies in institutions and organisations out of dynamic evolutionary processes, other forms of complexity become relevant. One is clearly that of hierarchical complexity as discussed by Simon (1962), but Simon was also concerned with problems of computational complexity arising from his study of bounded rationality, which led him to become an important figure in the development of artificial intelligence in computer science (Rosser and Rosser 2015). Part of the problem in this hierarchical structure of institutions is that there is also a hierarchy of information, but with the possibility of asymmetries of information arising between the different levels of the hierarchy.

The issue of hierarchy also appears within computational complexity in the appearance of different levels of computational complexity, with the apparent gap between those that can be solved in polynomial time versus those in non-polynomial time such as exponential time being a central gap. Highest of all are those that cannot be solved due to a halting problem, which may arise from paradoxical self-referencing (Koppl and Rosser 2002; Rosser 2004). Such issues may imply some sort of mechanism to move to a higher level beyond where the problem is posed in order for some sort of solution to be achieved. Whereas most of this kind of discussion is done in terms of the functioning of Turing machines, when this is placed in the world of institutions, problems of the inability of memes to solve problems may well instigate the emergence of a higher-level institution to resolve or manage the question at hand.

Indeed in the world of market institutions, we have seen the development and emergence of higher-order systems initially appearing to resolve problems arising at lower levels. Spot markets led to futures markets, which in turn led to options markets and to ever higher-order derivatives markets, with the higher orders dominating the lower orders. This sort of process has been argued for by Mirowski (2007) in his theory of *markomata*. In this view, markets are fundamentally information systems, and institutions are ways to manage these systems. They develop these hierarchical structures, but these structures in turn compete with each other in a manner that involves natural selection between them. They change by random mutations, and those that survive pass on their structures to future systems. This theory of markomata becomes a theory of higher-order institutional evolution fundamentally based on information systems and their computationally complex hierarchies. In this concept, the ideas of dynamic and computational complexity come together as a foundation for a more advanced theory of behavioural institutional evolution.

## 4.11 Conclusions

Central to understanding the complex behavioural institutional evolution is understanding the ideas of the founder of evolutionary institutional economics, Thorstein Veblen, and the founder of behavioural economics and hierarchical complexity, Herbert Simon. For Veblen his formulation of the concept of cumulative causation was important, later taken up more prominently by such figures as Young, Myrdal and Kaldor. This links to modern dynamic complexity theory through increasing returns, which leads to multiple equilibria and complex disequilibrium dynamics. Veblen's vision was thoroughly Darwinian in that he did not propose any directed teleological evolution in the way that favoured more by fellow institutional economist, John R. Commons.

Arising from Veblen's ideas of institutional evolution is also the possibility of complex emergence of higher orders of institutions based on cooperation, linking to ideas of Herbert Simon, as well as drawing on the theory of multilevel evolution developed by biologists such as Crow, Hamilton and Price. The existence and competition between hierarchical economic institutions also implies problems of computational complexity, again with no definite direction or outcome a likely result. Evolution itself is a profoundly complex process, and so it is when it happens to economic institutions.

## References

R.H. Abraham, L. Gardini, C. Mira, *Chaos in Discrete Dynamical Systems: A Visual Introduction in 2 Dimensions* (Springer, New York, 1997)

E. Ahmed, J.B. Rosser Jr., J.Y. Uppal, Are there nonlinear speculative bubbles in commodities prices? J. Post Keyne. Econ. **36**, 415–438 (2014)

A.A. Alchian, Uncertainty, evolution, and economic theory. J. Polit. Econ. **58**, 211–222 (1950)

W.B. Arthur, *Increasing Returns and Path Dependence in the Economy* (University of Michigan Press, Ann Arbor, 1994)

W.B. Arthur, S.N. Durlauf, D.A. Lane, Introduction, in *The Economy as an Evolving Complex System II*, ed. by W.B. Arthur, S.N. Durlauf, D.A. Lane (Addison-Wesley, Reading, 1997), pp. 1–14

W.B. Arthur, Y. Ermoliev, Y. Kaniovski, Adaptive growth processes modeled by urn schemes. Cybernetics **23**, 776–789 (1987)

R.L. Axtell, The complexity of exchange. Econ. J. **115**, F193–F210 (2005)

W.J. Baumol, Speculation, profitability, and stability. Rev. Econ. Stat. **39**, 263–271 (1957)

A.A. Bogdanov, *Tektologia: vseobshchaya organizatsionnay nauka*, 3rd edn, 3 vols (Kniga, Leningrad-Moscow, 1925–1929)

J.R. Commons, *The Legal Foundations of Capitalism* (Macmillan, New York, 1924)

J.R. Commons, *Institutional Economics* (Macmillan, New York, 1934)

J. Conlisk, Why bounded rationality? J. Econ. Lit. **34**, 1–64 (1996)

J.F. Crow, General theory of population genetics: Synthesis. Cold Spring Harb. Symp. Quant. Biol. **20**, 54–59 (1955)

J.P. Crutchfield, The calculi of emergence: Computation, dynamics and induction. Physica D **75**, 11–54 (1994)

J.P. Crutchfield, When Evolution Is Revolution – Origins of Innovation, in *Evolutionary Dynamics: Exploring the Interplay of Selection, Accident, Neutrality, and Function*, ed. by J.P. Crutchfield, P. Schuster (Oxford University Press, Oxford, 2003), pp. 101–133

R.M. Cyert, J.G. March, *A Behavioral Theory of the Firm* (Prentice-Hall, Englewood Cliffs, 1963)

C. Darwin, *On the Origin of Species by Means of Natural Selection or the Preservation of Favored Races in the Struggle for Life* (John Murray, London, 1859)

R. Dawkins, *The Selfish Gene* (Oxford University Press, Oxford, 1976)

R.H. Day, *Complex Economic Dynamics: An Introduction to Dynamical Systems and Market Mechanisms: Volume I* (MIT Press, Cambridge, 1994)

K. Dopfer, J. Foster, J. Potts, Micro-meso-macro. J. Evol. Econ. **14**, 263–279 (2004)

R.A. Easterlin, Does Economic Growth Improve the Human Lot? Some Empirical Evidence, in *Nations and Households in Economic Growth: Essays in Honor of Moses Abramovitz*, ed. by P.A. David, M.W. Reder, (Academic, New York, 1974), pp. 89–125

M. Eigen, P. Schuster, *The Hypercycle: A Principle of Natural Self-Organization* (Springer, Berlin)

N.J. Eldredge, S.J. Gould, Punctuated equilibria, in *Models of Paleobiology*, ed. by D.J.M. Schopf (Freeman, Cooper, San Francisco, 1972), pp. 82–115

R.A. Fisher, *The Genetical Theory of Natural Selection* (Oxford University Press, Oxford, 1930)

J.W. Forrester, *Industrial Dynamics* (MIT Press, Cambridge, 1961)

S.J. Gould, *The Structure of Evolutionary Theory* (Belknap Press of Harvard University Press, Cambridge, 2002)

S.J. Guastello, Discontinuities and Catastrophes with Polynomial Regression, in *Nonlinear Dynamical Systems Analysis for the Behavioral Sciences Using Real Data*, ed. by S.J. Guastello, R.A.M. Gregson (CRC Press, Boca Raton, 2009), pp. 251–280

J.B.S. Haldane, *The Causes of Evolution* (Longmans & Green, London, 1932)

W.D. Hamilton, Altruism and related phenomena, mainly in the social insects. Annu. Rev. Ecol. Syst. **3**, 192–232 (1972)

G.C. Hartmann, O. Rössler, Coupled flare attractors- a discrete prototype for economic modelling. Discret. Dyn. Nat. Soc. **2**, 153–159 (1998)

F.A. Hayek, *The Sensory Order: An Inquiry into the Foundations of Theoretical Psychology* (University of Chicago Press, Chicago, 1952)

F.A. Hayek, The Theory of Complex Phenomena, in *Studies in Philosophy, Politics, and Economics*, (University of Chicago Press, Chicago, 1967), pp. 22–42

F.A. Hayek, *The Fatal Conceit: The Errors of Socialism* (University of Chicago Press, Chicago, 1988)

J. Henrich, Cultural group selection, coevolutionary processes, and large-scale cooperation. J. Econ. Behav. Organ. **53**, 3–35 (2004)

G.M. Hodgson, Institutional economics: Surveying the 'old' and the 'new'. Metroeconomica **44**, 1–28 (1993)

G.M. Hodgson, T. Knutsen, Why we need a generalized Darwinism and why generalized Darwinism is not enough. J. Econ. Behav. Organ. **61**, 1–19 (2006)
J. Horgan, *The end of Science: Facing the Limits of Knowledge in the Twilight of the Scientific age* (Broadway Books, New York, 1997)
G. Israel, The science of complexity: Epistemological implications and perspectives. Sci. Context. **18**, 1–31 (2005)
N. Kaldor, The irrelevance of equilibrium economics. Econ. J. **82**, 1237–1255 (1972)
S.A. Kauffman, *The Origins of Order: Self-Organization and Selection in Evolution* (Oxford University Press, Oxford, 1993)
C.P. Kindleberger, *Manias, Panics, and Crashes: A History of Financial Crises*, 4th edn. (Wiley, New York, 2001)
R. Koppl, J.B. Rosser Jr., All I have to say has already crossed your mind. Metroeconomica **53**, 339–360 (2002)
G.H. Lewes, *Problems of Life and Mind* (Trübner, London, 1875)
P. Lewis, Emergent properties in the work of Friedrich Hayek. J. Econ. Behav. Organ. **82**, 368–375 (2012)
E.N. Lorenz, Deterministic non-periodic flow. J. Atmos. Sci. **20**, 130–141 (1963)
H.-W. Lorenz, Multiple Attractors, Complex Basin Boundaries, and Transient Motion in Deterministic Economic Systems, in *Dynamic Economic Models and Optimal Control*, ed. by G. Feichtinger (Amsterdam, North-Holland, 1982), pp. 311–430
C. MacKay, *Memoirs of Extraordinary Delusions and the Madness of Crowds* (Office of the National Illustrated Library, London, 1852)
R.M. May, Simple mathematical models with very complicated dynamics. Nature **261**, 549–467 (1976)
J.L. McCauley, Making Mathematics Effective in Economics, in *Computability, Complexity and Constructivity in Economics*, ed. by K. V. Velupillai (Blackwell, Victoria, 2005), pp. 51–84
D.H. Meadows, D.L. Meadows, J. Randers, W.W. Behrens III, *The Limits to Growth* (Universe Books, New York, 1972)
C. Menger, *Grundsätze der volkswirtschaftslehre, zweite auflag* (Holden-Pichler-Tempsky, Vienna, 1923)
J.S. Mill, *A System of Logic: Ratiocinative and Inductive* (Longmans Green, London, 1843)
M. Minniti, Membership has its privileges: Old and new mafia organizations. Comp. Econ. Stud. **37**, 31–47 (1995)
P. Mirowski, Markets come to bits: Evolution, computation, and markomata in economic science. J. Econ. Behav. Organ. **63**, 209–242 (2007)
C. Moore, Undecidability and unpredictability in dynamical systems. Phys. Rev. Lett. **64**, 2354–2357 (1990)
C.L. Morgan, *Emergent Evolution* (Williams and Norgate, London, 1923)
B. Mueller *Beliefs and the Persistence of Inefficient Institutions*. University of Brasilia Working Paper (2014), http://extranet.sioe.org/uploads/isnie2015/mueller.pdf
G. Myrdal, *Economic Theory and Underdeveloped Regions* (Methuen, London, 1957)
R.R. Nelson, S.G. Winter, *An Evolutionary Theory of Economic Change* (Harvard University Press, Cambridge, 1982)
Y.-K. Ng, *Mesoeconomics: A Micro-Macro Analysis* (St. Martins, New York, 1986)
D.C. North, *Institutions, Institutional Change, and Economic Progress* (Cambridge University Press, Cambridge, 1990)
E. Ostrom, *Governing the Commons* (Cambridge University Press, Cambridge, 1990)
T. Papageorgiou, I. Katselides, P.G. Michaelides, Schumpeter, commons, and Veblen on institutions. Am. J. Econ. Sociol. **72**, 1232–1254 (2013)
W. Paley, *Natural Theology: Or Evidences of the Existence and Attributes of the Deity, Collected from the Appearances of Nature* (R. Faulder, London, 1802)
M. Pingle, R.H. Day, Modes of economizing behavior: Experimental evidence. J. Econ. Behav. Organ. **29**, 191–209 (1996)

H. Poincaré, Sur les équations de la dynamique et le problème de trois corps. Acta Mathematica **13**, 1–270 (1890)
G. Price, Selection and covariance. Nature **227**, 520–521 (1970)
M.J. Radzicki, Institutional dynamics, deterministic chaos, and self-organizing systems. J. Econ. Issues **24**, 57–102 (1990)
J.B. Rosser Jr., *From Catastrophe to Chaos: A General Theory of Economic Discontinuities* (Kluwer, Boston, 1991)
J.B. Rosser Jr., The dialogue between the economic and ecologic theories of evolution. J. Econ. Behav. Organ. **17**, 195–215 (1992)
J.B. Rosser Jr., Speculations on nonlinear speculative bubbles. Nonlinear Dynamics Psychol. Life Sci. **1**, 275–300 (1997)
J.B. Rosser Jr., On the complexities of complex economic dynamics. J. Econ. Perspect. **13**, 169–192 (1999)
J.B. Rosser Jr., Epistemological implications of economic complexity. Ann. Japan Assoc. Philos. Sci. **16**, 485–499 (2004)
J.B. Rosser Jr., Computational and Dynamic Complexity in Economics, in *Handbook on Research in Complexity*, ed. by J. B. Rosser Jr. (Edward Elgar, Cheltenham, 2009), pp. 22–35
J.B. Rosser Jr., E. Ahmed, G.C. Hartmann, Volatility via social flaring. J. Econ. Behav. Organ. **50**, 77–87 (2003a)
J.B. Rosser, J. Rosser, M.V. Ahmed, E., Multiple unofficial economy equilibria and income distribution dynamics in systemic transition. J. Post Keynes. Econ. **25**, 425–447 (2003b)
J.B. Rosser Jr., M.V. Rosser, Institutional evolution of environmental management under economic growth. J. Econ. Issues **40**, 421–429 (2006)
Rosser, J.B., Jr. (2007). The rise and fall of catastrophe theory applications in economics: Was the baby thrown out with the bathwater? J Econ Dyn Control 31: 3255–3280.
J.B. Rosser Jr., M.V. Rosser, M. Gallegati, A Minsky-Kindleberger perspective on the financial crisis. J. Econ. Issues **45**, 449–458 (2012)
J.B. Rosser Jr., M.V. Rosser, Complexity and behavioral economics. Nonlinear Dynamics Psychol. Life Sci. **19**, 201–226 (2015)
J.B. Rosser Jr.; Rosser, M.V. (2016) Complexity and institutional evolution. Evolut Inst Econ Rev. https://doi.org/10.1007/s40844-06-0060-3.
T.C. Schelling, Dynamics models of segregation. J. Math. Sociol. **1**, 609–681 (1971)
J.A. Schumpeter, *Theorie der wirtschaftlichen entwicklung* (Dunker & Humblot, Leipzig, 1912)
J.A. Schumpeter, *The History of Economic Analysis* (Oxford University Press, New York, 1954)
R. Sethi, E. Somanathan, The evolution of social norms in common property use. Am. Econ. Rev. **80**, 766–788 (1996)
H.A. Simon, *Administrative Behavior* (Macmillan, New York, 1947)
H.A. Simon, A behavioral model of rational choice. Q. J. Econ. **69**, 99–118 (1955)
H.A. Simon, *Models of man* (Wiley, New York, 1957)
H.A. Simon, The architecture of complexity. Proc. Am. Philos. Soc. **106**, 467–482 (1962)
H.A. Simon, *The Sciences of the Artificial* (MIT Press, Cambridge, 1969)
H.A. Simon, From Substantive to Procedural Rationality, in *Method and Appraisal in Economics*, ed. by S. J. Latsis (Cambridge University Press, Cambridge, 1976), pp. 129–148
B.F. Skinner, *The Behavior of Organisms: An Experimental Analysis* (B.F. Skinner Foundation, Cambridge, 1938)
A. Smith, *The Theory of Moral Sentiments* (Miller, Kincaid, & Bell, London, 1759)
V.L. Smith, G.L. Suchanek, A.W. Williams, Bubbles, crashes, and endogenous expectations. Econometrica **56**, 1119–1151 (1988)
G.J. Stigler, The economics of information. J. Polit. Econ. **69**, 213–225 (1961)
R. Thom, *Stabilité structurelle et morphogenèse: essai d'une théorie générale des modèles* (Benjamin, New York, 1972)
D.W. Thompson, *On Growth and Form* (Cambridge University Press, Cambridge, 1917)
A. M. Turing, The chemical basis of morphogenesis. Philos. Trans. R. Soc. Lond. Ser. B Biol. Sci. **237**, 37–72 (1952)

V.J. Vanberg, Institutional evolution through purposeful selection: The constitutional political economy of John R. Commons. Constit. Polit. Econ. **8**, 105–122 (1997)

T.B. Veblen, Why is economics not an evolutionary science? Q. J. Econ. **12**, 373–397 (1898)

T.B. Veblen, *The Theory of the Leisure Class: And Economic Study of Institutions* (Macmillan, New York, 1899)

T.B. Veblen, *The Place of Science in Modern Civilization and Other Essays* (Huebsch, New York, 1919)

K.V. Velupillai, *Computable Economics* (Oxford University Press, Oxford, 2000)

L. Von Bertalanffy, An outline of general systems theory. Br. J. Philos. Sci. **1**, 114–129 (1950)

J. Von Neumann, *Theory of Self-Reproducing Automata* (University of Illinois Press, Urbana, 1966)

N. Wiener, *Cybernetics: Or Control and Communication in the Animal and the Machine* (MIT Press, Cambridge, 1948)

G.C. Williams, *Adaptation and Natural Selection* (Princeton University Press, Princeton, 1966)

O.E. Williamson, The new institutional economics: Taking stock, looking forward. J. Econ. Lit. **38**, 596–613 (2000)

E.O. Wilson, *The Social Conquest of Earth* (W.W. Norton, New York, 2012)

S. Wright, The roles of mutation, inbreeding, crossbreeding, and selection in evolution. Proc. Int. Congr. Genet. **1**, 356–366 (1932)

S. Wright, The genetical theory of populations. Ann. Eugenics **15**, 323–354 (1951)

A.A. Young, Increasing returns and economic progress. Econ. J. **38**, 527–542 (1928)

E.C. Zeeman, On the unstable behavior of the stock exchanges. J. Math. Econ. **1**, 39–44 (1974)

# Chapter 5
# Agent-Based Models and Their Development Through the Lens of Networks

**Shu-Heng Chen and Ragupathy Venkatachalam**

**Abstract** In this paper we attempt to provide a historical overview of the development of agent-based models from a network perspective. We trace two distinct stages in the development of agent-based models – first-generation models (prior to the 1990s) and second-generation models (since the late 1990s), with a salient break in the middle. We highlight the nexus and synergies between these two approaches. We show that the early developments of agent-based models in the cellular automata tradition contained crucial ingredients concerning networks in them even before they were applied to the social sciences. We argue that networks or interconnectivity (coupling) is a fundamental feature that has enabled agent-based modelers to develop a unique version of complexity science, both in terms of ontology and epistemology.

## 5.1 Introduction

Individuals in a society routinely interact, exchange information, cooperate, learn from each other, and also depend on each other. The nature of interdependence and interactions among individuals are increasingly seen as central or even indispensable to understanding many social phenomena. Agent-based models can be broadly seen as an approach that is a bottom-up approach to understanding a variety of social phenomena in fields such as economics, sociology, ecology, and epidemiology. It is an approach often concerned with understanding how (often unexpected) macro-level phenomena *emerge* from microlevel interactions among different agents. Network science on the other hand is broadly concerned with the nature and

---

S.-H. Chen (✉)
Department of Economics, AI-ECON Research Center, National Chengchi University, Taipei, Taiwan
e-mail: chen.shuheng@gmail.com

R. Venkatachalam
Institute of Management Studies, Goldsmiths, University of London, London, UK
e-mail: rpathy@gmail.com

 
Y. Aruka, A. Kirman (eds.), *Economic Foundations for Social Complexity Science*, Evolutionary Economics and Social Complexity Science 9,
DOI 10.1007/978-981-10-5705-2_5

structure of links between individuals or entities (such as banks) and how these structural characteristics constitute a potential explanation for various phenomena.

Even at this broader level, one can see a common element among these two fields, viz., the possibility of interaction. Despite this, the intimate connections between these two fields or approaches have not been well understood. Agent-based modelers are often thought to have embraced network-related phenomena in their pursuits only after the prolific developments around the turn of the century that led to the formal advent of network science. However, this narrative can be challenged once we consider the origins of agent-based modeling, where network structures play an important role.

In this chapter, we provide a review of the development of agent-based modeling in the social sciences from the perspective of networks. By tracing the historical origins of agent-based modeling, we argue that the approach begins by having networks as one of its essential ingredients, even before its introduction to the social sciences. In fact, it is the idea of networks and interdependence that enabled agent-based modelers to develop a unique version of complexity science, both in terms of the ontology and epistemology of complexity. The unique nature of complexity that is advanced by agent-based modeling, in our opinion, is best understood through two early examples in agent-based modeling, although they were then referred to by a more familiar name as cellular automata. The two models are John Conway's *Game of Life* and Stephen Wolfram's *Elementary Cellular Automata* (Wolfram 1983, 1994, 2002). In a nutshell, this notion of complexity can be described as follows: *programs based on simple rules do not always produce simple behavior*. We refer to this unique notion of complexity as the cellular automata tradition.

In tracing the development of agent-based models from a network perspective, we find three distinct periods. The first stage is a period roughly before the mid-1990s, during which some of the earliest agent-based models of social phenomena were developed. These models were squarely in the cellular automata tradition, and we refer to them as the Mark I or first-generation network-based agent-based models. These models relied on checkerboard-type models to invoke interdependencies in decision processes. The second stage saw a proliferation of agent-based models in the social sciences. A series of agent-based economic models, both macroeconomic models and financial market models, were introduced. However, agent-based models that were developed during this time, particularly those in economics, were largely network-free. Instead of the network structure, the focus of these models has been on developing more sophisticated functionalities for artificial agents, by applying many interesting insights from *computational intelligence*. Although it was burgeoning with interesting developments, we refer to this period as a *salient break*. The third stage of development has occurred since the late 1990s, after this so-called salient break. As a result of several important breakthroughs that unleashed network science as an independent area of research, the models of this era are referred to as ABM Mark III or the second-generation network-based agent-based models.

By choosing to look through the lens of networks, albeit only ex post, we hope to identify and illustrate the profound links that exist between agent-based models and networks since the former's inception and also in its modern incarnations. It is also our belief that the importance of spatial and network features is not going to diminish at least in the near future. The rest of this chapter is structured as follows: Sect. 5.2 traces the first-generation agent-based models and discusses the unique nature of the complexity that they encapsulate. Section 5.3 describes the second stage which marks a salient break and attempts to resolve the puzzle concerning the absence of networks. Section 5.4 outlines the third stage of development after the birth of network science and how it has contributed to the investigations of agent-based modelers. The final section provides some concluding remarks.

## 5.2 ABM Mark I: Cellular Automata Tradition

The initial developments in the field of agent-based models had explicit network characteristics even since its nascent stage. The origins of agent-based models, at least one of the origins certainly, can be traced back to biophysical computational models known as *cellular automata*. Foundational work on cellular automata was carried out by John von Neumann in the 1950s, where he was interested in self-replicating or self-reproducing automata as a pathway toward understanding biological evolution and self-organizing capabilities of the human brain. Cellular automata (CA) are discrete mathematical structures (both in space and time), and local interaction (neighborhood) is a key feature of them. The evolution of the cellular automata is based both on the rules of interaction and the structure of the neighborhood (network). The idea of a network is implicit, and it is related to the notion of physical proximity (neighborhood) in this class of models. For instance, in a two-dimensional square lattice, the number of cells that constitute the neighborhood (network) varies depending on how a step or distance is defined. If it is defined so as to include cells (or agents) within a unit Chebyshev distance, the resulting neighborhood will have eight neighbors (a Moore neighborhood). On the other hand, if it is defined based on Manhattan distance, there will be four neighbors (a von Neumann neighborhood).

The presence of a network structure in the early development of agent-based models can be best understood by examining two early models in the cellular automata tradition: (a) John Conway's Game of Life and (b) Stephen Wolfram's analysis of *elementary cellular automata* (Wolfram 1983, 1994, 2002). These precursors to modern day agent-based models had network structures as essential ingredients even before they were applied to the social sciences. These network characteristics, which are highly indispensable, have in fact enabled agent-based modelers to develop a unique version of complexity science both in terms of the ontology and epistemology of complexity. We will formally define cellular automata and outline the unique features concerning complexity of what we refer to as the *cellular automata tradition* in the following subsection.

### 5.2.1 Complexity and Agent-Based Models

CA models are capable of generating a wide variety of complex behavior from simple, local rules of interaction between different parts (cells) as seen in famous models such as John Conway's *Game of Life* and the elementary cellular automata of Stephen Wolfram. A *neighborhood-based decision rule* relies on a set of neighbors and a set of states and can be represented as follows:

$$X_{i,t+1} = f(X(N_{i,t})) \tag{5.1}$$

where $N_{i,t}$ represents the set of *neighbors*[1] of the cell (or agent) $i$ and $X(N_{i,t}$ is the *state* of the neighbors at time $i$. The neighborhood of agent $i$ can in principle be defined based on physical or social distance. In the elementary cellular automata studies by Wolfram, they are defined based on physical distance. From the decision rule, we can infer that the decisions of agent $i$ at time $t+1$ depend on the decisions made by the neighbors at (or up to) time $t$. Let us consider the simplest case of a one-dimensional lattice, where an agent has two neighbors – one on his left and one on his right. The agent has to make a binary choice 0 or 1. The decision rule of the agent is then a mapping from a binary string of length 3 (neighborhood state) to a binary variable (choice).

$$X_{i,t+1} = f : \{1,0\} \times \{1,0\} \times \{1,0\} \rightarrow \{1,0\} \tag{5.2}$$

The triplets refer to the decisions made by the neighbor to the right, agent $i$, and the neighbor to the left, respectively, at time $t$. In this simple case, there are only a total of $2^3$ possible combinations of decisions at a given time $t$. An eight-bit binary string can be then taken to represent a general form of the decision rule. Since each bit has two possible states (0 or 1), a total of 256 distinct possibilities are available for the decision rules. Some of these decision rules exhibit complex patterns of evolution over time and space.

Stephen Wolfram studied these 256 rules systematically and generated their time-space patterns (Wolfram 2002). Based on the patterns generated, these 256 rules were classified into four classes, which represent a hierarchy of complexity. This four-class classification also establishes the link between one-dimensional cellular automata, dynamical systems theory, formal language theory, and the theory of computation. In the parlance of dynamical systems theory, these four classes are fixed points, limit cycles (periodic cycles), chaos (pseudo-randomness, strange attractors), and on the edge of chaos (complex patterns):

- Class I: Nearly all initial patterns converge to a uniform homogeneous final state (fixed point), with all agents being in state 1 or 0

[1]This neighborhood can include agent $i$ as well.

- Class II: Nearly all initial patterns quickly converge to a stable structure or a continually recurring repetition of short cycles
- Class III: Nearly all initial patterns evolve in a pseudo-random, chaotic, or apparently irregular manner
- Class IV: Nearly all initial patterns evolve into localized structures that interact in a complex and interesting manner. They can exhibit behavior that is not completely pseudo-random for the entire duration. Patterns can emerge, persist for an unknown duration and dissolve into chaotic configurations. Some rules in this class are shown to exhibit computational universality.

The unique notion of complexity that can be synthesized from the cellular automata studies by Conway and Wolfram is that even the programs based on simple rules do not always produce simple behavior. Complexity or randomness is inherent in the embedded structure (network) on which the simple operational rules are defined. It is this notion of complexity that we find to be unique that the agent-based models in this tradition possess. Here, complexity or randomness is inherent in the embedded structure (network) in which the simple operating rules are defined. It is important to note that minor changes in the decision rules, which in turn are dependent on the state of the network (neighborhood), can lead to remarkable changes in the complexity classes to which the decision rule belongs. The emergence of complex structures through local interactions has enabled agent-based modelers to develop a unique explanation for complex phenomena in many domains. As these models indicate, the resulting structures cannot be predicted a priori in all cases, despite having simple rules. Some of these rules are shown to exhibit computational irreducibility which means that the complete evolution of the system cannot be guessed using shortcuts and can only be known through simulations. Therefore, the structure or network dependence is crucial in rendering a unique variant of complexity to be generated and studied.

The first-generation of network-oriented agent-based models was in the CA tradition and was characterized by the use of such simple local rules (neighborhood-based behavior) in the one- or two-dimensional checkerboard or lattice models. These are sometimes extended to irregularly spaced grids as well. Some of the earliest models in this strand of the literature exploited the power of cellular automata or checkerboard models to capture the essence of complex patterns of self-organization and cooperative behavior that are often observed in actual social systems. They captured the essence of social economics – i.e., an agent's preferences and decisions are interdependent and influenced by what other agents prefer and decide – using *neighborhood-based decision rules*. We will review two important models in this tradition, namely, segregation models and spatial games.

### 5.2.2 Schelling's Segregation Model

Thomas Schelling and James Sakoda employed neighborhood-based decision rules where agents' decisions (preferences) are under the influence of their neighbors in order to study patterns of segregation (Schelling 1971; Sakoda 1971), in residential choices, for instance. These models examined how unexpected and undesirable macro-level patterns can result from the interactions of several individuals under neighborhood-based decision rules in a rational choice theoretic framework. The decision rule underpinning the Schelling model is a migration rule or a residential choice rule. Agents have to decide whether or not to migrate to a different place. In Schelling's model, this decision depends on the ethnicity of an agent's neighbors. He considered two ethnic groups, and the migration rule is quite simple: if agent $i$ in his community (neighborhood) $\mathbf{N_i}$ considers himself to be in a "minority," then agent $i$ will decide to migrate to the nearest satisfactory place; otherwise, he will decide to remain. Notice that here the definition of majority or minority is subjective and determined by agents' perceptions. The network-based decision rule is therefore parameterized by a threshold, the migration threshold or tolerance capacity, denoted by $\theta$. If more than $\theta\%$ of his neighbors belong to a different ethnic group, the agent will decide to migrate.

$$a_i(t) = \begin{cases} 1 \text{ (migrate)}, & \text{if } f_{\mathbf{N_i}}(t) > \theta_i, \\ 0 \text{ (stay put)}, & \text{otherwise}, \end{cases} \tag{5.3}$$

where

$$f_{\mathbf{N_i}}(t) = \frac{\mid \{j : j \in \mathbf{N_i}(t), j(x) \neq i(x)\} \mid}{\mid \mathbf{N_i}(t) \mid} \tag{5.4}$$

In Eq. (5.4), $i(x)$ represents the ethnic group of agent $i$, and "$j(x) \neq i(x)$" implies that agents $i$ and $j$ have different ethnic groups. Hence, $f_{\mathbf{N_i}}(t)$ is the fraction of agent $i$'s neighbors who do not share the same ethnic group as agent $i$.

Even if we start from a well-integrated society, the resulting dynamics governed by neighborhood-based decision rules can produce a highly segregated society. Agents may or may not have homogeneous thresholds; some may have lower tolerance, while others have a higher tolerance capacity. Depending on the distribution of the threshold values, one may identify a *tipping point* or *critical point*, at which a morphogenetic change in demographics can occur (Schelling 1972). It should be noted that neither Schelling nor Sakoda explicitly referred to or identified their checkerboard models with formal cellular automata when they first used them in the social science context. Peter Albin (1975) was among the first to classify these checkerboard models under the cellular automata framework, and checkerboard models came to be classified as a special kind of cellular automata.

The Schelling-Sakoda model is a pioneering agent-based model that has been extensively studied in the literature and has also been extended in many directions.

It still continues to generate a lot of interest in understanding the micro-macro divide that often characterizes various social phenomena (Raub et al. 2011). The model shows that a system has the properties that the individuals do not have. These properties are surprising in the sense that they cannot be easily inferred or deduced and could therefore be referred to as *emergent properties*.The results of the original model were reinforced by Pancs and Vriend (2007), who modified the conditional preferences of the agent, i.e., preferences for the proportion of "others" conditioned on the agent remaining in the majority. Despite agents holding conditional preferences that favor higher proportions of members from other ethnicities (more tolerant, in other words), the emergence of segregation persists. Rogers and McKane (2011) have proposed a generalized analytical Schelling model to encompass a number of variants of the Schelling model. Both Rogers and McKane (2011) and Banos (2012) provide an extensive list of studies along the lines initiated by Thomas Schelling.

### 5.2.3 Spatial Games

The second important set of models that witnessed the explicit use of neighborhood or lattice models in the first generation of agent-based models are those concerned with *spatial games*. This class of games can be viewed as a predecessor to what came to be known as *network games*. The origins of spatial games can be traced back to the works of Robert Axelrod in the 1980s (Axelrod 1984; Matsuo 1985). One part of the origins of the spatial game can also be traced back to Thomas Schelling's multi-person games. Although Schelling's original setting of the multi-person game did not have a network, Albin (1992) transformed Schelling's multi-person game into a two-person game with multi-persons (neighbors). Spatial games were extensively pursued by Martin Nowak and Robert May (1992, 1993).

They were largely concerned with models with a structure akin to the prisoner's dilemma, and the game is played by many agents who are distributed across space (a lattice). Despite defection being the equilibrium strategy in many prisoner's dilemma-like situations, we often find that people do in fact cooperate. The key question that underpins this line of research is the following: why do people choose to cooperate, to be kind, even in the presence of a strong temptation to defect? The pioneers in this area were interested in understanding the emergence and sustenance of cooperation and its coexistence with other defection strategies.

Let $\mathbf{A}$ be the action or strategy space and $\mathbf{A} = \{1, 0\}$, where "1" denotes the action "cooperation" and "0" denotes the action "defection." In addition, let $\mathbf{X}(t)$ be the set of choices that each agent makes at time $t$: $\mathbf{X}(t) = \{a_i(t)\}$, $a_i(t) \in \mathbf{A}$. Let $A_i(t)$ be the history of the actions taken by agent $i$ up to the time $t$, i.e., $A_i(t) \equiv \{a_i(t), a_i(t-1), \ldots a_i(1)\}$. Let $A_i^k(t)$ be the subset of $A_i(t)$, which retains the most recent $k$ periods. Agent $i$'s strategy can then be defined as follows:

$$f_i : \{A_j^k(t)\}_{j \in \{i\} \cup \mathbf{N}_i} \rightarrow \mathbf{A}. \tag{5.5}$$

For instance, if agent $i$ has four neighbors (the von Neumann neighborhood) and assuming that $k = 1$, then, including agent $i$, there are a total of $2^5$ possible combinations of actions (scenarios) that $i$'s decision can be contingent upon. Since the agent's decision is a binary choice – either "1" or "0" – there are a total of $2^{2^5}$ ($=2^{32}$) possible strategies (the more general case contains $2^{2^{5\times k}}$ strategies) for a von Neumann neighborhood.

In Axelrod (1984, Chapter 8), the spatial game is introduced in the form of an imposed social structure. Axelrod explores the consequence of a social structure (network topology) on the overall pro-social behavior. He shows that the imposed social structure or a specific network topology can in fact help protect clusters of cooperators from the invasion of defectors. He assumes that all agents in a regular lattice use the tit-for-tat strategy and all agents start out as cooperators in the initial time period. This initial state, which is in fact is a steady state, is then perturbed by the introduction of an intruder at the center of the lattice. This intruder always defects, instead of adopting the tit-for-tat strategy that others use. Given the simple imitation dynamics wherein an agent copies the strategy of the most successful neighbor in the von Neumann neighborhood, the "always-defect" strategy is shown to spread from the center to the entire space in a snowflake-like pattern, with defectors bypassing islands of cooperators.

Thus, Axelrod's version of the spatial game demonstrates that, by imposing a social structure (network topology), the coexistence of both cooperators and defectors can be explained. Hence, the social structure (network topology) is sufficient to sustain the presence of cooperative behavior in such a society. The line of research initiated by Axelrod (1984) and extended by others has been surveyed by Nowak and Sigmund (2000).

> The main message so far is that neighborhood structure seems to offer a promising way out of the prisoner's dilemma toward the emergence of cooperation. There are many alternative explanations of the prevalence of cooperation, but, arguably, none require less sophistication on the part of the individual agents than those with spatial structure. The latter need no foresight, no memory, and no family structure. *Viscosity suffices.* (Ibid, p. 145; Italics added)

However, Axelrod did not use formal automata theory to further analyze the complex patterns emerging from his formulation of spatial games. Strategies of a spatial game (as in Eq. 5.5) can also be formulated like those in Schelling (1978). This has been carried out by Albin (1992), who modifies Schelling's formulation of multi-person games. In his rendition, only the local aggregate (the number of cooperative neighbors in the preceding period) matters. With this aggregate information, agent $i$ no longer cares about the action taken by each of his neighbors; hence, the number of possible scenarios can be drastically reduced. After such a reduction, it is easy to code this kind of strategy by a binary string with $N_i + 1$ bits. Such a formulation of the decision rules directly connects spatial games to *cellular automata*. Around the same time as Albin's paper was published, a formal research program on spatial games along the lines broached by Axelrod had already begun. The *Nowak-May model*, which remains a pioneering work, has shaped the study of the spatial game explicitly within the framework of cellular automata (Nowak and May 1992, 1993).

Let $\pi_i(a_i, a_j)$ be the payoff of a two-person game that agent $i$ plays with his opponent agent $j$, when agent $i$ takes the action $a_i$ ($a_i \in \mathbf{A}_i$) and agent $j$ takes the action $a_j$ ($a_j \in \mathbf{A}_j$). In the prisoner's dilemma game, $\mathbf{A}_i = \mathbf{A}_j = \{1(C), 0(D)\}$. Since the game is symmetric, $\pi_j(a_j, a_i) = \pi_i(a_i, a_j)$, we shall omit the subscript appearing in the payoff function and let $\Pi$ be the payoff matrix:

$$\Pi = \begin{bmatrix} \pi_{11} & \pi_{10} \\ \pi_{01} & \pi_{00} \end{bmatrix}, \tag{5.6}$$

where $\pi_{11} \equiv \pi(1,1)$, $\pi_{10} \equiv \pi(1,0)$, and so on. As a prisoner's dilemma game, the following inequality should be satisfied: $\pi_{01} > \pi_{11} > \pi_{00} > \pi_{10}$. What (Nowak and May 1992) did was to "normalize" the game by setting $\pi_{11} = 1$ and $\pi_{00} = \pi_{10} = 0$ and leave only one parameter free, i.e., $\pi_{01}$. Under this normalization, the game can be characterized by a single parameter, namely,

$$\text{temptation to defect} \equiv \frac{\pi_{01}}{\pi_{11}} = \pi_{01}.$$

Based on the value of $\pi_{01}$, one can delineate the boundary of the cluster of cooperators and the dynamics of its advancement. Various possible spreading patterns are shown to occur, wherein complex temporal-spatial patterns may emerge with the coexistence of the cooperators and defectors. Thus, the first generation of agent-based models was highly reliant on network structures to explain different social phenomena. It can be seen from the above example that the interdependence or network underpinning is what enables agent-based modelers to generate and explain complex patterns.

## 5.3 ABM Mark II: Salient Break

We have so far traced the development of agent-based modeling in its initial stages, largely prior to the mid-1990s (ABM Mark I models as we referred to above), and have shown that they had a distinct network flavor. Agent-based models witnessed rapid developments and applications to the social sciences around the mid-1990s, and this wave has been seen particularly in the field of economics. This may be partly attributed to the availability of cheaper and faster computing facilities to researchers in the social sciences. We refer to this as the burgeoning stage, which also marks a salient break with the models that came at the end of the 1990s and in the first decade of the millennium.

Within economics, agent-based modeling has been increasingly applied to study a variety of problems, and in particular, we identify three domains in which it has been prominently applied:

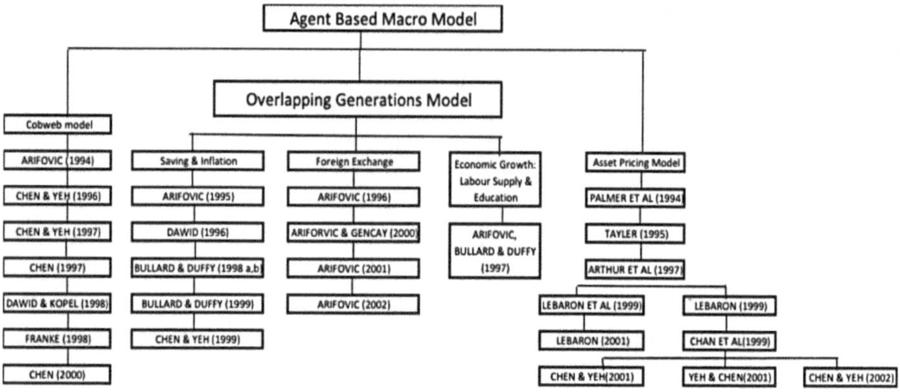


**Fig. 5.1** Agent-based models in macroeconomics (stage, ABM Mark II): This figure presents a summary of research studies in macroeconomics that employed agent-based models. These models belong to the second stage of development, which was characterized by a marked absence of explicit network features

- Double auction markets (Gode and Sunder 1993)
- Financial markets (Arthur 1992; Kirman 1993; Lux 2010)
- Macroeconomics (Arifovic 1994; Bullard and Duffy 1998c)

Although this stage witnessed the rapid growth of agent-based modeling, by being applied to newer domains and making interesting advances, the developments were largely devoid of network ingredients. In other words, these studies hardly exploited or made use of network concepts or measures. For instance, let us consider the case of agent-based models in macroeconomics during this stage. Within this area, agent-based models were predominantly applied in studies that employed cobweb models, overlapping generation models, and asset pricing models. Figure 5.1 provides a summary of the important research studies within three lines of research that used agent-based models. All these models belong to the second stage of development that occurred around or after the mid- or late 1990s before the advent of formal network theory. All the models that are listed in Fig. 5.1 are characterized by a marked absence of explicit network features. The conspicuous absence of network features is indeed puzzling. Why did these studies disregard the network aspects despite their crucial importance in the first-generation models? We attempt to resolve this puzzle in the following subsection.

### *5.3.1 Missing Networks: A Puzzle?*

The absence of prominent network elements in the studies during this salient break phase can be understood by considering the specific problems that the agent-based models were employed to study and also the developments that were happening outside the scope of agent-based models. We find three important aspects that can shed light on this puzzle:

- **Markets and trading mechanisms**: First, most of the agent-based models in this time period focused largely on the operations and aspects concerning markets. In doing so, they need to specify trading mechanisms that governed the interaction between different market participants. The underpinning trading mechanism in these artificial markets was the (clearing house) auction. This meant that there was no decentralized trade that was considered, which would have necessitated the application of network concepts. Under such auction mechanisms, network concepts, such as locality, neighborhood, or geographical specificity, play a negligible role at best. Therefore, it was understandable that the network factor is naturally assumed away.
- **Absence of network formation algorithms**: Second, the studies during this phase also lacked explicit network formation algorithms. However, this should not be taken to mean that network-oriented thinking was not introduced or absent in the broader context of the market mechanism. In fact, on the contrary, the literature in economics during that time had already pointed toward the need to reflect upon the relation between networks and markets. This was undertaken by scholars in two major streams of the literature: first, studies on the labor market (Rees 1966; Granovetter 1973; Boorman 1975; Montgomery 1991) and, second, studies investigating buyer-seller markets (Uzzi 1996; Kranton and Minehart 2001). Why, then, had the development of agent-based models pertaining to markets ignored this connection? Our view is that the first-generation models available to the scholars could not be immediately transformed or related to this need. The social structure or networks are exogenously specified in the checkerboard-type models, and they had very little empirical relevance. Given this context, it may be easier to understand why the significance of networks to market modeling could not have been immediately clear during that time.
- **Computational intelligence and autonomous agents**: Third, compared to the tools for network modeling, tools concerning the modeling of artificial agents were advancing rapidly during the early 1990s. Since agent-based models were focusing on complex, adaptive systems, agents had to learn from the past and make good, novel decisions for the future. Developments in computational intelligence and learning, such as reinforcement learning, swarm intelligence, neural networks, and evolutionary computation, were squarely addressing such needs and were being widely adopted. During this stage, the attention of the economists has been drawn to the *functioning* of autonomous artificial agents; hence, the modeling strategy that emerged was one that focused on *complex* agents (in terms of their functionality) embedded in *simple* or no networks, instead of simple agents being placed in elaborate or complex networks.

  In other words, the relative focus was on the functions that agents perform and learning, which assumed a higher priority than the network element in this stage. For example, the well-known *El-Farol Bar Problem* was originally introduced by Brian Arthur (1994) as an illustration of bounded rationality and learning. However, it is a clear instance of a situation where many agents decide simultaneously and network structure has an important role to play. It was only two decades later that the role and significance of network topologies were shown to be important in this problem (Chen and Gostoli 2017).

## 5.4 ABM Mark III: Network Science Influence

Although the idea and implicit awareness of networks have existed for a very long time, the topic itself was given a fresh lease of life during the late 1990s. This period witnessed a number of major, path-breaking achievements, which in turn made the study of networks become an independent discipline, i.e., *network science* (Lewis 2011).

A series of breakthroughs, for instance, Watts and Strogatz's *small-world networks* (Watts and Strogatz 1998) and Barabasi and Albert's *scale-free networks* (Barabási and Albert 1999), addressed some important questions concerning network generation. Developments in game-theoretic approaches, both cooperative and noncooperative, to network formation broke new ground in relating the emergence of different network structures to strategic, rational economic actions of interacting individuals (Jackson and Wolinsky 1996). However, the informational demands of these game-theoretic models concerning the costs of links and the value associated with networks were rather restrictive. By contrast, agent-based models provided an alternative path to understanding network formation.

These developments led to two major breakthroughs that are relevant to this context: First, these studies provided a set of behavioral heuristics that presented procedures to generate networks, in other words, *network formation algorithms*. Second, some essential characteristics of the real-world networks were capable of being replicated by these algorithms, such as clustering coefficients, diameter (degree of separation), and the power-law distribution of degree (super stars). These studies thus provided a framework to formalize research questions about the formation, dynamics, efficiency, stability, and evolution of networks.

The impact of these pioneering studies on subsequent research is fourfold:

- Networks again become a more relevant part of model building; economic issues which used to be network-free earlier were later reformulated to allow networks to play a part.
- Research that existed till then on the network effects, such as spatial games, were furthered and reexamined with various graph-oriented networks that went beyond cellular automata.
- Perhaps the most illuminating aspect for researchers is that it presented the possibility to use the characterizations of network topologies as possible economic explanatory variables. Hence, a series of new questions regarding the economic consequences of network aspects such as the clustering coefficient, centrality, and degree distribution were posed.

- Watts and Strogatz (1998) and Barabási and Albert (1999) motivated researchers to think about behavioral heuristics to *grow* networks. This is of particular relevance to agent-based models. Since agent-based modeling is behaviorally oriented, this naturally led to a research area known as *agent-based modeling of networks* being born. This area ties together the work on network dynamics and agent-based modeling. This new approach to network dynamics can also be seen as the behavioral extension of the game-theoretic approach to network formation (network dynamics) populated by Matthew Jackson and Asher Wolinsky (1996).

These breakthroughs essentially led economists to utilize the kind of sociometric data, initially collected and analyzed by Jacob Moreno (1889–1974), the founder of *sociometry*, in their agent-based models. These aspects that take into account social networks are a distinct feature that characterizes the second generation of the network-based agent-based models. It is interesting to note that even though *paths through us* were already seen in Moreno's pioneering work, a more imaginative study of these paths had not been pursued until Ithiel de Sola Pool (1917–1984) and Manfred Kochen (1928–1989) investigated this phenomenon in the 1950s, and it was later further investigated by Stanley Milgram (1933–1984) and Jeffrey Travers in the late 1960s. These later developments were known as the *small-world problem* and the *small-world experiment* and have gained prominence as they are being drastically reshaped by the ICT revolution.

The second-generation network-based agent-based model is very much a consequence of the advent of network science. It is characterized by the use of network formation algorithms (graphs), such as small-world networks, instead of the checkerboard model as in the case of ABM Mark I models, as the interaction platform. The economic and social consequences of different network topologies are addressed in light of various network characteristics, such as degree, degree distribution, diameter, clustering coefficient, centrality, etc. In the following subsections, we briefly outline two important applications that illustrate the nature of questions pursued by ABM Mark III models.

### 5.4.1 Networks and Market Efficiency

An essential issue that is frequently raised and debated in economics is the extent to which we rely on the market mechanism to solve different problems and to identify the areas where markets are likely to fail. Even though this issue has been extensively studied in economics over many decades, very few actually paid attention to the role that network topology played before the advent of network science. As we mentioned earlier, economists in general did not or still do not consider the checkerboard models as being empirically relevant and saw them as thought experiments. However, unlike the checkerboard model, the small-world networks (the small-world phenomena), which are highly clustered and interconnected where a given node in the network is only a few steps from any other node in the network, are quite ubiquitous. Given that they are so prevalent in reality, one cannot simply ignore their existence and their potential role in the market mechanism.

In economics, Wilhite (2001) was one of the first to address this question, and he found that by considering search intensity and search costs, a small-world network (where not everyone is connected with each other, but nevertheless each individual is only a few steps away from any given person) happens to be the most efficient network topology, even more efficient than the fully connected network. He also found that economic gains resulting from trading have a tendency to have a higher concentration on the centrality nodes. Agents with high betweenness centrality play two different roles: they help create economic gains, but they also help increase economic inequality. This mode of reasoning provides a flavor of the models that were being developed in the second-generation agent-based models.

### *5.4.2 Networks and Cooperation*

The second illustration involves cooperation and the role of network structure. This issue has been a subject of interest to researchers in the first-generation models as well. Naturally, when the new ideas of network formation began to spread with the breakthroughs mentioned earlier, this topic (spatial games) was pursued in a new light (network games). Building upon the pioneering work of Martin Nowak and Robert May on lattice models, studies were interested in whether the results about network reciprocity are robust with respect to different network topologies and whether or not this result is sensitive to different games (payoffs). In other words, the quest was to understand whether cooperative behavior can be shown to be universal even when there is a strong temptation to defect.

The initial attempts in this direction have shown that the Nowak-May results are in fact sensitive to both the payoff matrix (i.e., are game specific) and, most importantly, are sensitive to network topologies (Abramson and Kuperman 2001; Hauert and Doebeli 2004). Some of the earlier results were built upon the test with small-world networks, but the later trials using the scale-free networks seem to have overthrown these counterarguments (Santos and Pacheco 2005; Gómez-Gardeñes et al. 2007; Santos et al. 2008).

Hence, the research focus can be seen as increasingly shifting toward the scale-free network, which allows for the formation of hubs and hence highly heterogeneous graphs. Properties of the scale-free networks are also examined in terms of their contribution to the emergence of a cooperative structure, such as a degree distribution, degree-degree correlations (assortative mixing), enhanced clustering coefficients (where the cluster coefficient conditional is on associative mixing) and so on (Pusch et al. 2008; Assenza et al. 2008).

## 5.5 Concluding Remarks

In this chapter, we have traced the evolution of agent-based models from the perspective of networks. We have argued that the network element was inherent in agent-based models right from the outset. We also argued that the unique nature of complexity that was developed by agent-based modeling – the cellular automata tradition – owes it to the structure of interdependencies that manages to generate complex phenomena from simple, neighborhood-dependent rules of interaction. Despite the salient break, network elements have again become prominent in agent-based models after the network science revolution. The behavioral aspect of network science has continued to promote the use of agent-based modeling as a network formation algorithm (ch. 4, Namatame and Chen 2016). The models that were network-free agent-based models in the past have focused mainly on agent functionality (complex agents). This is in sharp contrast to the cellular automata tradition, where the network is an indispensable aspect mainly to model complex phenomena using agents with simple design. Given that the networks can be modeled as complex structures, the need for highly complex agents can be questioned. Can human decision-making become truly simpler or easier when the society becomes increasingly complex and, in a sense, "smarter"? The answer to this question does not seem obvious and warrants efforts in this direction.

**Acknowledgements** The first and second authors are grateful for the research support in the form of Ministry of Science and Technology (MOST) Grants, MOST 103-2410-H-004-009-MY3 and MOST 104-2811-H-004-003, respectively.

## References

G. Abramson, M. Kuperman, Social games in a social network. Phys. Rev. E **63**(3), 030901 (2001)
P. Albin, *The Analysis of Complex Socioeconomic Systems* (Lexington Books, Lexington, 1975)
P. Albin, Approximations of cooperative equilibria in multi-person prisoners' dilemma played by cellular automata. Math. Soc. Sci. **24**(2), 293–319 (1992)
J. Arifovic, Genetic algorithm learning and the cobweb model. J. Econ. Dyn. Control. **18**, 3–28 (1994)
J. Arifovic, Genetic algorithms and inflationary economies. J. Monet. Econ. **36**(1), 219–243 (1995)
J. Arifovic, The behavior of the exchange rate in the genetic algorithm and experimental economies. J. Polit. Econ. **104**(3), 510–541 (1996)
J. Arifovic, Evolutionary dynamics of currency substitution. J. Econ. Dyn. Control. **25**, 395–417 (2001)
J. Arifovic, Exchange rate volatility in the artificial foreign exchange market, in *Evolutionary Computation in Economics and Finance*, ed. by S.H. Chen (Physica-Verlag, Heidelberg, 2002), pp. 125–136
J. Arifovic, R. Gencay, Statistical properties of genetic learning in a model of exchange rate. J. Econ. Dyn. Control. **24**, 981–1005 (2000)
J. Arifovic, J. Bullard, J. Duffy, The transition from stagnation to growth: an adaptive learning approach. J. Econ. Growth **2**(2), 185–209 (1997)
W.B. Arthur, On learning and adaptation in the economy. SFI working paper, 92-07-038 (1992)

W.B. Arthur, Complexity in economic theory: inductive reasoning and bounded rationality. Am. Econ. Rev. **82**(2), 406–411 (1994)

W.B. Arthur, J. Holland, B. LeBaron, R. Palmer, P. Tayler, Asset pricing under endogenous expectations in an artificial stock market, in *The Economy as an Evolving Complex System II*, ed. by W.B. Arthur, S. Durlauf, D. Lane (Addison-Wesley, Reading, 1997), pp. 15–44

S. Assenza, J. Gómez-Gardeñes, V. Latora, Enhancement of cooperation in highly clustered scale-free networks. Phys. Rev. E **78**(1), 017101 (2008)

R. Axelrod, *The Evolution of Cooperation* (Basic Books, New York, 1984)

R. Axelrod, The dissemination of culture: a model with local convergence and global polarization. J. Confl. Resolut. **41**(2), 203–226 (1997)

A. Banos, Network effects in Schelling's model of segregation: new evidences from agent-based simulation. Environ. Plann. B. Plann. Des. **39**(2), 393–405 (2012)

A.L. Barabási, R. Albert, Emergence of scaling in random networks. Science **286**, 509–512 (1999)

S.A. Boorman, A combinatorial optimization model for transmission of job information through contact networks. Bell J. Econ. **6**(1), 216–249 (1975)

J. Bullard, J. Duffy, A model of learning and emulation with artificial adaptive agents. J. Econ. Dyn. Control. **22**, 179–207 (1998)

J. Bullard, J. Duffy, Learning and the stability of cycles. Macroecon. Dyn. **2**(1), 22–48 (1998)

J. Bullard, J. Duffy, A model of learning and emulation with artificial adaptive agents. J. Econ. Dyn. Control **22**, 179–207 (1998c)

J. Bullard, J. Duffy, Using genetic algorithms to model the evolution of heterogeneous beliefs. Comput. Econ. **13**(1), 41–60 (1999)

N.T. Chan, B. LeBaron, A.W. Lo, T. Poggio, Agent-based models of financial markets: a comparison with experimental markets. Unpublished working paper, MIT Artificial Markets Project, MIT, MA (1999)

S.H. Chen, On the artificial life of the general economic system (I): the role of selection pressure, in *Proceedings of the International Symposium on System Life*, ed. by F. Hara, K. Yoshida, pp. 233–240 (1997)

S.H. Chen, Toward an agent-based computational modeling of bargaining strategies in double auction markets with genetic programming, in *Intelligent Data Engineering and Automated Learning- IDEAL 2000: Data Mining, Financial Engineering, and Intelligent Agents*, ed. by K.S. Leung, L.-W. Chan, H. Meng. Lecture Notes in Computer Sciences (Springer, Berlin, 2000), pp. 517–531

S.H. Chen, U. Gostoli, Coordination in the El Farol Bar problem: the role of social preferences and social networks. J. Econ. Interac. Coord. **12**(1), 59–93 (2017)

S.H. Chen, C.H. Yeh, Genetic programming learning and the cobweb model, in *Advances in Genetic Programming*, ed. by P. Angeline, vol. 2, Chap. 22 (MIT Press, Cambridge, 1996), pp. 443–466

S.H. Chen, C.H. Yeh, Modeling speculators with genetic programming, in *Evolutionary Programming VI*, ed. by P. Angeline, R.G. Reynolds, J.R. McDonnell, R. Eberhart. Lecture Notes in Computer Science, vol. 1213 (Springer, Berlin, 1997), pp. 137–147

S.H. Chen, C.H. Yeh, Modeling the expectations of inflation in the OLG model with genetic programming. Soft. Comput. **3**(2), 53–62 (1999)

S.H. Chen, C.H. Yeh, Evolving traders and the business school with genetic programming: a new architecture of the agent-based artificial stock market. J. Econ. Dyn. Control **25**, 363–393 (2001)

S.H. Chen, C.H. Yeh, On the emergent properties of artificial stock markets: the efficient market hypothesis and the rational expectations hypothesis. J. Econ. Behav. Organ. **49**(2), 217–239 (2002)

H. Dawid, Learning of cycles and sunspot equilibria by genetic algorithms. J. Evol. Ecol. **6**(4), 361–373 (1996)

H. Dawid, M. Kopel, On economic applications of the genetic algorithm: a model of the cobweb type. J. Evol. Econ. **8**(3), 297–315 (1998)

R. Franke, Coevolution and stable adjustments in the cobweb model. J. Evol. Econ. **8**(4), 383–406 (1998)

D.K. Gode, S. Sunder, Allocative efficiency of markets with zero-intelligence traders: market as a partial substitute for individual rationality. J. Polit. Econ. **101**(1), 119–137 (1993)

J. Gómez-Gardeñes, M. Campillo, L.M. Floría, Y. Moreno, Dynamical organization of cooperation in complex topologies. Phys. Rev. Lett. **98**(10), 108103 (2007)

M. Granovetter, The strength of weak ties. Am. J. Sociol. **78**(6), 1360–1380 (1973)

C. Hauert, M. Doebeli, Spatial structure often inhibits the evolution of cooperation in the snowdrift game. Nature **428**(6983), 643–646 (2004)

R. Hegselmann, A. Flache, Understanding complex social dynamics: a plea for cellular automata based modelling. J. Artif. Soc. Soc. Simul. **1**(3) (1998). http://www.soc.surrey.ac.uk/JASSS/1/3/1.html

M.O. Jackson, A. Wolinsky, A strategic model of social and economic networks. J. Econ. Theory **71**, 44–74 (1996)

A. Kirman, Ants, rationality, and recruitment. Q. J. Econ. **108**, 137–156 (1993)

R.E. Kranton, D.F. Minehart, A theory of buyer-seller networks. Am. Econ. Rev. **91**(3), 485–508 (2001)

B. LeBaron, Building financial markets with artificial agents: desired goals and present techniques, in *Computational Markets*, ed. by G. Karakoulas (MIT Press, Cambridge, 1999)

B. LeBaron, Evolution and time horizons in an agent based stock market. Macroecon. Dyn. **5**, 225–254 (2001)

B. LeBaron, W.B. Arthur, R. Palmer, Time series properties of an artificial stock market. J. Econ. Dyn. Control. **23**, 1487–1516 (1999)

T.G. Lewis, *Network Science: Theory and Applications* (John Wiley and Sons, Hoboken, 2011)

T. Lux, Herd behaviour, bubbles and crashes. Econ. J. **105**(431), 881–896 (1995)

K. Matsuo, Ecological characteristics of strategic groups in a 'dilemmatic world', in *Proceedings of the IEEE International Conference on Systems and Cybernetics*, Tucson (1985), pp. 1071–1075

J.D. Montgomery, Social networks and labor-market outcomes: toward an economic analysis. Am. Econ. Rev. **81**(5), 1408–1418 (1991)

A. Namatame, S.H. Chen, *Agent Based Modelling and Network Dynamics* (Oxford University Press, Oxford, 2016)

M.A. Nowak, *Evolutionary Dynamics* (Harvard University Press, Cambridge, 2006)

M.A. Nowak, R.M. May, Evolutionary games and spatial chaos. Nature **359**(6398), 826–829 (1992)

M.A. Nowak, R.M. May, The spatial dilemmas of evolution. Int. J. Bifurcation Chaos **3**, 35–78 (1993)

M.A. Nowak, K. Sigmund, Games on grids, in *The Geometry of Ecological Interactions: Simplifying Spatial Complexity*, ed. by U. Dieckmann, R. Law, J.A.J. Metz (Cambridge University Press, New York, 2000), pp. 135–150

R.G. Palmer, W.B. Arthur, J.H. Holland, B. LeBaron, P. Tayler, Artificial economic life: a simple model of a stock market. Physica D **75**, 264–274 (1994)

R. Pancs, N. Vriend, Schelling's spatial proximity model of segregation revisited. J. Public Econ. **91**, 1–24 (2007)

A. Pusch, S. Weber, M. Porto, Impact of topology on the dynamical organization of cooperation in the prisoner's dilemma game. Phys. Rev. E **77**(3), 036120 (2008)

W. Raub, V. Buskens, M.A. Van Assen, Micro-macro links and microfoundations in sociology. J. Math. Sociol. **35**(1–3), 1–25 (2011)

A. Rees, Information networks in labor markets. Am. Econ. Rev. **56**(1/2), 559–566 (1966)

T. Rogers, A.J. McKane, A unified framework for Schelling's model of segregation. J. Stat. Mech. Theory Exp. **2011**(07), P07006 (2011)

J. Sakoda, The checkerboard model of social interaction. J. Math. Sociol. **1**, 119–132 (1971)

F.C. Santos, J.M. Pacheco, Scale-free networks provide a unifying framework for the emergence of cooperation. Phys. Rev. Lett. **95**(9), 098104 (2005)

F.C. Santos, M.D. Santos, J.M. Pacheco, Social diversity promotes the emergence of cooperation in public goods games. Nature **454**(7201), 213–216 (2008)

T. Schelling, Dynamic models of segregation. J. Math. Sociol. **1**, 143–186 (1971)

T. Schelling, A process of residential segregation: neighborhood tipping, in *Racial Discrimination in Economic Life*, ed. by A. Pascal (D.C. Heath, Lexington, 1972), pp. 157–84

T. Schelling, *Micromotives and Macrobehavior* (W.W. Norton and Company, New York/London, 1978)

P. Tayler, Modelling artificial stock markets using genetic algorithms, in *Intelligent Systems for Finance and Business*, ed. by S. Goonatilake, P. Treleaven (Wiley, New York, 1995), pp. 271–287

B. Uzzi, The sources and consequences of embeddedness for the economic performance of organizations: the network effect. Am. Sociol. Rev. **61**(4), 674–698 (1996)

D. Watts, S. Strogatz, Collective dynamics of 'small-world' networks. Nature **393**, 440–442 (1998)

A. Wilhite, Bilateral trade and 'small-world' networks. Comput. Econ. **18**, 49–64 (2001)

S. Wolfram, Statistical mechanics of cellular automata. Rev. Mod. Phys. **55**(3), 601–644 (1983)

S. Wolfram, *Cellular Automata and Complexity: Collected Papers* (Westview Press, Boulder, 1994)

S. Wolfram, *A New Kind of Science* (Wolfram Media, Champaign, 2002)

C.H. Yeh, S.H. Chen, Market diversity and market efficiency: the approach based on genetic programming. J. Artif. Simul. Adapt. Behav. **1**(1), 147–165 (2001)

# Chapter 6
# Calculus-Based Econophysics with Applications to the Japanese Economy

**Jürgen Mimkes**

**Abstract** In engineering calculus of closed Stokes integrals leads to the Carnot circuit of motors or refrigerators with two levels: hot and cold. In economics calculus of closed Stokes integrals leads to capitalistic circuits with two levels: capital and labour in production and buy cheap and sell expensive in trade. There are many parallels between motors and economies: A heat pump can extract heat from a cold river and heat up a warm house. Production, trade and banking may extract capital from a poor population and make a rich population richer! A running motor gets hotter, and the efficiency, the difference in temperatures, grows with time. By foreign trade as well as in internal markets, capitalistic economies like the USA, China and others get richer, and the efficiency, the difference between rich and poor, grows with time.

However, Japan is different: If the cooling of a motor breaks down, the lower temperature rises too high, the motor runs hot and stops. In foreign trade Japanese wages have risen too close to US levels, and Japanese automobile exports to the USA have stagnated. Japanese wages need to cool down. At a refrigerator the upper temperature (outside) stays constant, and the inner temperature cools down. The Japanese internal market presently follows this refrigerator effect: incomes of the lower classes cool down at constant incomes of the upper class. Again, the difference between rich and poor in Japan also grows with time. The cooling down of lower incomes in Japan will make foreign trade with the USA profitable again and stop stagnation.

## 6.1 Introduction

Social and natural sciences have developed independently in the last three centuries with little interaction. Social sciences are of philosophical nature based on economic principles. Natural sciences rely on mathematical equations and on experiments. Only 30 years ago, scientists have started to discover the close relationship

---

J. Mimkes (✉)
Physics Department, Paderborn University, Warburger Str. 100, Paderborn, D-33098, Germany
e-mail: juergen.mimkes@upb.de

 
Y. Aruka, A. Kirman (eds.), *Economic Foundations for Social Complexity Science*, Evolutionary Economics and Social Complexity Science 9,
DOI 10.1007/978-981-10-5705-2_6

between different fields of social and natural sciences (Mantegna and Stanley 2000; Yakovenko and Rosser 2009; Mimkes 2010; Mimkes 2012; Richmond et al. 2013). The basic idea of the present paper is to introduce mathematics as a common basis of social and natural sciences. It will be easier to understand natural or capitalistic production by looking at motors, refrigerators or heat pumps. The convergence of economic and natural sciences occurs due to the common mathematical basis of calculus and probability. However, a full convergence of the complex fields of socioeconomics and natural sciences may only be expected after students in both fields are given a similar mathematical foundation in calculus and probability theory.

### *6.1.1 Ex Ante–Ex Post*

Is it possible to predict the outcome of investments like the flight of an arrow? Of course, not! Economic terms are divided into terms that can be predicted, ex ante, and terms that can only be answered afterwards, ex post. Investments are definitely ex post terms. But economic theory does not provide mathematical means to classify economic terms as ex ante or as ex post. This leads to grave errors in neoclassical theory.

*Example* The Solow model of economic growth is based on the relationship, $Y = F(K, L)$; economic output is a function of capital and labour. However, this law cannot be correct. Income ($Y$) is an ex post term; we can file our tax returns only at the end of the fiscal year. And functions are ex ante terms; we may calculate a function $F(K, L)$ at any time, before or after we earned the money. Accordingly, we find $Y \neq F(K, L)$; income cannot be a function, and ex post cannot be equal to ex ante!

Neoclassical theory introduces a factor $A$ (advancement of technology, human capital, etc.) in the production function. But this approach still leaves $Y \neq F(A, K, L)$ as an ex ante function, but output ($Y$) is still ex post!

Does this mean a production function ($F$) cannot exist? The answer is the production function ($F$) exists, but it is coupled to output ($Y$) in a different way.

## 6.2 Calculus in Economics

In order to understand the important problem of ex ante and ex post, we have to introduce calculus to economic theory!

### *6.2.1 Exact–Not Exact*

Economic theory depends on two production factors, capital and labour. In two dimensions we have two variables x and y and two differential forms:

1. Exact differential forms $\mathrm{d}f = a\mathrm{d}x + b\mathrm{d}y$ with $\partial b/\partial x = \partial a/\partial y$ have a stem function and correspond to ex ante economic terms.

2. Not exact differential forms $\delta g = a\mathrm{d}x + b\mathrm{d}y$ with $\partial b/\partial x \neq \partial a/\partial y$ do not have a stem function and correspond to ex post economic terms.
3. A not exact differential form may be transformed into an exact differential form by an integrating factor ($\lambda$): $\delta g = \lambda \mathrm{d}f$.

*Example* This law is the solution to the problem of the Solow growth model: $\delta Y = \lambda \mathrm{dF}$.

Output as an ex post term is represented by a not exact differential ($\delta Y$), which may be linked to an exact differential form (d$F$) or ex ante term by an integrating factor ($\lambda$). Production function ($F$) and integrating factor ($\lambda$) always exist. More details for $\lambda$ and $F$ will be discussed, below.

### 6.2.2 Riemann–Stokes

1. Exact differential forms lead to Riemann integrals; they depend on the limits ($A$) and ($B$) and are path independent. Riemann integrals are well known in economic theory. The closed Riemann integral is always zero; the integral with path A B is equal to the integral with path B A. The closed path corresponds to a ring.
2. Not exact differential forms lead to Stokes integrals; they depend on the limits ($A$) and ($B$) and on the path of integration! A different path of integration leads to a different result!

The closed Stokes integral is never zero; it is positive or negative, as the path A B differs from the path B A. The closed path corresponds to a spiral going up or down.

*Example* Surplus is a closed Stokes integral and an ex post term that depends on the path of integration, which means: A different way of production with the same production factors ($A$, $K$, $L$) leads to a different surplus! Accordingly, surplus is either positive or negative, spiraling up or down.

### 6.2.3 Stokes Integrals

Before we start applying Stokes integrals to economic theory, it may be helpful to discuss some examples. Figure 6.1 shows a closed Stokes integral from A to B and back from B to A.

The closed Stokes integral may be interpreted as a monetary balance of households: Income $\mathrm{Y_H}$ is paid from (A: industry) to (B: households), and consumption costs $\mathrm{C_H}$ are paid from (B: households) to (A: industry). The Stokes integral is path dependent; $\mathrm{Y_H}$ and $\mathrm{C_H}$ differ! This balance may be calculated as an excel sheet and represents a spiral: a positive balance goes up, and a negative balance goes down.

Figure 6.2 shows Stokes integrals from various scientific fields.

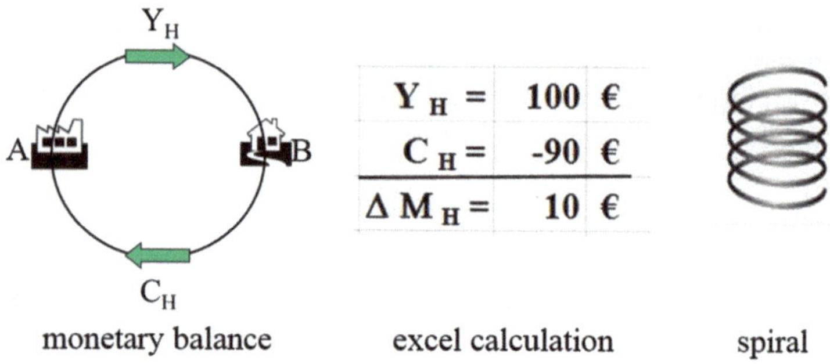


**Fig. 6.1** A Stokes integral may be interpreted as an economic balance, as an excel calculation or as a spiral winding up or down

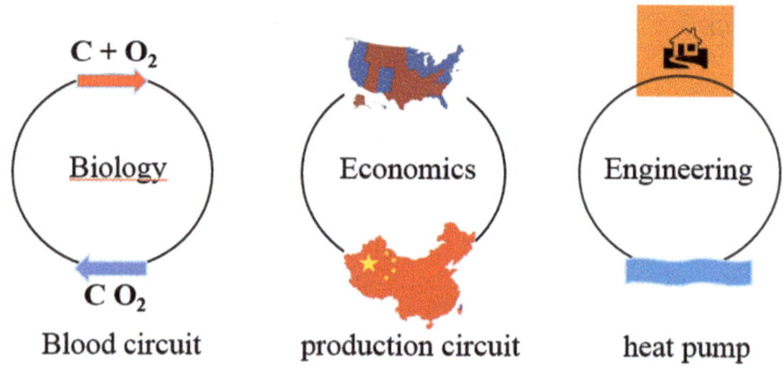


**Fig. 6.2** Three examples of a closed Stokes integral: the blood circuit in biology, the production circuit in economics and the circuit of a heat pump in engineering

In biology the blood circuit is a typical example of a closed Stokes integral. The arteries carry carbon and oxygen and potential energy of combustion from the lungs to the muscles. On the way back the veins carry carbon dioxide without energy from muscles to lungs. In each cycle the heat of combustion is the energy used by the muscles.

In economics the production circuit is a closed Stokes integral. Commodities may be produced at low wages in China, and they may be sold afterwards at a high price in the USA. Each cycle leads to a profit, the difference between prices and costs.

In engineering the closed Stokes integral is the basis of heat pumps. The pump may extract heat with little effort from a cold river. Afterwards the pump will heat up a warm room with high efficiency. All three models are equivalent and will be used in economics.

## 6.3 The Laws of Economics in Integral Form

We will now apply calculus to economic theory of natural and modern production.

### 6.3.1 *Natural Production (F. Quesnay)*

The natural production circuit is a further example of a closed Stokes integral. One of the first economists, the French medical doctor Françoise Quesnay (1694–1774), was inspired by the human blood circuit to look at natural production not as a linear process, but as a production circuit: Every day again a number ($N$) of workers go to work from households into the fields, the capital ($K$) of the village, and in return bring home consumption goods from the fields, Fig. 6.3.

The balance of production is an ex post term and cannot be predicted in advance; the natural productive circuit ($\delta P$) is a closed Stokes integral,

$$\oint \delta P = \int_{\text{Field}}^{\text{H}} \delta P_1 + \int_{\text{H}}^{\text{Field}} \delta P_2 = \text{G}_\text{H} - \text{W}_\text{H} = \Delta \text{P}_\text{H} \tag{6.1}$$

Household goods ($G_\text{H}$) and work input ($W_\text{H}$) are Stokes line integrals of the same closed production circuit, and ($\Delta P_\text{H}$) is the production surplus of households. After each circuit the surplus rises and spirals up by ($\Delta P_\text{H}$). Labour is not just the number of labourers ($N$), but real physical work. Work and consumption goods are part of the same production circuit and can be measured in units of energy per circuit, in Joules, kWh or kcal. The length of the cycle depends on the system: A normal person requires a meal at least once a day. Blue-collar workers often get paid by the week, white-collar workers by the month and a farmer needs to harvest at least once a year. The number of circuits is dimensionless.

The energy output of food from the fields ($\Delta P_\text{H}$) must be equal or larger than the energy input by work. But in order to survive in critical situations without

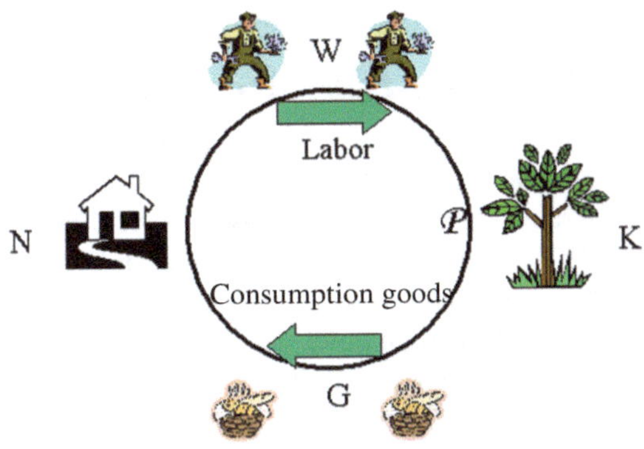


**Fig. 6.3** Shows the model of a closed natural production circuit (P). Labourers from households work (W) in the fields, the capital (K) of the village, and consumption goods (G) are brought back from agriculture to households

production (illness, winter, war, etc.), the energy balance must be positive for most cycles. And additional energy must be provided for children, for population growth. Only a positive balance leads to survival and growth, a negative balance to destruction and death. This law of survival leads to permanent growth rather than sustainable production in most economies.

### 6.3.2 Modern Production

In modern times people work in industry rather than on a field, and they obtain wages to buy food in stores. Let us look at a typical situation:

*Example A household (H) works in industry (In) earning 100 € per day and spending 90 € for food and goods. The surplus is 10 € per day.*

Modern production is a two-circuit process with a monetary and a productive circuit. The Stokes integral of the productive circuit,

$$\oint \delta P = \int_{\text{Field}}^{\text{H}} \delta P_1 + \int_{\text{H}}^{\text{Field}} \delta P_2 = \text{G}_\text{H} - \text{W}_\text{H} = \Delta P_\text{H} \tag{6.1a}$$

The productive circuit ($\delta P$) is given by commodities ($G_\text{H}$) and labour input ($W_\text{H}$) and is measured in energy units, like J, kWh and kcal.

The monetary balance ($\delta M$) of income ($Y_\text{H} = 100€$), costs ($C_\text{H} = 90€$) and surplus ($\Delta M_\text{H} = 10€$) may again be represented by a closed Stokes integral:

$$\oint \delta M = \int_{In}^{\text{H}} \delta M_1 + \int_{\text{H}}^{In} \delta\, M_2 = \text{Y}_\text{H} - \text{C}_\text{H} = \Delta \text{M}_\text{H} \tag{6.2}$$

Income and costs are Stokes line integrals of the same closed circuit measured in monetary units, like \$, €, £ and ¥. If the closed Stokes integral is positive, the spiral goes up after each round by $\Delta M_\text{H}$. If the Stokes integral is negative, the spiral goes down after each round by $\Delta M_\text{H}$. Again industry as the capital of society is not part of these equations.

### 6.3.3 Double-Entry Accounting

We are now ready to start looking for a basic law of economics, from which modern production and trade as well as all other economic activities may be derived. In physical sciences every field has a certain natural basis: Mechanics is ruled by Newton's law, from which all movements of apples, atoms or stars may be derived. Electrodynamics is governed by Maxwell's equations, and all modern communication systems like radio, TV, computer and smart phone are based on these equations. And the first and second law of thermodynamics may be applied to modern engineering; to motors, generators, heat pumps and refrigerators; to chemistry; to metallurgy; or to the weather. And a similar law must now be found to solve the problems of economics! This law is called "double-entry accounting" and is applied to professional production and management in all modern states.

According to Wikipedia "Luca Pacioli, a Franciscan friar, first codified the system in his mathematics textbook *Summa de arithmetica, geometria, proportioni et proportionalità* published in Venice in 1494. Economist Benedetto Cotrugli's 1458 treatise *Della mercatura e del mercante perfetto* contained the earliest known manuscript of a double-entry bookkeeping system; however, Cotrugli's manuscript was only officially published in 1573."

We may apply again the above example of a household ($H$) working in industry (In) earning 100€ per day and spending 90€ for food and goods. The surplus is 10 € per day. The excel calculation for the monetary and the productive circuits is given by:

$$\begin{array}{rr@{\qquad}rr} Y_{\mathrm{H}} = & 100\,€ & W_{\mathrm{H}} = & -100\,€ \\ C_{\mathrm{H}} = & -90\,€ & G_{\mathrm{H}} = & 90\,€ \\ \hline \Delta M_{\mathrm{H}} = & 10\,€ & \Delta P_{\mathrm{H}} = & -10\,€ \end{array}$$

Monetary account + Productive account = 0

## 6.3.4 The Fundamental Law of Economics in Integral Form

The sum of the monetary ($\delta M$) and the productive ($\delta P$) circuit is zero; this is the basic law of double-entry accounting of money-based societies. According to calculus, we may write the monetary ($\delta M$) and productive ($\delta P$) circuits as closed Stokes integrals:

$$\oint \delta M + \oint \delta P = 0 \tag{6.3}$$

This integral equation of the double-entry balance may be considered as the basic law of macro- and microeconomics! It may also be written as

$$\oint \delta M = -\oint \delta P \tag{6.3a}$$

The monetary circuit ($\delta M$) measures the production ($\delta P$) circuit in monetary units. Economics in integral forms leads to the integrals of production ($P$), goods ($G$) and labour ($W$) and to the integrals of surplus or profit ($M$), of income ($Y$) and costs (C), measured in US \$, €, £ or ¥. But before we can apply the fundamental law of economics in Eq. (6.3a) to production (Chap. 4), we have to look at the laws of economics in differential forms.

## 6.4 The Laws of Economics in Differential Forms

The differential laws of economics may be derived from the integral law (6.3a).

### 6.4.1 The First Law of Economics

The first law of economics in differential forms is

$$\delta \mathrm{M} = \mathrm{d}\, \mathrm{K} - \delta\, \mathrm{P} \tag{6.4}$$

Integrating Eq. (6.4) by a closed integral leads back to the fundamental law, Eq. (6.3a). The new term (d*K*) is an exact differential form, and the closed integral of (d*K*) will be zero.

The first law contains three differential forms with the common dimension *money*:

$\delta P$ is a not exact differential and refers according to Eq. (6.1) to the *ex post* term of production, to goods ($\delta G$) and labour ($\delta W$).

$\delta M$ is inexact and refers according to Eq. (6.2) to the *ex post* terms income ($\delta Y$), costs ($\delta C$) and surplus.

d*K* is exact and (*K*) a function. This is the reason why capital does not appear in the Stokes integrals of monetary and productive circuits. The meaning of (d*K*) is capital, the fields of the farmer, the company of the entrepreneur, the industries of an economy and the earth of all beings. Capital (*K*) includes all unperishable resources like property, firms, houses and money in cash that may be calculated by a function (*K*). This will be discussed in more detail, below.

The first law is the differential balance of every economic system: profits ($\delta M$) depend on capital (d*K*) and labour ($\delta P$). This result is well known in economics, but so far, it has not been stated in a proper mathematical form by a differential equation.

### 6.4.2 The Second Law of Economics

According to calculus (Chapter 6.5), a not exact differential form ($\delta M$) may be transformed into an exact differential form (d*F*) by an integrating factor $\lambda$,

$$\delta \mathrm{M} = \lambda\, \mathrm{d}\, \mathrm{F} \tag{6.5}$$

We may call this the second law of economics. ($\delta M$) is not exact or *ex post*. For positive values ($\delta M$) refers to production output, profits or income ($\delta Y$). At negative values ($\delta M$) refers to losses or costs ($\delta C$). For positive values ($\delta M$) may be replaced by income ($\delta Y$): $\delta Y = \lambda \mathrm{d}F$. The second law replaces the erratic Solow equation $Y = F\,(K, N)$.

These are the new laws of economics in differential form.

### 6.4.3 Economics and Thermodynamics

The differential laws of calculus-based economics have the same mathematical structure as the laws of thermodynamics:

$$\delta\,\mathrm{Q} = \mathrm{d}\,\mathrm{E}–\delta\,\mathrm{W} \tag{6.6}$$

In thermodynamics heat ($\delta Q$) and work ($\delta W$) are not exact differential forms; they are measured in energy units. The not exact differential of heat ($\delta Q$) may be linked to an exact differential form of entropy (d$S$) by an integrating factor ($T$):

$$\delta\,\mathrm{Q} = \mathrm{T}\,\mathrm{d}\,\mathrm{S} \tag{6.7}$$

The formal equivalence of Eqs. (6.4 and 6.5) and (6.6 and 6.7) seems striking. But this is not a formal coincidence, but rather an identity:

The productive circuit is originally measured in energy units. In double-entry accounting, the monetary circuit measures the productive circuit in monetary units. By applying the oil price fixing, this law may be reversed: the productive circuit measures the monetary circuit in energy units! The laws of thermodynamics are the laws of economics in energy units! This leads to an identity of terms in economics and thermodynamics, as shown in Table 6.1.

Table 6.1 is an indication of the convergence of economics and social sciences. And Table 6.1 makes it easy to identify the new economic terms of the first and second law in Eqs. (6.4) and (6.5) by the thermodynamic terms in Eqs. (6.6) and (6.7)): Surplus, income and costs correspond to heat, capital to energy and labour to physical work. But it is more difficult to identify the integrating factor ($\lambda$) and the new production function ($F$).

**Table 6.1** Corresponding functions of economics and thermodynamics

| Symbol | Economics | Unit | | Symbol | Thermodynamics | Unit |
|---|---|---|---|---|---|---|
| **M** | Income, costs | €, $, £, ¥ | ↔ | **Q** | Heat | kWh |
| **K** | Capital | €, $, £, ¥ | ↔ | **E** | Energy | kWh |
| **P** | Production, labour | €, $, £, ¥ | ↔ | **W** | Work | kWh |
| **λ** | Mean capital | €, $, £, ¥ | ↔ | **T** | Mean energy | kWh |
| **F** | Production function | – | ↔ | **S** | Entropy | – |
| **N** | Number | – | ↔ | **N** | Number | – |

### 6.4.4 Standard of Living ($\lambda$) as Economic Temperature

The integrating factor ($\lambda$) exists in all economic systems, in production, in markets and finance. In interacting economic systems and efficient markets, we always find a common economic level ($\lambda$). A market will lead to a common price level ($\lambda$) for each commodity; an economy will have a common mean standard of living ($\lambda$). The integrating factor $\lambda$ is the mean capital and equivalent to temperature ($T$), the mean energy of an atomic system.

The equivalence of ($T$) and ($\lambda$) is shown in Fig. 6.4: GDP per capita and energy consumption per capita are equivalent for most of the 126 largest countries in the world. Both mean capital and mean energy consumption follow the same line.

The standard of living ($\lambda$) may be defined by the mean capital per capita,

$$\lambda = \mathrm{K/c\,N}. \tag{6.8a}$$

It is proportional to the social temperature or energy consumption per capita,

$$\mathrm{T} = \mathrm{E/c\,N}. \tag{6.9}$$

The constant c may be called specific income and reflects the degrees of freedom to obtain income from work, stocks, etc. It corresponds to the specific heat or degrees of freedom in thermodynamics.

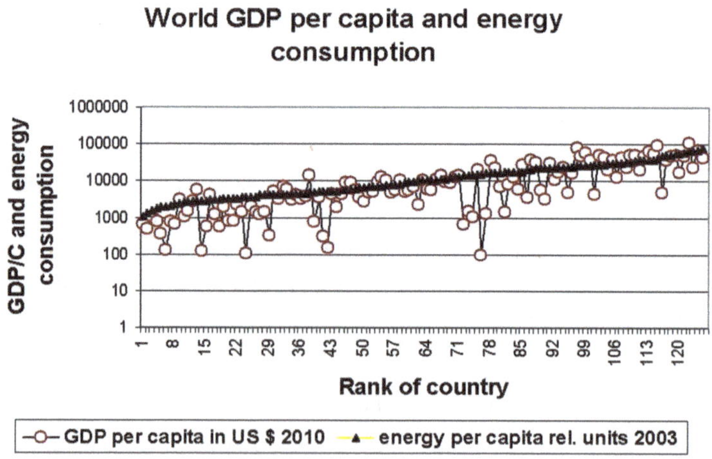


**Fig. 6.4** GDP per capita and energy consumption per capita follow the same line for most of the 126 largest countries in the world. Both mean capital and mean energy consumption are equivalent. Economics and thermodynamics are converging

### 6.4.5 Capital

According to the first law, capital is a monetary term represented by an exact differential form (d K). Capital corresponds to energy in thermodynamics.

The difference between capital (*ex ante*) and income (*ex post*) is clear for a boy that goes out to work at a restaurant. He has five dollars in his pocket, which will be his capital tonight. But he does not know how much he will earn from tips at the restaurant. He will count his income afterwards. For the boy capital is the cash money in his pocket.

According to Eqs. (6.10 and 6.11), capital is equivalent to goods that do not lose in value, $\delta M = 0$:

$$\delta \mathrm{M} = \mathrm{d\,K} - \delta \mathrm{P} = 0 \quad (6.10)$$

$$\mathrm{d\,K} = \mathrm{d\,P} \quad (6.11)$$

Capital (d$K$) is equivalent to production means (d$P$) of permanent value, $\delta M = 0$. The capital of a farmer is given by the fields; they are the basis for production and profit. The capital of a company is the production plant. The capital of countries is the resources like water, air, land and oil. But capital is also knowledge, technology, industry and education. Noble metals like gold or platinum do not rust and are regarded as capital. And money is also capital as long as we have no inflation, $\delta M = 0$.

### 6.4.6 Entropy as the New Production Function

The most important result is the equivalence of economic production function ($F$) and natural entropy ($S$), $F = S$,

$$\mathrm{S} = \ln\,\mathrm{P}\,(\mathrm{N}) \quad (6.12)$$

*P (N)* is the probability of arrangement of the $N$ elements (atoms) in the thermodynamic system. In natural science entropy is linking thermodynamics to statistical mechanics; in economics entropy is linking macroeconomics to microeconomics:

In a system with (i) different elements (atoms), the entropy may be calculated from the relative number $x_i = N_i/N$ of atoms; this is called the Shannon entropy:

$$\mathrm{F} \equiv \mathrm{S} = \mathrm{N}\,\Sigma \mathrm{x_i} \ln\,\mathrm{x_i} \quad (6.13)$$

Accordingly, the production function of an economic system of (i) different elements (goods, people, etc.) may be calculated from the relative number $x_i = N_i/N$ of people, goods, etc. The function is shown in Fig. 6.5.

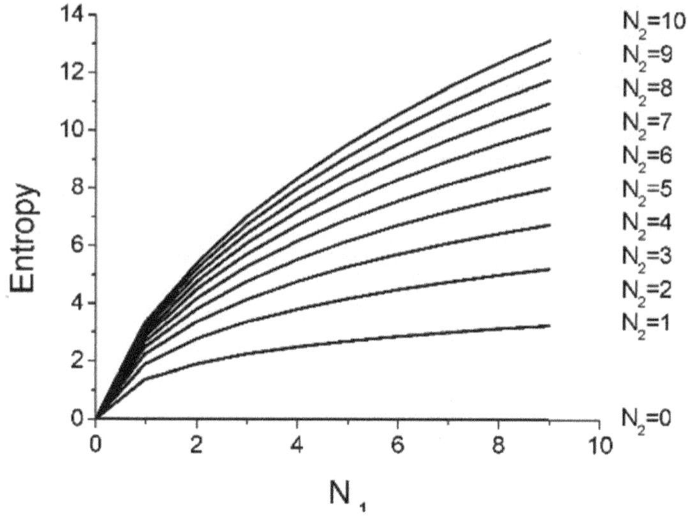


**Fig. 6.5** The entropy production function $F(N_1, N_2)$ in Eq. (6.13) is plotted versus $N_1$ in the range from 0 to 10. The parameter $N_2$ is in the range from 0 to 10

Entropy replaces the Cobb–Douglas production function of neoclassical economics,

$$F = -N\,(x_i)^{\alpha}\left(x_j\right)^{\beta} \tag{6.14}$$

Both functions are very similar, but according to Figs. 6.5 and 6.6, entropy is larger by a factor of 1,4. Entropy is the natural and optimal production function.

Entropy in Fig. 6.5 is the natural production function, which is always larger than the Cobb–Douglas function in Fig. 6.6. In addition there is no elasticity in calculus-based economics, which is used in standard economics to adjust to real data. Beyond numerical values, entropy has a very general meaning, as will now be discussed.

### 6.4.7 Entropy and Work

The new production function entropy is a measure of disorder. This leads to a new understanding of entropy in thermodynamics and economics:

1. *Thermodynamics: Combining the first and second law of thermodynamics, we obtain*

$$T\,d\,S. = d\,E – d\,W \tag{6.15}$$

*A light breeze in a park with the energy (dE) will easily empty a paper basket and generate more and more disorder (dS > 0). The paper will distribute throughout*

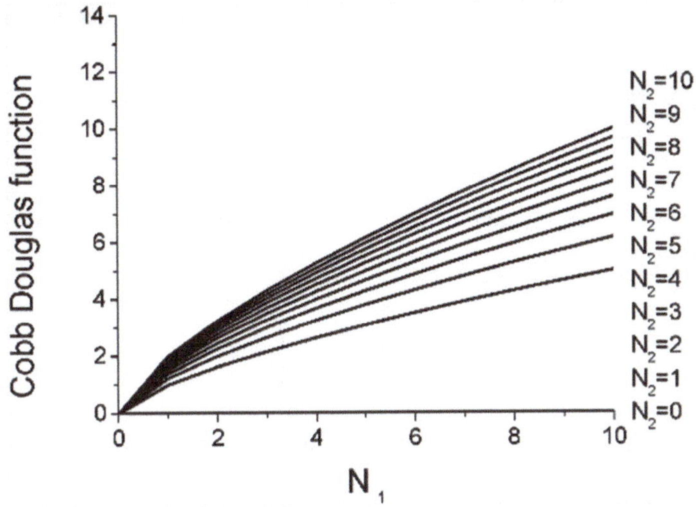


**Fig. 6.6** The Cobb–Douglas production function $F(N_2) = AN_1^{\alpha}N_2^{1-\alpha}$ in Eq. (6.14) is plotted versus $N_1$ in the range from 0 to 10. The parameter is N $_2$ in the range from 0 to 10. The parameters are $N = 1$ and $\alpha = 0{,}7$

**Fig. 6.7** Production of automobiles requires the ordering of many parts: screws, nuts, bolts, wires, tires, wheels, etc. All parts have to be placed in the correct position and in the right order; hence entropy d$F < 0$

*the park and never come back into the basket. Positive entropy means creating disorder or distributing items.*

*But a janitor may work ($\delta W$) and sweep the paper together and put it back into the basket.*

*Work reduces disorder: (d$S$ < 0). Negative entropy means reducing disorder or ordering, collecting items.*

2. *Economics: In the same way we may combine the first and second law of economics,*

$$\lambda \, \mathrm{d}\, \mathrm{F} = \mathrm{d}\, \mathrm{K} - \delta\, \mathrm{P} \tag{6.16}$$

*The capital (d$K$) you pay for a snack will easily empty your purse and distribute (dF > 0) the money to the shop keeper. The money will never come back into the* *purse. But in the afternoon you may work (δP) in the office, and the money will come back into your purse.*

We may solve Eq. (6.16) for ($\delta P$):

$$\delta \mathrm{P} = \mathrm{d\,K} - \lambda\, \mathrm{d\,F} \tag{6.17}$$

Labour (d$P$) increases capital (d$K$) and reduces disorder (−d$F$). Work means ordering:

Entropy reduction also applies to mental work. Mental work orders ideas like in a puzzle:

$$\text{Brain work}: \quad \mathrm{g+i+r+r+n+o+d+e} \quad \rightarrow \quad \text{ordering}$$

Medical doctors order deficiencies within a body; teachers order or develop the minds of young people. Housewives have known for long times that making order is hard (unpaid) work! This may be one reason that today most women prefer to work as professionals outside of the house.

### *6.4.8 Production Costs*

According to Eq. (6.17), production costs ($\delta P$) depend on capital or energy costs (d$K$), on the amount of ordering (−d$F$), and on the standard of living ($\lambda$) of the producing country. Many companies produce cars with the same energy costs and parts in the same order (d$F$) in Europe and China. However, wages are much lower in China than in Europe due to the lower standard of living ($\lambda$) in China.

## 6.5 The Mechanism of Production and Trade

We will now go back to double-entry accounting to discuss the mechanism of production and trade. Modern economic production contains two circuits, the monetary circuit ($\delta M$) and the production circuit ($\delta P$), Fig. 6.8; the sum according to double-entry accounting in Eq. (6.3) is zero!

$$\oint \delta M + \oint \delta P = 0 \tag{6.3}$$

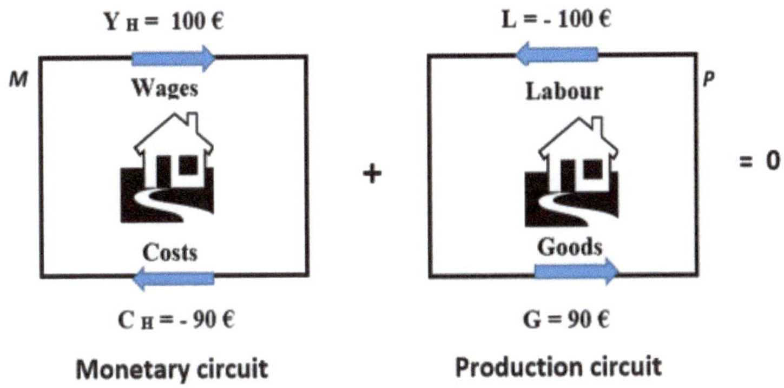


**Fig. 6.8** The production process contains two balances or two circuits, the monetary circuit ($\delta M$) and the production circuit ($\delta P$). The sum of both is zero. The balances are closely tied to a specific economic system, to a person, a household or an industry. This has to be marked by a proper index, like $Y_{John}$, $C_H$ or $M_{Ind}$

### 6.5.1 *Production Factors: λ and F*

We will now investigate the proper production factors or variables for a coordinate system of the economic circuits ($\delta M$) and ($\delta P$). For this we go back to the second law of economics:

$$\delta \, M = \lambda \, d \, F \tag{6.5}$$

Income and costs are part of the same closed integral of the monetary circuit ($\delta M$), which depends on the integrating factor or standard of living ($\lambda$) and the production function or entropy ($F$). In a two-dimensional system like economics, we may choose two independent variables or production factors. In the present frame, it is most useful to introduce the standard of living ($\lambda$) and the production function or entropy ($F$) as production factors. Other production factors will be discussed in a later chapter.

### 6.5.2 *Mechanism of Cyclic Production*

We will now discuss permanent (cyclic) production and trade. Within one cycle general costs and prices are constant by contract or by short cycle times. Monetary and productive circuits are presented in a $\lambda$–$F$ coordinate system, Fig. 6.9.

*Example: Europe importing coal from South Africa* In Fig. 6.9 coal import–export is presented in a $\lambda$–$F$ coordinate system. The y-axis is given by ($\lambda$) representing the mean capital or price of coal, which is low in South Africa (SA) and more expensive in Europe (EU). The x-axis is given by entropy ($F$): growing $F$, $dF > 0$, means

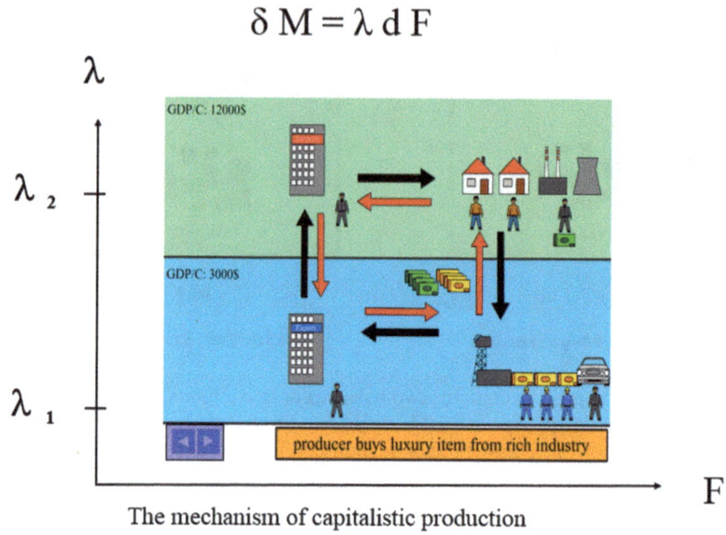


**Fig. 6.9** Europe (EU) buying coal from South Africa (SA): The $\lambda$-axis represents the price of coal, which is low ($\lambda_1$) in SA and high in the EU ($\lambda_2$). The F-axis is given by entropy: growing entropy, $dF > 0$, means distributing of coal or money; $dF < 0$ means collecting of coal or money (see text)

distribution (coal, money). Reducing $F$, $dF < 0$, means collecting (coal, money) or buying. The productive circuit discusses collecting and distributing of coal; the monetary circuit discusses collecting and distributing money.

Figure 6.9 shows Europe (EU) buying coal from South Africa (SA) in a productive and a monetary circuit: In the productive circuit, workers are collecting cheap coal from the mines in SA bringing it to the harbour of Cape Town. From there cheap coal ($\lambda_1$) is shipped to Rotterdam, where coal is more expensive ($\lambda_2$). From Rotterdam the expensive coal is distributed to the consumers.

In the monetary circuit, the importer in Rotterdam collects the money from the consumers in the EU and sends it to the exporter in Cape Town. Of course, he keeps some of the money for himself. The exporter in Cape Town distributes the wages to the workers, but he also keeps some money for himself.

In the productive circuit, coal is collected (bought) at a low price and distributed (sold) at a higher price. In the monetary circuit, much money is collected (earned) in the EU, and less money is distributed (spent) in SA. The difference between both levels is the surplus, profit or output ($\Delta M$) of this trade, which is distributed between both sides: the importer in the EU and the exporter in SA. Both sides have gained and raised their standards of living ($\lambda$). The next trade will occur at a higher level ($\lambda$) for both sides.

This is the mechanism of all economic actions. Production and trade in shops, companies, banks and economies follow this two-level mechanism: collecting (buying) at a low price level and distributing (selling) at a high price level. At each level the exchange of money and goods is without profit. Nobody would go to a market, where he feels being robbed! The workers in SA get the normal wages for their work; the customers in the EU pay the normal price for coal. The surplus is generated by the two-level trade in the EU and in SA.

Every trader has to buy cheap and sell expensive; trade needs two price levels. But two price levels are a problem in neighbourhoods. For this reason traders used to travel long distances to sell their products at a profitable price. Spices, which were cheap in the Indies, would bring wealth to the traders in Europe. Today export still is a way to generate price differences. Within a country or town, trade is possible between producers of different products: an apple farmer cannot sell his apples to another apple farmer, but to a farmer of pears or plums. We cannot get richer by cutting each other's hair, but a tailor can sell his products to a baker and vice versa. The two-level system is the basis of production and trade.

### *6.5.3 Two-Level Systems*

In Fig. 6.2 the Stokes integral has been applied to the production circuit of economics and to the Carnot circuit of motor and heat pump engineering. Both fields have similar properties:

In engineering all machines are two-level systems: in motors, generators, heat pumps and refrigerators, we always have two temperatures, hot and cold. If the two temperatures are the same, the machine will stop: if the motor is either cold or is running hot, it cannot run. If the door is open, a refrigerator cannot run.

In economics all actions are two-level systems: in production we have capital and labour, in trade we have buy and sell, in banking we have savers and investors and in societies we have rich and poor. If the price is the same in buying and selling, trade will stop.

### *6.5.4 Efficiency*

The efficiency ($\eta$) of a Carnot motor is proportional to the difference of the two levels,

$$\eta_{\text{ideal}} = \frac{Y_2 - Y_1}{Y_1} = \frac{\lambda_2 - \lambda_1}{\lambda_1} \tag{6.24a}$$

In economics the efficiency is given by the difference of prices in producing or buying and selling. If the levels are the same, the efficiency is zero.

## 6.6 Economic Growth

In motors the heat of combustion is distributed to the hot and cold side of the motor. This leads to a uniform high temperature that will lead to zero efficiency. In order to prevent a stop of the motor, the outside of the motors is cooled by water or air. This keeps a motor running and raises the efficiency.

Economic growth depends on the distribution of the surplus of a production circuit to the higher and the lower levels. In companies this distribution is generally negotiated between capital and labour unions or enforced by strikes. Economic growth may be calculated from the levels $\lambda_1$ and $\lambda_2$ of capital and labour. The cycle time is chosen to be short enough to keep the value of ($\lambda$) constant. According to the second law (6.5), income is now: $Y = \lambda \Delta F$. In each cycle the production process ($\Delta F$) remains constant. Income ($Y$) is proportional to ($\lambda$), and we may calculate the growth of income ($Y$) of any production process ($\Delta F$):

$$\mathrm{d}Y_1(t) = p\,(Y_2 - Y_1)\,\mathrm{d}\omega t \tag{6.18}$$

$$\mathrm{d}Y_2(t) = (1-p)\,(Y_2 - Y_1)\,\mathrm{d}\omega t \tag{6.19}$$

$Y_1$ is the income of labour and $Y_2$ the income of capital owners. The factor p (in %) is the percentage of the surplus (or profit) after each circuit given to the lower side (labour). The factor ω is the frequency of circuits within a given time (t), like a year.

For p ≠ ½ the solution of this set of differential equations is given by:

$$Y_1(t) = Y_{10} + p\ \ [Y_{20} - Y_{10}]\,[\exp\ ((1-2p)\,\omega t) - 1]\,/\,(1-2p) \tag{6.20}$$

$$Y_2(t) = Y_{20} + (1-p)\,[Y_{20} - Y_{10}]\,[\exp\,((1-2p)\,\omega t) - 1]\,/\,(1-2p) \tag{6.21}$$

For $p = ½$ the solution is given by

$$Y_1(t) = Y_{10} + \frac{1}{2}\ [Y_{20} - Y_{10}]\ \omega t \tag{6.22}$$

$$Y_2(t) = Y_{20} + \frac{1}{2}\ [Y_{20} - Y_{10}]\,\omega t \tag{6.23}$$

The equations are presented in Fig. 6.10 and may be applied to all economic systems; to internal trade with labour and capital, unions and industry; or to international trade.

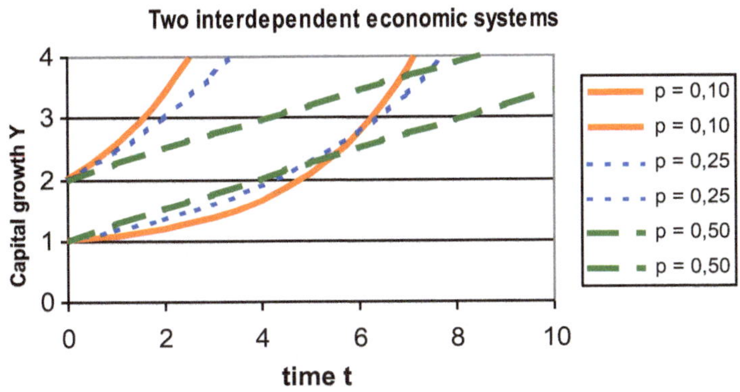


**Fig. 6.10** Economic growth depends on the proportion (p) of surplus after each circuit given to the lower side (labour), and $(1 - p)$ given to the higher side (capital). Calculus-based economic theory so far does not depend on time. But closed Stokes integrals lead to economic circuits and a number (ω) of cycles within a given time. The number of cycles is n = ω t. In this way time may be introduced as a new variable in economics

### *6.6.1 National Markets Between Capitalism and Communism*

We may classify economic systems by the different values of p, the way surplus is divided between the two levels ($Y_1$) and ($Y_2$), rich and poor and capital and labour. The value of $p$ is only a guide line, not an accurate value:

| | | | |
|---|---|---|---|
| (a) | $p = 0{,}00$: Capitalism | $\eta \to$ Max! | (Very high prices, very low wages) strong market |
| (b) | $p = 0{,}10$: Capitalistic | $\eta \to$ Max! | (High prices, low wages) strong market |
| (c) | $p = 0{,}25$: Socialism | $\eta \to$ Min! | (Lower prices, higher wages) weak market |
| (d) | $p = 0{,}50$: Fair deal | $\eta \to$ Constant! | (Prices and wages nearly equal) |
| (e) | $p = 0{,}75$: Stagnation | $\eta \to$ Decreasing! | (Low prices, too high wages) weak market |
| (f) | $p = 1{,}00$: Communism | $\eta \to 0$! | (One class system) the market will break down |

Capitalism, $p = 0$, $\eta \to$ Max!: Capital takes all surplus of the production circuit; labour receives nothing. This leads to exponential growth of capital and zero growth for labour. The parameter $p = 0$ is the pure form of capitalism with no chance for growth for labour. This market has been enforced by many countries in the time of the industrial revolution.

Capitalistic, $p = 0{,}10$, $\eta \to$ Max!: In many countries capital will take a very large amount of surplus, perhaps 90–98% of a production circuit, and labour will receive perhaps only 2–10%, as indicated by the solid line in Fig. 6.10. The efficiency ($\eta$) is maximal; capital and industry prefer high prices and low wages. But in contrast to pure capitalism, wages will indeed grow exponentially after some time due to the coupling to an exponentially growing industry, Eqs. (6.18 and 6.19). In addition, a lower rise in wages is often more easily obtained by negotiations; accordingly, in countries with lower wage rises, we observe fewer strikes. Lower wage rises and fewer strikes will strengthen industry, but with taxation also the economy.

The factor $p = 10\%$ (solid lines in Fig. 6.10) is—surprisingly—in the long run more profitable for workers than the higher value $p = 25\%$ (dotted lines in Fig. 6.10): At $p = 10\%$ the strong growing industry after some time pulls up the working class, as well, Eqs. (6.18 and 6.19). This effect has been called the German/Japanese economic miracle after World War II. Both countries had very low wages, and industry was exporting to the USA with high profits. Wages only rose many years later. This has been the basis for the present high standard of living in both countries. A similar development is observed today for countries like China.

Typical capitalistic countries in the last years have been the USA, Japan, Germany and China.

Socialism, $p = 0{,}25$, $\eta \rightarrow$ Min!: In countries following the ideas of socialism, workers will receive a higher percentage of annual surplus. These countries have strong unions, which prefer low prices and high wages and will enforce their ideas by strikes. Strikes weaken industry and the economy; the efficiency ($\eta$) in socialistic countries is minimal. Economic growth will not be as strong as in capitalistic production. The effect of the economic miracle is too vague; socialist people prefer present-day good life to future expectations.

A typical socialist country may be France and Italy.

Summary, $p < 0{,}5$, $\eta \rightarrow$ growing!: As long as capital receives more of the surplus of each cycle than labour, economic growth will be exponential for both, capital and labour. But economic growth will always be larger for capital than for labour. This leads to a rising difference between capital and labour and between rich and poor. This has already been observed experimentally by Piketty (2014), Fig. 6.11:

The rising difference between capital and labour and between rich and poor leads to a growing efficiency,

$$\eta \sim (\lambda_2 - \lambda_1) \rightarrow \max! \tag{6.24b}$$

In a well-running economy, the difference between rich and poor grows permanently. This fact is often described as opening of a scissor, Fig. 6.12.

A permanent growing efficiency is also observed in a starting motor. In the beginning the inside and outside of the motor have the same temperature ($T$); the efficiency is zero. Only after stimulating the motor with a crank or a battery, the temperatures start to differ and the efficiency grows; the motor keeps getting hotter and stronger.

Fair deal, $p = 0{,}5$: Surprisingly, the fair deal, the even distribution of profits between capital and labour, leads to linear growth. In times of dwindling resources, the fair deal is a solution to stop exponential growth. Capital and labour would have the same (linear) growth rate, as required by Piketty (2014), and the world would come closer to sustainable production. However, many economists argue with

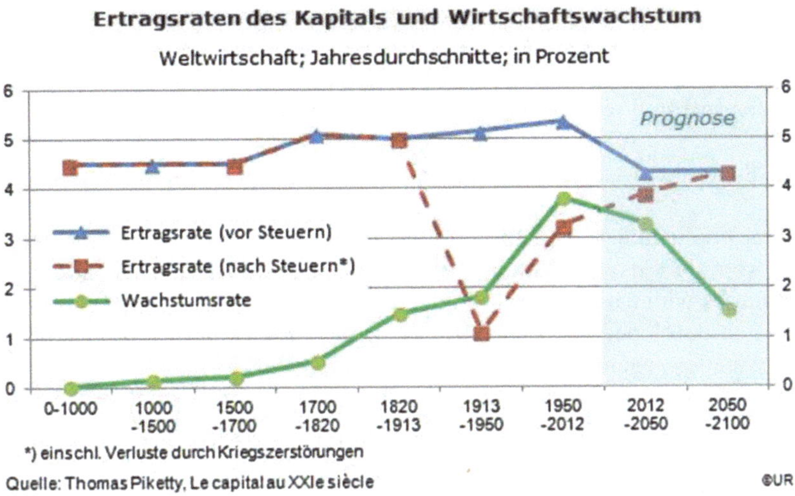


**Fig. 6.11** Growth of world economy since 1000 A.D. In the last 1000 years, capital has always grown by 4–6%, whereas labour has only grown by 0–4%. Since 1820 capital growth rates have been reduced by taxation. But presently, capital is again taxed less than labour

**Fig. 6.12** The open scissor reflects the growing difference between rich and poor, which is always observed for $p < 0{,}5$, when the poor side (labour) gets less than half after each production cycle and the rich side (capital) gets more than half

Julian Simon (1932–1998) that there is no need for economic restrictions, as human intelligence will always find new resources to survive. This view is supported by Darwin's law, the survival of the fittest, and by evolution: dinosaurs as symbols of growth have died out, but other species have survived.

Stagnation, $p = 0{,}75$, $\eta \to 0$: If the productivity of a country is reduced, it is politically very difficult to reduce wages or raise prices. Accordingly, labour may get a too high share of the annual surplus, ($p = 0{,}75$). In unstable democracies this is often supported by corruption and money disappearing in secret deposits like offshore tax heavens. This leads to a weak market; the efficiency of inner trade is decreasing.

Typical stagnating countries are Argentina and Greece.

Communism, $p = 1{,}00$: $\eta = 0$!: Communism is a very honest vision by Karl Marx, who observed the mechanism of production very thoroughly in his famous book the *Capital*. He wanted to avoid the exploitation of workers by the capitalists by abolishing the capitalistic class. The capital should be owned by the proletariat, by labour. Labour should get 100% of the surplus of production. However, according to the Carnot process of motors, which is equivalent to the production process, communism is like opening the door of the refrigerator and mixing hot and cold. The refrigerator will stop working. The efficiency with only one level is zero, $\eta \sim (\lambda_1 - \lambda_1) = 0$!

Typical communist countries have been USSR, German Democratic Republic, etc.

Natural production, $p = 1$, $\eta = 0$: In addition to economic systems that have evolved from the Eqs. (6.18 and 6.19) of modern (industrial) production, we still have countries with natural (rural) production and little trade as a one class system. In these non-developed and generally non-democratic countries, productivity and efficiency are low, again.

Typical rural countries are Zimbabwe and some African and East Asian countries.

### 6.6.2 *Income and Fertility*

So far we have discussed economic growth and the political status of different countries. But capital growth also includes population growth. According to Eq. (6.8a)), we have

$$K = c\,\lambda\,N \tag{6.8b}$$

The total capital ($K$) of a country depends on standard of living ($\lambda$), the total number of people ($N$) and a specific constant ($c$) that reflects different types of capital.

$$dK = c\,N\,d\lambda + c\,\lambda\,dN \tag{6.25}$$

A change of capital is a change in the standard of living ($\lambda$) and a change in population ($N$). The standard of living grows according to Eqs. (6.18 and 6.19) above; populations generally grow exponentially with fertility ($f$), the average number of children per woman in a country. Different countries grow different in GDP per capita and in population, Fig. 6.13.

According to Fig. 6.13, the world population is divided into two classes: (a) rich and democratic and (b) poor and non-democratic. About 1/3 of the largest countries in the world are democratic and industrialized. The structure of governments is between capitalistic and socialistic, and the GDP per capita is between 50.000 and 5.000 US $ per year. People invest in stocks or bonds, and they have less than three children per woman.

About 2/3 of the largest countries in the world are non-democratic, hierarchic. People live in rural, less industrialized areas. The structure of governments is between socialistic, communistic or military rule, and the GDP per capita is between

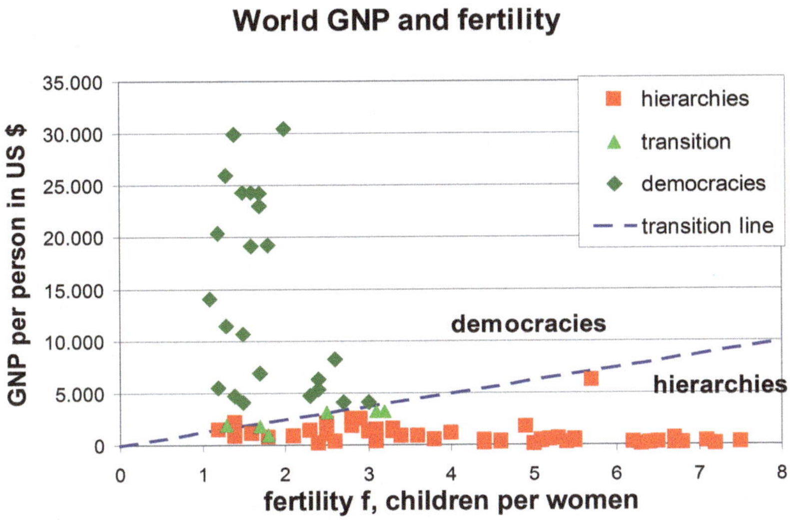


**Fig. 6.13** GDP per capita as a function of fertility for the 90 biggest countries in the world. Diamonds correspond to democratic countries, squares to non-democratic countries and triangles to countries in transition

5.000 and 500 US $ per year. People have no money to invest in stocks or bonds, they invest in children and the fertility is between three and eight children per woman.

The industrial revolution has turned the poor, rural and hierarchic western world of the nineteenth century into industrial democracies with one to two children per family. Improving poverty today in less privileged countries of the second and third world can only be achieved again by investments in industrial production and modern energy facilities in these countries, as already indicated by Fig. 6.4. Production will reduce poverty and fertility, raise the GDP per capita and install democracy.

### *6.6.3 International Trade*

Figure 6.14 shows economic growth in China and the USA between 1990 and 2005 according to the calculations in Eqs. (6.20 and 6.21). The large difference in GDP per capita ($Y_1$) in China and ($Y_2$) in the USA is the basis for the strong economic motor between China and western countries (the USA).

The efficiency ($\eta$) of the US–Chinese trade relations for labour intensive products corresponds to the efficiency of heat pumps and is determined by the standard of living in the USA ($\lambda_2$) and in China ($\lambda_1$). In the USA the GDP per capita is $\lambda_2 = 43.000$ US $ per capita, and in China it is about 3.000 US $ in 2010. The ideal efficiency is about $\eta_{\text{ideal}} = 40/3 = 13$ dollar for each dollar invested into Chinese labour intensive products. The real efficiency—like in heat pumps—is much lower, but still way above one, $\eta >> 1$.

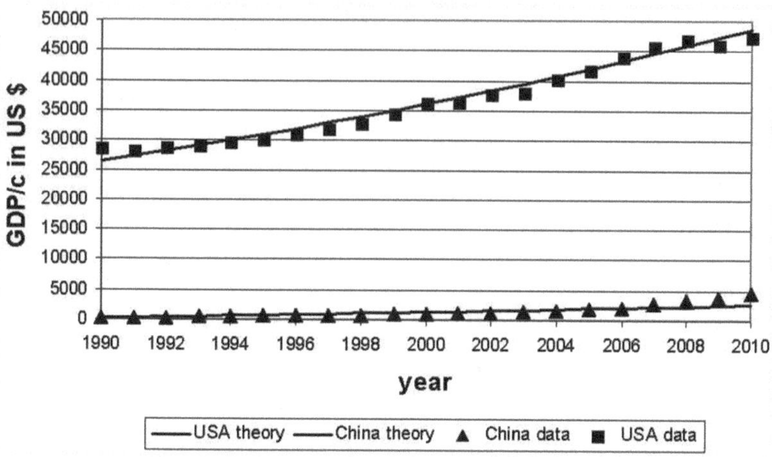


**Fig. 6.14** Exponential growth of US–China GDP per capita between 1990 and 2010. Data points from Wikipedia, lines according to calculation, Eqs. (6.20 and 6.21)

We may also discuss the scissor effect for the US–China trade in Fig. 6.14: both countries grow exponentially; China has doubled the GDP/capita from 2000 US $ per capita to 4000 US $ per capita. The USA has only raised the GDP/capita by a factor of 1,7 from 27.000 to 45.000 US $ per capita. But in absolute numbers, the rich country the USA has increased the personal income by 18.000 US $/C, while the poor China only by 2000 US $ per capita.

## 6.7 Limited Growth in Japan

### *6.7.1 Japan–US Trade*

A distribution of surplus of US–Japan trade mainly to the lower side (Japan) leads to limited growth and finally to stagnation of the US–Japan trade circuit; the efficiency goes to zero, like the motor without cooling. Both levels of the motor will soon have the same temperature; the motor runs hot and stops!

Figure 6.15 shows the trade relations between the USA and Japan between 1990 and 2011.

Japanese economic data are different from data in other countries of the Western world. What has happened? After World War II, Japan was a country with low wages and the US–Japan trade at that time looked like the US–China trade today, Fig. 6.10. After 40 years of technological improvement, wages and the GDP per capita in Japan have risen to the level of the USA. The US–Japan trade motor has run hot and stagnates. It is not any more profitable to export cars from Japan to the USA. Japanese cars on the US market are now produced in the USA. And the production

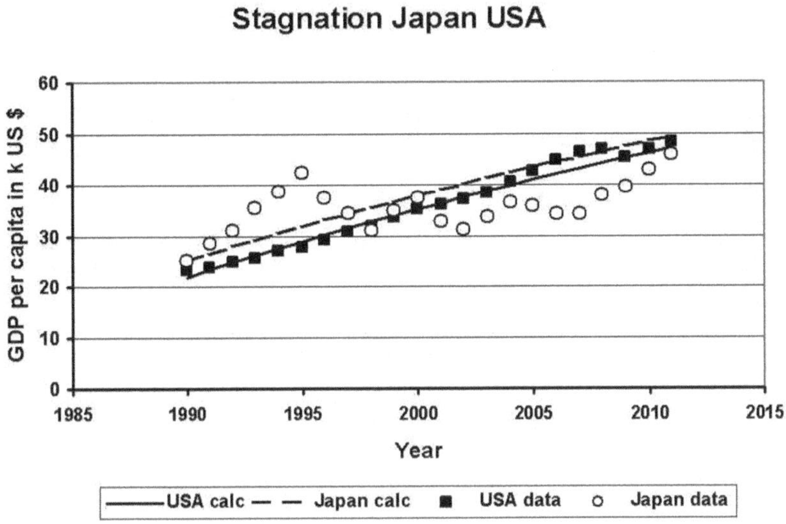


**Fig. 6.15** The growth of GDP per capita in US–Japan trade relations between 1990 and 2011. Both countries show the same values of GDP per capita; trade shows stagnation. Variations of Japan data are due to variations in the exchange rates of dollar and yen

of new technologies like computers or iPhones has shifted from Japan to China, where wages are much lower than in Japan. Germany has suffered a similar fate; wages and the GDP per capita have also risen tremendously, and German exports were slowing down. Only after Chancellor Schroeder had lowered production costs by reducing payments to the unemployed, productivity of German industry has risen, again. On the other hand trade between equal partners and equal wages is also possible, if different goods are exchanged. Germany has rich neighbours like France, and machines or chemicals are traded, e.g. for whine and for cheese.

In contrast Japan has only few equal neighbours like Taiwan, a country that also exports technology. Japan has only few resources. The only important resource is excellent labour, which has now become very expensive for exports.

### 6.7.2 *Japanese Internal Trade*

Japanese internal production and trade also depend on capital and labour, the income levels of the upper percentiles (industry, capital) and the income of the lower percentiles (labour, unemployed). The annual Japanese income growth is given in Fig. 6.16 for the upper and lower percentiles for the decades 1960–1979 and for the decades 2000–2010.

In the decades 1960–1970, the Japanese economy was growing, automobile exports to the USA were flourishing and incomes of the upper percentiles (capital)

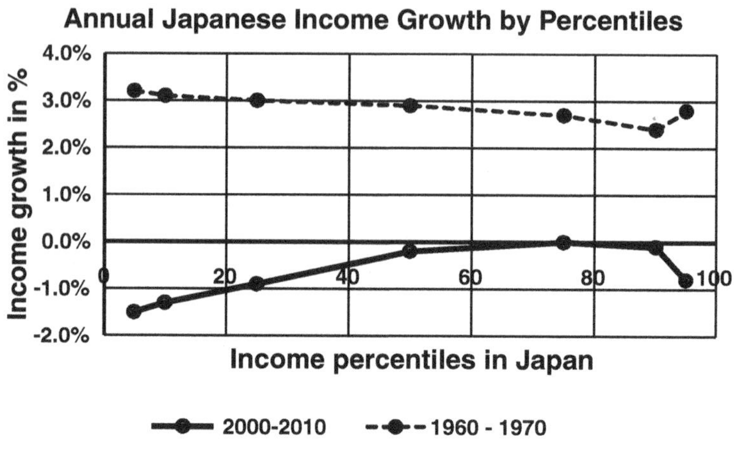


**Fig. 6.16** The annual Japanese income growth in % for the upper and lower percentiles in the decades 1960–1970 and in the decades 2000–2010

and the lower percentiles (labour) were both rising by nearly 3% per annum. Due to the growing wages compared to the USA, the export-dependent economy started to stagnate. In the decades 2000–2010, the income of the upper percentiles (capital) arrived at zero growth. However, the lowest percentiles (labour, unemployed) have suffered annual losses by 1–1,5%; in 10 years this has summed up to 10–15%. This may be called the refrigerator effect. In a refrigerator the upper temperature is the constant outside room temperature. The inside of a running fridge gets colder with time; the difference in temperatures and the efficiency grows. The same effect is observed in internal trade. Lower wages lead to a more efficient internal market. The mechanism is similar to growth: in order to obtain growth, the largest portion of surplus has to be given to the capital side. The refrigerator effect in economics is obtained by giving the largest part of obligations to labour, to the poor side. The Schroeder reforms have lowered wages suddenly in Germany, which became competitive again quickly compared to its neighbours France, Great Britain and Italy. In Japan this refrigerator effect will step by step make the Japanese economy more profitable again. This will certainly not overthrow the Chinese competition, but it will help to sell Japanese products like cars on the US market with higher profits.

## 6.8 Conclusion

Calculus-based econophysics has shown to lead to very reasonable results in general theory as well as in applications to real data. This approach may be extended to microeconomics, finance and to applications in social sciences and brings economics and social sciences closer to natural sciences and engineering.

## References

R.N. Mantegna, H.E. Stanley, *Introduction to Econophysics: Correlations and Complexity in Finance* (Cambridge University Press, Cambridge, 2000)

J. Mimkes, Stokes integral of economic growth. Calculus and the Solow model. Physica A **389**, 1665–1676 (2010)

J. Mimkes, Introduction to macro-econophysics and finance. Contin. Mechan. Thermodyn. Springer **24**(4–6), 731–737 (2012)

T. Piketty, *Capital in the Twenty-First Century* (Harvard University Press, Cambridge, MA, 2014)

P. Richmond, J. Mimkes, S. Hutzler, *Econophysics and Physical Economics* (Oxford University Press, Oxford, 2013)

V.M. Yakovenko, J.B. Rosser, Colloquium: Statistical mechanics of money, wealth and income, arXiv:0905.1518. Rev. Mod. Phys. **81**, 1703 (2009)

# Chapter 7
# A Stylised Model for Wealth Distribution

**Bertram Düring, Nicos Georgiou, and Enrico Scalas**

**Abstract** The recent book by T. Piketty (Capital in the Twenty-First Century) promoted the important issue of wealth inequality. In the last twenty years, physicists and mathematicians developed models to derive the wealth distribution using discrete and continuous stochastic processes (random exchange models) as well as related Boltzmann-type kinetic equations. In this literature, the usual concept of equilibrium in economics is either replaced or completed by statistical equilibrium.

In order to illustrate this activity with a concrete example, we present a stylised random exchange model for the distribution of wealth. We first discuss a fully discrete version (a Markov chain with finite state space). We then study its discrete-time continuous-state-space version, and we prove the existence of the equilibrium distribution. Finally, we discuss the connection of these models with Boltzmann-like kinetic equations for the marginal distribution of wealth. This paper shows in practice how it is possible to start from a finitary description and connect it to continuous models following Boltzmann's original research programme.

**Mathematics Subject Classification (2000)** 60J05 · 60J10 · 60J20 · 82B31 · 82B40

## 7.1 Introduction

The recent book by T. Piketty (2013) shifted the general attention as well as the attention of economists towards the important issue of wealth inequality. The question "Why is there wealth inequality?" has attracted the attention of a diverse set of researchers, including economists, physicists and mathematicians. Particularly, during the last 20 years, physicists and mathematicians developed models to theoretically derive the wealth distribution using tools of statistical physics and

B. Düring • N. Georgiou • E. Scalas (✉)
Department of Mathematics, University of Sussex, Brighton, UK
e-mail: b.during@sussex.ac.uk; n.georgiou@sussex.ac.uk; e.scalas@sussex.ac.uk



Y. Aruka, A. Kirman (eds.), *Economic Foundations for Social Complexity Science*,
Evolutionary Economics and Social Complexity Science 9,
DOI 10.1007/978-981-10-5705-2_7

probability theory: discrete and continuous stochastic processes (random exchange models) as well as related Boltzmann-type kinetic equations. In this framework, the usual concept of equilibrium in economics is complemented or replaced by statistical equilibrium (Garibaldi and Scalas 2010).

The original work of Pareto concerned the distribution of income (Pareto 1897). Pareto observed a skewed distribution with power-law tail. However, he also dealt with the distribution of wealth, for which he wrote:

> La répartition de la richesse peut dépendre de la nature des hommes dont se compose la societé, de l'organisation de celle-ci, et aussi, en partie, du *hasard* (les *conjonctures* de Lassalle), [...]
>
> The distribution of wealth can depend on the nature of those who make up society, on the social organization and, also, in part, on *chance*, (the *conjunctures* of Lassalle), [...]

More recently, Champernowne (1952), Simon (1955), Wold and Whittle (1957) as well as Mandelbrot (1961) used random processes to derive distributions for income and wealth. Starting from the late 1980s and publishing in the sociological literature, Angle introduced the so-called inequality process, a continuous-space discrete-time Markov chain for the distribution of wealth based on the surplus theory of social stratification (Angle 1986). However, the interest of physicists and mathematicians was triggered by a paper written by Drăgulescu and Yakovenko (2000) and explicitly relating random exchange models with statistical physics. Among other things, they discussed a simple random exchange model already published in Italian by Bennati (1988). An exact solution of that model was published in Scalas et al. (2006). Lux wrote an early review of the statistical physics literature up to 2005 (Lux 2005). An extensive review was written by Chakrabarti and Chakrabarti in 2010. Boltzmann-like kinetic equations for the marginal distribution of wealth were studied in Cordier et al. (2005) and several other works; we refer to the review article by Düring et al. (2009) and the book by Pareschi and Toscani (2014) and the references therein.

We will focus on the essentials of random modelling for wealth distributions, and we will explicitly show how continuous-space Markov chains can be derived from discrete-space (actually finite-space) chains. We will then focus on the stability properties of these chains, and, finally, we will review the mathematical literature on kinetic equations while studying the kinetic equation related to the Markov chains. In doing so, we will deal with a stylised model for the time evolution of wealth (a *stock*) and not of income (a *flow*).

Distributional problems in economics can be presented in a rather general form. Assume one has $N$ economic agents, each one endowed with his/her stock (for instance, wealth) $w_i \geq 0$. Let $W = \sum_{i=1}^{N} w_i$ be the total wealth of the set of agents. Consider the random variable $W_i$, i.e. the stock of agent $i$. One is interested in the distribution of the vector $(W_1, \ldots, W_N)$ as well as in the marginal distribution $W_1$ if all agents are on a par (exchangeable). The transformation

$$X_i = \frac{W_i}{W}, \tag{7.1}$$

normalises the total wealth of the system to be equal to one since

$$\sum_{i=1}^{N} X_i = 1 \tag{7.2}$$

and the vector $(X_1, \ldots, X_N)$ is a finite random partition of the interval $(0, 1)$. The $X_i$s are called *spacings* of the partition.

The following remarks are useful and justify this simplified modelling of wealth distribution.

1. If the stock $w_i$ represents wealth, it can be negative due to indebtedness. In this case, one can always shift the wealth to non-negative values by subtracting the negative wealth with largest absolute value.
2. A mass partition is an infinite sequence $\mathbf{s} = (s_1, s_2, \ldots)$ such that $s_1 \geq s_2 \geq \ldots \geq 0$ and $\sum_{i=1}^{\infty} s_i \leq 1$.
3. Finite random interval partitions can be mapped into mass partitions, just by ranking the spacings and adding an infinite sequence of 0s.

The vector $\mathbf{X} = (X_1, \ldots, X_N)$ lives on the $N-1$ dimensional simplex $\Delta_{N-1}$, defined by

**Definition 7.1 (The simplex $\Delta_{N-1}$)**

$$\Delta_{N-1} = \left\{ \mathbf{x} = (x_1, \ldots, x_N) : x_i \geq 0 \text{ for all } i = 1, \ldots, N \text{ and } \sum_{i=1}^{N} x_i = 1 \right\}. \tag{7.3}$$

There are two natural questions that immediately arise from defining such a model.

1. Which is the distribution of the vector $(X_1, \ldots, X_N)$ with $X_i$ given by (7.1) at a given time?
2. Which is the distribution of the random variable $X_1$, the proportion of the wealth of a single individual?

One well-studied probabilistic example is to set the vector $(W_1, \ldots, W_N)$ of i.i.d. random variables such that $W_i \sim \text{gamma}(\alpha_i, \lambda)$. Then $W = \sum_{i=1}^{N} W_i \sim \text{gamma}\left(\sum_{i=1}^{N} \alpha_i, \lambda\right)$. In this case the mass function of $(X_1, \ldots, X_N)$ is the Dirichlet distribution, given by

$$f_{\mathbf{X}}(\mathbf{x}) = \frac{\Gamma(\alpha_1 + \cdots + \alpha_N)}{\Gamma(\alpha_1) \cdots \Gamma(\alpha_N)} x_1^{\alpha_1 - 1} \cdots x_N^{\alpha_N - 1}, \quad \mathbf{x} = (x_1, \ldots, x_N) \in \Delta_{N-1}. \tag{7.4}$$

We say $\mathbf{X} \sim \text{Dir}_{N-1}(\alpha_1, \ldots, \alpha_N)$ and the parameters $\alpha_1, \ldots, \alpha_N$ are assumed strictly positive as they can be interpreted as the shapes of gamma random variables. A particular case is when $\alpha_1 = \cdots = \alpha_n = \alpha$. Then the Dirichlet distribution is called symmetric. The symmetric Dirichlet distribution with $\alpha = 1$ is uniform on the simplex $\Delta_{N-1}$.

One can now answer the two questions above using the following proposition which we present in its simplest form.

**Proposition 7.1** *Let* $(W_1, \ldots, W_N)$ *be i.i.d. random variables such that* $W_i \sim \exp(1)$. *Then* $W = \sum_{i=1}^{N} W_i \sim \text{gamma}(N, 1)$. *Define* $X_i = W_i/W$*; then the vector* $\mathbf{X} = (X_1, \ldots, X_N)$ *has the uniform distribution on the simplex* $\Delta_{N-1}$ *and one-dimensional marginals* $X_1 \sim \text{beta}(1, N-1)$, *namely,*

$$f_{X_1}(x) = \frac{(1-x)^{N-2}}{B(1, N-1)}, \tag{7.5}$$

*where, for* $a, b > 0$,

$$B(a, b) = \frac{\Gamma(a)\Gamma(b)}{\Gamma(a+b)}. \tag{7.6}$$

The proof of this proposition can be found in several textbooks of probability and statistics including Devroye's book (2003). Specifically, the part of Proposition 7.1 concerning the uniform distribution is a corollary of Theorem 4.1 in Devroye (2003). Equation (7.5) is a direct consequence of the aggregation property of the Dirichlet distribution.

In this chapter, we define three related models that incorporate a stochastic time evolution for the agent wealth distribution. The models increase in mathematical complexity in the order they are presented.

The first one is a discrete-time discrete-space (DD) Markov chain with a Pólya limiting invariant distribution. We keep the dynamics as simple as possible so in fact the invariant distribution will be uniform (not a generic Pólya distribution), but the ideas and techniques are the same for more complicated versions. The Markov chain of the DD model is then generalised to a discrete-time continuous-space (DC) Markov chain. The extension is natural in the sense that the dynamics, irreducibility and the invariant distribution of the DC model can be viewed as limiting case of the DD model. In the process, we effectively prove that Monte Carlo algorithms will approximate the DC model well. Finally, we present a continuous-continuous-space (CC) model for which the temporal evolution of the (random) wealth of a single individual is governed by a Boltzmann-type equation.

## 7.2 Random Dynamics on the Simplex

In order to define our simple models, we first introduce two types of moves on the simplex.

**Definition 7.2 (Coagulation)** By coagulation, we denote the aggregation of the stocks of two or more agents into a single stock. This can happen in mergers, acquisitions and so on.

**Definition 7.3 (Fragmentation)** By fragmentation, we denote the division of the stock of one agent into two or more stocks. This can happen in inheritance, default and so on.

### *7.2.1 Discrete-Time Continuous-Space Model: Coagulation-Fragmentation Dynamics*

Before introducing the DD model, let us define the main object of our study: The DC model.

At each event time, the state of the process $\mathbf{X} \in \Delta_{N-1}$ changes according to a composition of one coagulation and one fragmentation step.

To be precise, let $\mathbf{X} = \mathbf{x}$ be the current value of the random variable $\mathbf{X}$. For any ordered pair of indices $i,j$, $1 \leq i,j \leq N$, chosen uniformly at random, define the coagulation application $\text{coag}_{ij}(\mathbf{x}) : \Delta_{N-1} \rightarrow \Delta_{N-2}$ by creating a new agent with stock $x = x_i + x_j$ while the proportion of wealth for all others remain unchanged. Next enforce a random fragmentation application $\text{frag}(\mathbf{x}) : \Delta_{N-2} \rightarrow \Delta_{N-1}$ that takes $x$ defined above and splits it into two parts as follows. Given $u \in (0, 1)$ drawn from the uniform distribution $U[0, 1]$, set $x_i = ux$ and $x_j = (1 - u)x$.

The sequence of coagulation and fragmentation operators defines a time-homogeneous Markov chain on the simplex $\Delta_{N-1}$. Let $\mathbf{x}(t) = (x_1(t), \ldots, x_i(t), \ldots, x_j(t), \ldots, x_N(t))$ be the state of the chain at time $t$ with $i$ and $j$ denoting the selected indices. Then the state at time $t + 1$ is

$$\begin{aligned}\mathbf{x}(t+1) = (x_1(t), \ldots, x_i(t+1) &= u(x_i(t) + x_j(t)), \ldots, x_j(t+1) \\ &= (1-u)(x_i(t) + x_j(t)), \ldots, x_N(t)).\end{aligned}$$

The Markov kernel for this process is however degenerate because each step only affects a Lebesgue measure 0 of the simplex. To avoid this technical complication for the moment, we define the same dynamics on the discrete simplex, and we then analyse the DC model.

### *7.2.2 Discrete-Time Discrete-Space Model*

Let $N$ denote the number of categories (individuals) into which $n$ objects (coins or tokens) are classified (Garibaldi and Scalas 2010). In the frequency or statistical description of this system, a state is a list $\mathbf{n} = (n_1, \ldots, n_N)$ with $\sum_{i=1}^{N} n_i = n$ which gives the number of objects belonging to each category. In this framework, a coagulation move is defined by picking up a pair of ordered integers $i,j$ at random without replacement from $1 \leq \ldots \leq N$ and creating a new category with $n_i + n_j$ objects. A fragmentation move takes this category and splits it into two new categories relabeled $i$ and $j$ where $n_i'$ is a uniform random integer between 0 and $(n_i + n_j - 1) \vee 0$ and $n_j' = n_i + n_j - n_i'$. The state of the process at time $t \in \mathbb{N}_0$ is denoted by $\mathbf{X}(t)$, and its state space is the scaled integer simplex

$$S_{N-1}^{(n)} = n\Delta_{N-1} \cap \mathbb{Z}^N = \left\{ \mathbf{n} = (n_1, n_2, \dots, n_N) : 0 \le n_i \le n, \quad \sum_{i=1}^{N} n_i = n, \quad n_i \in \mathbb{N}_0 \right\}.$$

*Remark 7.1* Note that we have seemingly introduced a slight asymmetry; the agent picked first runs the risk of ending up with 0 fortune. The dynamics are overall not asymmetric, however, since we select $i$ before $j$ with the same probability as selecting $j$ before $i$. The reason for introducing the model in this way is to simplify the presentation and error estimate in the proof of the weak convergence of the finite dimensional marginals from the DD to the DC model.

Formally, with coagulation, we move from the state space $S_{N-1}^{(n)}$ to $S_{N-2}^{(n)}$, and then again with fragmentation, we come back to $S_{N-1}^{(n)}$. While it is interesting to actually study all stages of the procedure, we are only interested in the aggregated wealth, and therefore we can bypass the intermediate state space by defining the process only on $S_{N-1}^{(n)}$; it is straightforward to write down the transition probabilities for $\mathbf{X}(t)$

$$\begin{aligned}
&\mathbb{P}\{\mathbf{X}(t+1) = \mathbf{n}' | \mathbf{X}(t) = \mathbf{n}\} \\
&= \sum_{i,j:i \ne j} \left\{ \frac{1}{N} \frac{1}{N-1} \left( \frac{\mathbb{1}\{n_i + n_j \ge 1, n_j' \ge 1\}}{n_i + n_j} + \mathbb{1}\{n_i + n_j = 0\} \right) \right. \\
&\qquad \left. \times\, \delta_{n_i+n_j, n_i'+n_j'} \prod_{k \ne i,j} \delta_{n_k', n_k} \right\}. \qquad (7.7)
\end{aligned}$$

The notation is shorthand and implies that we are adding over all ordered pairs $(i,j), i \neq j$ where the first coordinate indicates the index $i$ that was selected first.

The chain is *time-homogeneous* as the transition (7.7) is independent of the time parameter $t$. It is also *aperiodic* since with positive probability, during each time step, the chain may coagulate and then fragment to the same state. To see this, consider any vector $(X_1, \dots, X_N) = (x_1, \dots, x_N)$ on the simplex. It must have at least one non-zero entry, say $x_1 > 0$. Select index $i = 1$ first (with probability $N^{-1}$), and then select any other index $j$. After that, fragment at precisely $x_1, x_j$ (with probability $1/(x_1 + x_j) > 0$). Finally, the chain is *irreducible*, since from any point $\mathbf{X} = (x_1, \dots, x_N)$, the chain can move with positive probability to any of the neighbouring $((x_1, \dots, x_N) \pm (e_i - e_j)) \cap S_{N-1}^{(n)}$, i.e. to any point in the simplex that is $\ell^1$-distance 2 away from the current state. Therefore, we can conclude that the chain $\{\mathbf{X}(t)\}_{t \in \mathbb{N}_0}$ has a unique equilibrium distribution $\boldsymbol{\pi}$ which we identify in the next proposition.

**Proposition 7.2** *The invariant distribution of this Markov chain* $\mathbf{X}(t)$ *is the uniform distribution on* $n\Delta_{N-1} \cap \mathbb{Z}^N$.

*Proof* Define

$$A_{i,j}(\mathbf{n}') = \left\{ \mathbf{n} : \mathbf{n} \overset{\text{coag} - \text{frag}_{i,j}}{\longrightarrow} \mathbf{n}' \right\}$$

to be the set of all simplex elements $\mathbf{n}$ that map to $\mathbf{n}'$ via a coagulation-fragmentation procedure in the $i, j$ indices ($i$ selected before $j$). This set is empty only when $n'_j = 0$ while $n'_i \geq 1$, but otherwise it contains at least one vector. Assuming that $A_{i,j}(\mathbf{n}')$ is not empty, we have that its cardinality is

$$\text{card}(A_{i,j}(\mathbf{n}')) = (n'_i + n'_j) \vee 1. \tag{7.8}$$

Using this notation, we may rewrite the transition probability in (7.7) as

$$\mathbb{P}\{\mathbf{X}(t+1) = \mathbf{n}' | \mathbf{X}(t) = \mathbf{n}\}$$

$$= \sum_{i,j:i \neq j} \left\{ \frac{1}{N} \frac{1}{N-1} \left( \frac{\mathbb{1}\{n_i + n_j \geq 1, n'_j \geq 1\}}{n_i + n_j} + \mathbb{1}\{n_i + n_j = 0\} \right) \delta_{n_i+n_j, n'_i+n'_j} \prod_{k \neq i,j} \delta_{n'_k, n_k} \right\}$$

$$= \sum_{i,j:i \neq j} \left\{ \frac{1}{N} \frac{1}{N-1} \left( \frac{\mathbb{1}\{n'_i + n'_j \geq 1, n'_j \geq 1\}}{n'_i + n'_j} + \mathbb{1}\{n'_i + n'_j = 0\} \right) \delta_{n_i+n_j, n'_i+n'_j} \prod_{k \neq i,j} \delta_{n'_k, n_k} \right\}$$

from the $\delta_{n_i+n_j, n'_i+n'_j}$ factor,

$$= \sum_{i,j:i \neq j} \left\{ \frac{1}{N} \frac{1}{N-1} \left( \frac{1}{(n'_i + n'_j) \vee 1} \right) \delta_{n_i+n_j, n'_i+n'_j} \prod_{k \neq i,j} \delta_{n'_k, n_k} \mathbb{1}\left\{ \mathbf{n} \in A_{i,j}(\mathbf{n}') \right\} \right\}$$

$$= \sum_{i,j:i \neq j} \left\{ \frac{1}{N} \frac{1}{N-1} \left( \frac{1}{\text{card}(A_{i,j}(\mathbf{n}'))} \right) \delta_{n_i+n_j, n'_i+n'_j} \prod_{k \neq i,j} \delta_{n'_k, n_k} \mathbb{1}\left\{ \mathbf{n} \in A_{i,j}(\mathbf{n}') \right\} \right\}$$

$$= \sum_{i,j:i \neq j} \left\{ \frac{1}{N} \frac{1}{N-1} \left( \frac{1}{\text{card}(A_{i,j}(\mathbf{n}'))} \right) \mathbb{1}\left\{ \mathbf{n} \in A_{i,j}(\mathbf{n}') \right\} \right\} \tag{7.9}$$

since the $\delta$ product is equivalent to the last indicator.

Now fix a $\mathbf{n}'$ and add up all the transition probabilities in (7.9) over $\mathbf{n}$. We get

$$\sum_{\mathbf{n}} \sum_{i,j:i\neq j} \left\{ \frac{1}{N}\frac{1}{N-1}\left(\frac{1}{\mathrm{card}(A_{i,j}(\mathbf{n}'))}\right) 1\!\!1\{\mathbf{n} \in A_{i,j}(\mathbf{n}')\}\right\}$$
$$= \sum_{i,j:i\neq j} \left\{ \frac{1}{N}\frac{1}{N-1}\left(\frac{1}{\mathrm{card}(A_{i,j}(\mathbf{n}'))}\right) \sum_{\mathbf{n}} 1\!\!1\{\mathbf{n} \in A_{i,j}(\mathbf{n}')\}\right\}$$
$$= \sum_{i,j:i\neq j} \left\{ \frac{1}{N}\frac{1}{N-1}\left(\frac{1}{\mathrm{card}(A_{i,j}(\mathbf{n}'))}\right) \mathrm{card}(A_{i,j}(\mathbf{n}'))\right\}$$
$$= 1.$$

Therefore, the transition matrix is doubly stochastic, and in particular the invariant distribution must be uniform.

### 7.2.3 *Convergence of the Finite Markov Chain as the Overall Wealth Increases*

Reaching a similar conclusion in the case of the DC model is slightly more complicated. The difficulty is related to the fact that time is changing in discrete steps and the chain cannot explore the whole available state space because real numbers cannot be put in 1-to-1 correspondence with integers. How can we be sure that the Markov chain with continuous state space can explore its state space uniformly? We begin our analysis by studying the convergence of the finite state-space Markov chain to the continuous-state-space Markov chain.

Let $\mathbf{X}^{(n)}$ be the DD Markov chain for wealth, when the wealth of the system is $n$, and let $\mathbf{X}^{(\infty)}$ be the chain for the DC model introduced in Sect. 7.2.1. We scale the state space of each process $\mathbf{X}^{(n)}$ so that it is a subset of $\Delta_{N-1}$ by defining a new, coupled process

$$\mathbf{Y}^{(n)} = n^{-1}\mathbf{X}^{(n)}.$$

The state space for the process $\mathbf{Y}^{(n)}$ is the simplex

$$\Delta_{N-1}(n) = \{(q_1, \ldots, q_d) : 0 \leq q_i \leq 1,\, q_1 + \ldots + q_d = 1,\, nq_i \in \mathbb{N}_0\} \subset \Delta_{N-1}.$$

It can be considered as partition of $\Delta_{N-1}$ with mesh $n^{-1}$, i.e. inversely proportional to the total wealth.

In this section we first prove weak convergence of the one-dimensional marginals

$$\mathrm{Y}_k^{(n)} \overset{n\to\infty}{\Longrightarrow} \mathrm{X}_k^{(\infty)}, \text{ for all } k \in \mathbb{N}$$

and then prove the existence of a unique invariant distribution for $X^{(\infty)}$ (the DC model) that we identify as the uniform distribution on $\Delta_{N-1}$.

Let $\mu_0^{(n)}$ the initial distribution of $Y_0^{(n)}$ and $\mu_0^{(\infty)}$ the initial distribution of $X_0^{\infty}$.

**Proposition 7.3** *Assume the weak convergence* $\mu_0^{(n)} \Longrightarrow \mu_0^{(\infty)}$ *as* $n \to \infty$. *Then for each* $k \in \mathbb{N}$, *we have weak convergence of the one-dimensional marginals*

$$Y_k^{(n)} \Longrightarrow X_k^{(\infty)} \quad \text{as } n \to \infty.$$

*Proof* We first show this for $k = 1$ and then show it in general with an inductive argument. Let $f$ be a bounded continuous function on $\Delta_{N-1}(n)$. Let $U$ be a uniform random variable on $[0, 1]$, and define the bounded and continuous $F_{i,j} : \Delta_{N-1} \to \mathbb{R}$ by

$$F_{i,j}(x_1, \dots, x_d) = \mathbb{E}^U(f(x_1, \dots, U(x_i+x_j), x_{i+1}, \dots, (1-U)(x_i+x_j), x_{j+1}, \dots, x_d)).$$

Pick an $\varepsilon > 0$. By compactness, we can find a $\delta = \delta(\varepsilon) > 0$ such that whenever $\|x - y\|_1 < \delta$ we have that

$$\sup_{\{i,j\}} |F_{ij}(x) - F_{ij}(y)| + |f(x) - f(y)| < \varepsilon.$$

From this relation, choose $n$ large enough so that the discrete simplex $\Delta_{N-1}(n)$ is fine enough, namely, two neighbouring points $x^{(n)}, y^{(n)}$ satisfy $\|x^{(n)} - y^{(n)}\|_1 < \delta$. In particular, this implies that $n > 2\delta^{-1}$.

The function $F_{i,j}$ evaluated on the partition points is

$$F_{i,j}(x^{(n)}) = \int_0^1 f(x_1^{(n)}, \dots, u(x_i^{(n)} + x_j^{(n)}), \dots, (1-u)(x_i^{(n)} + x_j^{(n)}), \dots, x_d)\, du$$

$$= \begin{cases} f(x^{(n)}), & x_i^{(n)} + x_j^{(n)} = 0 \\ \frac{1}{x_i^{(n)} + x_j^{(n)}} \int_0^{x_i^{(n)} + x_j^{(n)}} f(x_1^{(n)}, \dots, s, \dots, x_i^{(n)} + x_j^{(n)} - s, \dots, x_d)\, ds, & \text{otherwise.} \end{cases}$$

Focus on the integral of the second branch for a moment. We discretise the integral on $\Delta_{N-1}(n)$ with $s$-values $0, 1/n, \dots, x_i^{(n)} + x_j^{(n)} - 1/n$. Then

$$\Big| \int_{k/n}^{(k+1)/n} f(x_1^{(n)}, \dots, s, \dots, x_i^{(n)} + x_j^{(n)} - s, \dots, x_d)\, ds$$
$$- n^{-1} f(x_1^{(n)}, \dots, k/n, \dots, x_i^{(n)} + x_j^{(n)} - k/n, \dots, x_d) \Big| < \varepsilon/n.$$

Therefore, the overall error,

$$\left| F_{i,j}(x^{(n)}) - \sum_{k=0}^{n(x_i^{(n)}+x_j^{(n)})-1} \frac{f(x_1^{(n)},\ldots,k/n,\ldots,x_i^{(n)}+x_j^{(n)}-k/n,\ldots,x_d)}{n(x_i^{(n)}+x_j^{(n)})} \right| < \varepsilon. \tag{7.10}$$

Now we turn to prove the weak convergence:

$$\begin{aligned}
\mathbb{E}(f(Y_1^{(n)})) &= \sum_{x\in\Delta_{N-1}(n)} f(x)\mathbb{P}\{Y_1^{(n)}=x\} \\
&= \sum_{x\in\Delta_{N-1}(n)} \sum_{y\in\Delta_{N-1}(n)} f(x)\mathbb{P}\{Y_1^{(n)}=x|Y_0^{(n)}=y\}\mu_0^{(n)}(y) \\
&= \sum_{y\in\Delta_{N-1}(n)} \mu_0^{(n)}(y) \sum_{x\in\Delta_{N-1}(n)} f(x)\mathbb{P}\{X_1^{(n)}=x|X_0^{(n)}=y\} \\
&= \sum_{y\in\Delta_{N-1}(n)} \mu_0^{(n)}(y) \sum_{i,j:,i\neq j} \sum_{\substack{x_i+x_j=y_i+y_j \\ x_k=y_k}} f(x)\mathbb{P}\{Y_1^{(n)}=x|Y_0^{(n)}=y\} \\
&= \frac{1}{N(N-1)} \sum_{y\in\Delta_{N-1}(n)} \mu_0^{(n)}(y) \sum_{i,j:i\neq j} \sum_{\substack{x_i+x_j=y_i+y_j \\ x_k=y_k}} f(x_1,\ldots x_i,\ldots,x_j,\ldots,x_d) \\
&\quad \times \left( \frac{\mathbb{1}\{y_i+y_j\geq n^{-1}, x_j\geq n^{-1}\}}{n(y_i+y_j)} + \mathbb{1}\{y_i+y_j=0\} \right) \\
&= \frac{1}{N(N-1)} \sum_{y\in\Delta_{N-1}(n)} \mu_0^{(n)}(y) \\
&\quad \times \sum_{i,j:i\neq j} \left\{ \sum_{k=0}^{n(y_i+y_j)-1} f(y_1,\ldots,n^{-1}k,\ldots,y_i+y_j-n^{-1}k,\ldots,y_d)\frac{1}{n(y_i+y_j)} \right. \\
&\qquad \left. + f(y_1,\ldots,0,\ldots,0,\ldots,y_d)\mathbb{1}\{y_i+y_j=0\} \right\} \\
&= \frac{1}{N(N-1)} \sum_{y\in\Delta_{N-1}(n)} \mu_0^{(n)}(y) \\
&\quad \times \sum_{i,j:i\neq j} F_{i,j}(y)\mathbb{1}\{y_i+y_j>0\} + O(\varepsilon) + F_{i,j}(y)\mathbb{1}\{y_i+y_j=0\} \\
&= \frac{1}{N(N-1)} \sum_{y\in\Delta_{N-1}(n)} \mu_0^{(n)}(y) \sum_{i,j:i\neq j} F_{i,j}(y) + O(\varepsilon) \\
&= \frac{1}{N(N-1)} \sum_{i,j:i\neq j} \mathbb{E}^{\mu_0^{(n)}}(F_{i,j}) + O(\varepsilon).
\end{aligned}$$

Now let $n \to \infty$ and recall that $F_{i,j}$ is bounded and continuous to conclude that

$$- C\varepsilon \leq \underline{\lim}_{n\to\infty} \mathbb{E}(f(Y_1^{(n)})) - \frac{1}{N(N-1)} \sum_{i,j:i\neq j} \mathbb{E}^{\mu_0^{(\infty)}}(F_{i,j}) \leq \overline{\lim}_{n\to\infty} \mathbb{E}(f(Y_1^{(n)}))$$
$$- \frac{1}{N(N-1)} \sum_{i,j:i\neq j} \mathbb{E}^{\mu_0^{(\infty)}}(F_{i,j}) \leq C\varepsilon.$$

where $C$ is a constant independent of $n$ that comes from the error term. Let $\varepsilon \to 0$ to conclude the limit exists and observe that the definition of $F_{i,j}$ and the disintegration theorem imply that

$$\lim_{n\to\infty} \mathbb{E}(f(Y_1^{(n)})) = \frac{1}{N(N-1)} \sum_{i,j:i\neq j} \mathbb{E}^{\mu_0^{(\infty)}}(F_{i,j}) = \mathbb{E}(f(X_1^{(\infty)})).$$

Therefore, we have now shown that the $\mu_1^{(n)} \Longrightarrow \mu_1^{(\infty)}$ if $\mu_0^{(n)} \Longrightarrow \mu_0^{(\infty)}$. An inductive construction and the Markov property are enough to guarantee that all one-dimensional marginals converge.

### 7.2.4 *Irreducibility, Uniqueness of the Invariant Measure and Stability*

We can now proceed to study of irreducibility, of the uniqueness of the invariant measure and of the stability for the continuous-space Markov chain.

We begin with a proposition that will simplify the mathematical technicalities associated with general state-space discrete-time Markov chains.

**Proposition 7.4 (Duality of coagulation and fragmentation)** *Let* $\mathbf{X}(t)$ *denote the coagulation-fragmentation Markov chain defined in Sect.* 7.2.1. *If* $\mathbf{X}(t) \sim U[\Delta_{N-1}]$ *then* $\mathbf{X}(t+1) \sim U[\Delta_{N-1}]$, *as well.*

*Proof* See Bertoin (2006) chapter 2, corollary 2.1, page 77.
This proposition means that the uniform distribution on the simplex $\Delta_{N-1}$ is an invariant distribution for the coagulation-fragmentation chain.

What we prove in the sequel is that this is the unique invariant measure and the transition kernels converge to it in the total variation norm. With this goal in mind, we begin with some definitions.

**Definition 7.4 (Phi-irreducibility)** Let $(S, \mathscr{B}(S), \phi)$ be a measured Polish space. A discrete-time Markov chain $\mathbf{X}$ on $S$ is $\phi$-irreducible if and only if for any Borel set $A$, the following implication holds:

$$\phi(A) > 0 \Longrightarrow L(u, A) > 0, \qquad \text{for all } u \in S.$$

Above, we used notation

$$L(u, A) = P_u\{X_n \in A \text{ for some } n\} = P\{X_n \in A \text{ for some } n | X_0 = u\}.$$

This replaces the notion of irreducibility for discrete Markov chains and means that the chain is visiting any set of positive measure with positive probability.

The existence of a Foster-Lyapunov function V defined as

**Definition 7.5 (Foster-Lyapunov function)** For a *petite* set $C$, we can find a function $V \geq 0$ and a $\rho > 0$ so that for all $x \in S$

$$\int P(x, dy)V(y) \leq V(x) - 1 + \rho 1\!\!1_C(x), \tag{7.11}$$

implies convergence of the kernel $P$ of $\phi$-irreducible, aperiodic chain to a unique equilibrium measure $\pi$

$$\sup_{A \in \mathscr{B}(S)} |P^n(x, A) - \pi(A)| \to 0, \text{ as } n \to \infty. \tag{7.12}$$

(see Meyn and Tweedie 1993) for all $x$ for which $V(x) < \infty$. If we define $\tau_C$ to be the number of steps it takes the chain to return to the set $C$, the existence of a Foster-Lyapunov function (and therefore convergence to a unique equilibrium) is equivalent to $\tau_C$ having finite expectation, i.e.

$$\sup_{x \in C} \mathbf{E}_x(\tau_C) < M_C$$

which in turn is implied when $\tau_C$ has geometric tails. This is in fact what we prove in the following.

In our case, $\phi$ will be the Lebesgue measure, and the role of the petite set $C$ will be played by any set with positive Lebesgue measure. This useful simplification of the mathematical technicalities is an artefact of the compact state space ($\Delta_{N-1}$) and the fact that the uniform distribution on the simplex is invariant for the chain (Proposition 7.4).

**Proposition 7.5** *Let $t \in \mathbb{N}$. The discrete chain $\mathbf{X} = \{X_n\}_{n \in \mathbb{N}}$ as defined in Sect. 7.2.1 is $\phi$-irreducible, where $\phi \equiv \lambda_{N-1}$ is the Lebesgue measure on the simplex.*

At this point it is useful to explain the idea of the proof of Proposition 7.5 when we have deterministic dynamics. We do this in the (easy to visualise) case $N = 3$, while the proof is done generally, with Markov dynamics. For any pair $u, v \in \Delta_2^\circ$, there is a deterministic way to move from $u = (x_u, y_u, z_u)$ to $v = (x_v, y_v, z_v)$ in precisely two steps. The same happens in higher dimensions; on $\Delta_{N-1}^\circ$, we can move from any starting point to any target point using deterministic coagulation-fragmentation dynamics in precisely $N - 1$ steps.

Since the dynamics is symmetric with respect to the coordinates, we may assume without loss of generality that $z_u \leq 2/3$ and therefore there exists an entry in $v$, say

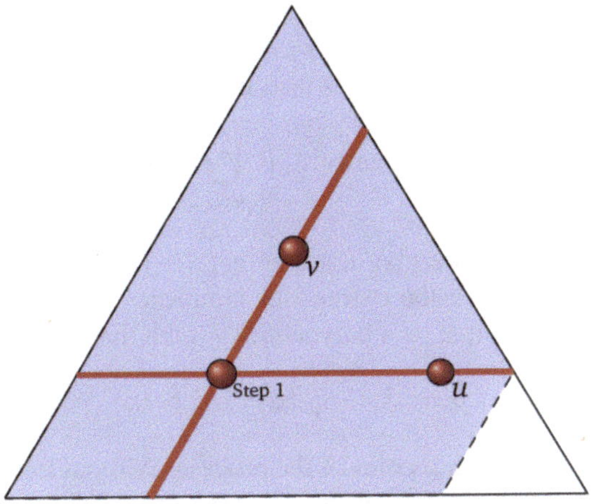


**Fig. 7.1** Schematic of a possible coagulation-fragmentation route from $u$ to $v$ in two steps. Starting from point $u \in \Delta_2$, fix $z_u$. Then on the line $x+y = 1-z_u$, pick the point $(x_v, 1-z_u-x_v, z_u)$. From there, fix $x_v$ and choose $(y_v, z_v)$ on the line $1-x_v = y_v + z_v$. The *shaded region* are all points $v$ that can be reached with this procedure from $u$, first by fixing $z_u$ and then by fixing $x_v$. Points in the *white region* can be reached from $u$ first by fixing $z_u$ and then $y_v$

$x_v$, such that $m_1 = \dfrac{x_v}{1-z_u} \leq 1$. Furthermore, $m_2 = \dfrac{x_v}{1-y_v} \leq 1$. Then consider the mapping

$$\begin{aligned} u =&(x_u, y_u, z_u) \mapsto (m_1(x_u+y_u), (1-m_1)(x_u+y_u), z_u) \\ &\mapsto (m_1(x_u+y_u), m_2[(1-m_1)(x_u+y_u)+z_u], (1-m_2)[(1-m_1)(x_u+y_u)+z_u]) \\ &= (m_1(1-z_u), m_2[(1-m_1)(1-z_u)+z_u], (1-m_2)[(1-m_1)(1-z_u)+z_u]) \\ &= (x_v, y_v, z_v) = v. \end{aligned} \tag{7.13}$$

This idea captures the proof of the Lebesgue irreducibility (see also Fig. 7.1).

*Proof (Proof of Proposition* 7.5*)* First observe that excluding one coordinate (say $x_1$) from the coagulation process in $\Delta_{N-1}$, we are merely restricting the dynamics to $(1-x_1)\Delta_{N-2}$. This observation is what allows us to proceed by way of induction.

*Base case* $N = 3$. We choose the base case $N = 3$ for purposes of clarity, in a way that can be immediately generalised to higher dimensions. We are working on $(\Delta_2, \mathscr{B}(\Delta_2), \lambda_2 \equiv \lambda \otimes \lambda)$.

Let $A$ be a Borel set and assume $\lambda_2(A) = \alpha > 0$. We will show that starting from any $x$, the probability of hitting $A$ in just two steps with the coagulation-fragmentation dynamics described above is strictly positive.

For any $\delta > 0$ and point $u$, let $\delta\Delta_2(u)$ denote the scaled simplex with length side $\delta\sqrt{2}$ and barycentre $u$.

Since $A$ has positive Lebesgue measure, for any $\varepsilon > 0$, we can find an open set $G_{A,\varepsilon} \supseteq A$ so that $\lambda_2(G_{A,\varepsilon} \setminus A) < \varepsilon$. Fix an $\varepsilon > 0$ and construct $G_{A,\varepsilon}$. Enumerate all rationals in $G_{A,\varepsilon}$ and find $\delta = \delta(A, \varepsilon) > 0$ so that

$$A \subseteq \bigcup_{q \in G_{A,\varepsilon}} (\delta\Delta_2(q)), \quad \text{and} \quad \lambda_2\Big(\bigcup_{q \in G_{A,\varepsilon}} (\delta\Delta_2(q))\Big) \leq \alpha + 2\varepsilon.$$

Without loss of generality, we may assume that $\delta\Delta_2(q) \cap A \neq \varnothing$ for all $q$ in the union; otherwise we remove the extraneous simplexes from the union. Since the union is countable, there must be a barycentre $q_0$ such that

$$\beta := \lambda_2(\delta\Delta_2(q_0) \cap A) > 0.$$

Let $u$ be an arbitrary starting point of the process. Without loss of generality, and by possibly decreasing our initial choice of $\delta$, assume

1. $z_u \leq 2/3$,
2. $u$ can be deterministically mapped at any point $v \in \delta\Delta_2(q_0) \cap A$ by first fixing $z_u$ and then $x_v$, as in calculation (7.13).

We denote the three corners $\delta\Delta_2(q_0)$ by $a = (x_a, y_a, z_a), b = (x_a - \delta, y_a + \delta, z_a)$ and $c = (x_a - \delta, y_a, z_a + \delta)$. Then,

$$\begin{aligned}
0 < \beta &= \int_{A \cap \delta\Delta_2(q_0)} d\lambda_2 = \int \int_{A \cap \delta\Delta_2(q_0)} d\lambda_1 \, d\lambda_1 \\
&= \int d\lambda_1 \Big( \mathbb{1}\{x_a - \delta < x < x_a\} \int d\lambda_1 \, \mathbb{1}\{A \cap \Delta_2(q_0) \cap \{z + y = 1 - x\}\}\Big) \\
&= \int d\lambda_1 \Big( \mathbb{1}\{x_a - \delta < x < x_a\} \int d\lambda_1 \, \mathbb{1}\{(x, y, z) \in A \cap \Delta_2(q_0) : y + z = 1 - x\}\Big).
\end{aligned}$$

Thus, for a positive $\lambda_1$ measure of $x \in (x_a - \delta, x_a)$, we can find positive $\lambda_1$ measure of the intersection between the set $A$ and the line $z + y = 1 - x$. Thus, we restrict to the measurable set $F = \{x \in [x_a - \delta, x_a] : \gamma_x > n^{-1}\}$ where

$$\gamma_x = \lambda_1\{A \cap \delta\Delta_2(q_0) \cap \{y + z = 1 - x\}\}.$$

Integer $n$ is chosen large enough so that $\lambda_1(F) > 0$. We have established the existence of a set $C$ so that

$$C = \{(x, y, z) : x \in F, (x, y, z) \in A \cap \delta\Delta_2(q_0)\}, \quad \lambda_2(C) > 0.$$

This is enough to finally complete the proof of the base case. Recall the starting point $u = (x_u, y_u, z_u)$ of the Markov chain. Define the projection set

$$F_{C,u} = \{(x, y, z) : z = z_u, \exists (x_0, y_0, z_0) \in C \text{ s.t. } y + z_u = y_0 + z_0 = 1 - x_0, x = x_0\},$$

which has positive measure as $\lambda_1(F) = \lambda_1(F_{C,u})$.

With strictly positive probability, we select to coagulate the first and second coordinate. Then with strictly positive probability, we fragment into the set $F_{C,u}$. This is because the coagulation-fragmentation process of two coordinates picks a uniformly distributed point on the line $x + y = 1 - z_u$ by virtue of construction. The uniform distribution is a scalar multiple of the Lebesgue measure, thus guaranteeing that the probability of selecting such a point is strictly positive. Then, given the chain's current position, with strictly positive probability, we coagulate the last two coordinates. For the same reason as before, with strictly positive probability, we terminate in the set $C \subseteq A$ since for any fixed $x \in F$, the fragmentation has probability no less than $1/n$ to pick up a point $(x, y_0, z_0) \in C$.

To conclude, in just two steps, we have a positive probability of hitting $A$ from any starting point $u$.

*Induction case:* Now consider simplex $\delta_{N-1}$, when $N \geq 4$, and assume that the proposition is true for all $k < N$. Let $A \subseteq \Delta_{N-1}$ be a Borel set of positive $\lambda_{N-1}$ measure. As for the base case, we can find a simplex $\delta\Delta_{N-1}(q_0)$ such that $\lambda_{N-1}(A \cap \delta\Delta_{N-1}(q_0)) > 0$ and with the same Fubini-Tonelli argument conclude that there exist a positive integer $n$ and a positive $\lambda_1$-measure set $F$ of $x$ values so that

$$\lambda_{N-2}(A \cap \delta\Delta_{N-1}(q) \cap \{x = x_0 \in F\}) > n^{-1}.$$

Without loss of generality, assume that from the starting point $u$, we can coagulate and fragment two coordinates, say $u_1$ and $u_2$, so that $\lambda_1\{x \in F, x < u_1 + u_2\} > 0$. Then, for the Markov chain, this implies

$$P_u\{X_1 \cdot e_1 \in F\} > 0. \tag{7.14}$$

Let

$$B = \{X_2, X_3, \ldots, X_{N-1} \text{ does not coagulate the first coordinate}\}.$$

Again,

$$P_u\{B|X_1 \cdot e_1 \in F\} = P_u\{B\} > 0. \tag{7.15}$$

Then it is immediate to compute

$$\begin{aligned} L(u, A) &= P_u\{X_\ell \in A \text{ for some } \ell\} \geq P_u\{X_{N-1} \in A\} \\ &\geq P_u\{X_{N-1} \in A, X_1 \cdot e_1 \in F, B\} \\ &\geq P_u\{X_{N-1} \in A|B, X_1 \cdot e_1 \in F\}P_u\{B|X_1 \cdot e_1 \in F\}P_u\{X_1 \cdot e_1 \in F\} > 0. \end{aligned}$$

Strict positivity of the last two factors follows from (7.14) and (7.15), while $P_u\{X_{N-1} \in A|B, X_1 \cdot e_1 \in F\}$ equals the probability that the $N-2$ dimensional fragmentation-coagulation process starting from a random point $u_0$ with $x_{u_0} \in F$ hits the set $A \cap \delta\Delta_{N-1}(q) \cap \{x = x_0\}$ in $N-2$ steps. By the induction hypothesis, this probability is strictly positive (given the starting point). By restricting the set $F$ so that its measure remains positive, we may further assume that these probabilities are uniformly bounded away from 0, independently of the starting point. This concludes the proof.

**Proposition 7.6 (Existence of a Foster-Lyapunov function)** *The return times $\tau_A$ to any set $A \in \mathscr{B}(\Delta_2)$ of positive measure have at most geometric tails, under $P_{x_0}$. As a consequence, the Foster-Lyapunov function exists.*

*Proof (Proof of Proposition* 7.6*)* Let $A$ be a positive Lebesgue measure set. By repeating the construction in the proof of Proposition 7.5 to all coordinates, we can find $\alpha_i > 0, n_i > 0, 1 \leq i \leq N, \delta > 0$ and a rational barycentre $q_0$ and a sequence of measurable sets

$$A \supseteq A_1 \supseteq A_2 \supseteq \ldots \supseteq A_N$$

of positive measure, $\lambda_{N-1}(A_N) = \eta > 0$ and a collection of one-dimensional measurable sets $F_1, \ldots, F_N$ with the following properties:

1. $\lambda_{N-2}\{A \cap \delta\Delta_2(q_0) \cap \{x_1 = x_1^* \in F_1\}\} = \gamma(x_1^*) > n_1^{-1}, \quad \lambda_1(F_1) \geq \alpha_1,$
   $A_1 = \{A \cap \delta\Delta_2(q_0), x_1 \in F_1\}$
2. $\lambda_{N-2}\{A_{k-1} \cap \delta\Delta_2(q_0) \cap \{x_k = x_k^* \in F_k\}\} = \gamma(x_k^*) > n_k^{-1}, \quad \lambda_1(F_k) \geq \alpha_k,$
   $A_k = \{A_{k-1} \cap \delta\Delta_2(q_0), x_k \in F_k\}, \quad k \geq 2.$

The basic property of $A_N$ is that it is accessible with positive probability (that depends on $A$), uniformly bounded from below from any point $u_0 \in \Delta_{N-1}$. Let $a = \min\{a_1, \ldots, a_N, 1\}$ and $n_0 = \max\{n_1, \ldots, n_N\}$. We bound above the probability that we do not hit $A_N$ in the first $N-1$ steps, i.e. $P_{u_0}\{\tau_{A_N} > N-1\}$. Suppose we hit in $N-1$ steps or less. Then there is at least one sequence of $N-1$ coagulation-fragmentation steps for which, if we follow it, we land in $A_N$. We select the appropriate pair of indices at each step with probability $1/N(N-1)$, and, given this, we fragment at an appropriate point with probability at least $a$. Therefore

$$\inf_{x \in \Delta_{N-1}} P_x\{\tau_{A_N} \leq N-1\} \geq \left(\frac{2a}{N(N-1)}\right)^{N-1} = \rho_A > 0.$$

Pick a starting point $u_0 \in A$. Then $P_{u_0}(\tau_A > M) \leq P_{u_0}(\tau_{A_N} > M)$. We will show that the larger tail is bounded above geometrically by an expression independent of $u_0$. We compute

$$\begin{aligned}P_x\{\tau_{A_N} > (N-1)M\} &= P_x\{X_1 \notin A_N, \ldots, X_{(N-1)M} \notin A_N\} \\ &\leq \left(\sup_{u \in \Delta_{N-1} \setminus A_N} P_u\{X_1 \notin A_N, \ldots, X_{N-1} \notin A_N\}\right)^M\end{aligned}$$

$$\leq \left( \sup_{u \in \Delta_{N-1} \setminus A_N} P_u\{\tau_{A_N} > N-1\} \right)^M$$
$$\leq (1-\rho_A)^M.$$

Finally, since for any $1 \leq k \leq N-1$ we have $\{\tau_{A_N} > (N-1)(M+1)\} \subseteq \{\tau_{A_N} > (N-1)M + k\} \subseteq \{\tau_{A_N} > (N-1)M\}$, we conclude that $\tau_{A_N}$ has geometric tails.

We assemble these propositions in the following theorem.

**Theorem 7.1** *Let $\mathbf{X}(t)$ denote the coagulation-fragmentation Markov chain defined in Sect. 7.2.1 and initial distribution $\mu_0$ on $\Delta_{N-1}$. Let $\mu_t$ denote the distribution of $\mathbf{X}(t)$ at time $t \in \mathbb{N}_0$. Then the uniform distribution on $\Delta_{N-1}$ is the unique invariant distribution that can be found as the weak limit of the sequence $\mu_t$.*

*Proof* From Proposition 7.4 we have that $U[\Delta_{N-1}]$ is an invariant distribution for the process. Since the chain is $\phi$-irreducible as shown in Proposition 7.5, uniqueness of the equilibrium follows from the existence of a Foster-Lyapunov function, proven in Proposition 7.6.

### 7.2.5 Kinetic Equation as Limit of the Agent System

Kinetic equations for the one-agent distribution function with a bilinear interaction term can be derived using mathematical techniques from the kinetic theory of rarefied gases (Cercignani 1988; Cercignani et al. 1994). In this section we discuss how in a time-continuous setting, where the stock (or wealth) of each agent is a continuous variable $w \in \mathcal{I} = [0, \infty)$, the exchange mechanism discussed above constitutes a special case of a kinetic model for wealth distribution, proposed by Cordier, Pareschi and Toscani in 2005. In this setting the microscopic dynamics leads to a homogeneous Boltzmann-type equation for the distribution function of wealth $f = f(w,t)$. One can study the moment evolution of the Boltzmann equation to obtain insight into the tail behaviour of the cumulative wealth distribution. We also discuss the grazing collision limit which yields a macroscopic Fokker-Planck-type equation.

Cordier, Pareschi and Toscani (2005) propose a kinetic model for wealth distribution where wealth is exchanged between individuals through pairwise (binary) interactions: when two individuals with pre-interaction wealth $v$ and $w$ meet, then their post-trade wealths $v^*$ and $w^*$ are given by

$$v^* = (1-\lambda)v + \lambda w + \tilde{\eta} v, \quad w^* = (1-\lambda)w + \lambda v + \eta w. \tag{7.16}$$

Herein, $\lambda \in (0,1)$ is a constant, the so-called *propensity to invest*. The quantities $\tilde{\eta}$ and $\eta$ are independent random variables with the same distribution (usually with mean zero and finite variance $\sigma^2$). They model randomness in the outcome of the interaction in a diffusive fashion. Note that to ensure that post-interaction wealths remain in the interval $\mathcal{I} = [0, \infty)$, additional assumptions need to be made. The discrete exchange dynamics considered in the previous sections find their continuous kinetic analogue when setting $\eta = \tilde{\eta} \equiv 0$ in (7.16).

With a fixed number $N$ of agents, the interaction (7.16) induces a discrete-time Markov process on $\mathbb{R}_+^N$ with $N$-particle joint probability distribution $P_N(w_1, w_2, \ldots, w_N, \tau)$. One can write a kinetic equation for the one-marginal distribution function

$$P_1(w, \tau) = \int P_N(w, w_2, \ldots, w_N, \tau)\, dw_2 \cdots dw_N,$$

using only one- and two-particle distribution functions (Cercignani 1988; Cercignani et al. 1994),

$$\begin{aligned} &P_1(w, \tau+1) - P_1(w, \tau) = \\ &\quad \Bigg\langle \frac{1}{N}\Bigg[\int P_2(w_i, w_j, \tau)\big(\delta_0(w - ((1-\lambda)w_i + \lambda w_j + \tilde{\eta} w_i)) \\ &\qquad + \delta_0(w - ((1-\lambda)w_j + \lambda w_i + \eta w_j))\big) dw_i\, dw_j - 2P_1(w, \tau)\Bigg]\Bigg\rangle. \end{aligned}$$

Here, $\langle \cdot \rangle$ denotes the mean operation with respect to the random variables $\eta, \tilde{\eta}$. This process can be continued to give a hierarchy of equations of so-called BBGKY type (Cercignani 1988; Cercignani et al. 1994), describing the dynamics of the system of a large number of interacting agents. A standard approximation is to neglect correlations between the wealth of agents and assume the factorisation

$$P_2(w_i, w_j, \tau) = P_1(w_i, \tau) P_1(w_j . \tau).$$

Standard methods of kinetic theory (Cercignani 1988; Cercignani et al. 1994) can be used to show that, scaling time as $t = 2\tau/N$ and taking the thermodynamical limit $N \to \infty$, one obtains that the time evolution of the one-agent distribution function is governed by a homogeneous Boltzmann-type equation of the form

$$\begin{aligned} &\frac{\partial}{\partial t} f(w, t) = \\ &\quad \frac{1}{2}\Bigg\langle \int f(w_i, t) f(w_j, t)\big(\delta_0(w - ((1-\lambda)w_i + \lambda w_j + \tilde{\eta} w_i)) \\ &\qquad + \delta_0(w - ((1-\lambda)w_j + \lambda w_i + \eta w_j))\big)\, dw_i\, dw_j \Bigg\rangle - f(w, t). \end{aligned} \tag{7.17}$$

Recalling the results from Düring et al. (2008) and Matthes and Toscani (2008), we have the following proposition.

**Proposition 7.7** *The distribution $f(w, t)$ tends to a steady-state distribution $f_\infty(w)$ with an exponential tail.*

*Proof* The results in Düring et al. (2008) and Matthes and Toscani (2008) imply that $f(w, t)$ tends to a steady-state distribution $f_\infty(w)$ which depends on the initial distribution only through the conserved mean wealth $M = \int_0^\infty wf(w, t)\,dw > 0$. As detailed in Düring et al. (2008) and Matthes and Toscani (2008), the long-time behaviour of the $s$-th moment $\int_0^\infty w^s f(w, t)\,dw$ is characterised by the function $\mathscr{S}(s) = (1-\lambda)^s + \lambda^s - 1$ which is negative for all $s > 1$; hence all $s$-th moments for $s > 1$ are bounded, and the tail of the steady-state distribution is exponential. □

In general, such equations like (7.17) are rather difficult to treat, and it is usual in kinetic theory to study certain asymptotic limits. In a suitable scaling limit, a partial differential equation of Fokker-Planck type can be derived for the distribution of wealth. Similar diffusion equations are also obtained in Slanina and Lavička (2003) as a mean-field limit of the Sznajd model (Sznajd-Weron and Sznajd 2000). Mathematically, the model is related to works in the kinetic theory of granular gases (Cercignani et al. 1994).

To this end, we study by formal asymptotics the so-called continuous trading limit ($\lambda \to 0$ while keeping $\sigma_\eta^2/\lambda = \gamma$ fixed).

Let us introduce some notation. First, consider test functions $\phi \in \mathscr{C}^{2,\delta}([0, \infty))$ for some $\delta > 0$. We use the usual Hölder norms

$$\|\phi\|_\delta = \sum_{|\alpha|\leq 2} \|D^\alpha \phi\|_{\mathscr{C}} + \sum_{\alpha=2} [D^\alpha \phi]_{\mathscr{C}^{0,\delta}},$$

where $[h]_{\mathscr{C}^{0,\delta}} = \sup_{v \neq w} |h(v) - h(w)|/|v - w|^\delta$. Denoting by $\mathscr{M}_0(A)$, $A \subset \mathbb{R}$, the space of probability measures on $A$, we define by

$$\mathscr{M}_p(A) = \left\{ \Theta \in \mathscr{M}_0 \,\middle|\, \int_A |\eta|^p d\Theta(\eta) < \infty,\ p \geq 0 \right\}$$

the space of measures with finite $p$th moment. In the following all our probability densities belong to $\mathscr{M}_{2+\delta}$, and we assume that the density $\Theta$ is obtained from a random variable $Y$ with zero mean and unit variance. We then obtain

$$\int_{\mathscr{I}} |\eta|^p \Theta(\eta)\, d\eta = \mathrm{E}[|\sigma_\eta Y|^p] = \sigma_\eta^p \mathrm{E}[|Y|^p], \tag{7.18}$$

where $\mathrm{E}[|Y|^p]$ is finite. The weak form of (7.17) is given by

$$\frac{d}{dt} \int_{\mathscr{I}} f(w, t)\phi(w)\, dw = \int_{\mathscr{I}} \mathscr{Q}(f, f)(w)\phi(w)\, dw \tag{7.19}$$

where

$$\int_{\mathcal{I}} \mathcal{Q}(f,f)(w)\phi(w)\,dw$$
$$= \frac{1}{2}\Big\langle \int_{\mathcal{I}^2} (\phi(w^*) + \phi(v^*) - \phi(w) - \phi(v))f(v)f(w)\,dv\,dw\Big\rangle.$$

Here, $\langle\cdot\rangle$ denotes the mean operation with respect to the random variables $\eta, \tilde{\eta}$. To study the situation for large times, i.e. close to the steady state, we introduce for $\lambda \ll 1$ the transformation $\tilde{t} = \lambda t,\ g(w,\tilde{t}) = f(w,t)$. This implies $f(w,0) = g(w,0)$, and the evolution of the scaled density $g(w,\tilde{t})$ follows (we immediately drop the tilde in the following)

$$\frac{d}{dt}\int_{\mathcal{I}} g(w,t)\phi(w)\,dw = \frac{1}{\lambda}\int_{\mathcal{I}} \mathcal{Q}(g,g)(w)\phi(w)\,dw. \tag{7.20}$$

Due to the interaction rule (7.16), it holds

$$w^* - w = \lambda(v - w) + \eta w.$$

A Taylor expansion of $\phi$ up to second order around $w$ of the right hand side of (7.20) leads to

$$\Big\langle \frac{1}{\lambda}\int_{\mathcal{I}^2} \phi'(w)\,[\lambda(v-w) + \eta w]\,g(w)g(v)\,dv\,dw\Big\rangle$$
$$+ \Big\langle \frac{1}{2\lambda}\int_{\mathcal{I}^2} \phi''(\tilde{w})\,[\lambda(v-w) + \eta w]^2\,g(w)g(v)\,dv\,dw\Big\rangle$$
$$= \Big\langle \frac{1}{\lambda}\int_{\mathcal{I}^2} \phi'(w)\,[\lambda(v-w) + \eta w]\,g(w)g(v)\,dv\,dw\Big\rangle$$
$$+ \Big\langle \frac{1}{2\lambda}\int_{\mathcal{I}^2} \phi''(w)\big[\lambda(v-w) + \eta w\big]^2 g(w)g(v)\,dv\,dw\Big\rangle + R(\lambda,\sigma_\eta)$$
$$= -\int_{\mathcal{I}^2} \phi'(w)(w-v)g(w)g(v)\,dv\,dw$$
$$+ \frac{1}{2\lambda}\int_{\mathcal{I}^2} \phi''(w)\big[\lambda^2(v-w)^2 + \lambda\gamma w^2\big]g(w)g(v)\,dv\,dw + R(\lambda,\sigma_\eta),$$

with $\tilde{w} = \kappa w^* + (1-\kappa)w$ for some $\kappa \in [0,1]$ and

$$R(\lambda,\sigma_\eta) = \Big\langle \frac{1}{2\lambda}\int_{\mathcal{I}^2} (\phi''(\tilde{w}) - \phi''(w))\,[\lambda(v-w) + \eta w]^2\,g(w)g(v)\,dv\,dw\Big\rangle.$$

Now we consider the limit $\lambda, \sigma_\eta \to 0$ while keeping $\gamma = \sigma_\eta^2/\lambda$ fixed. It can be seen that the remainder term $R(\gamma,\sigma_\eta)$ vanishes in this limit (see Cordier et al. (2005) for details). In the same limit, the term on the right-hand side of (7.20) converges to

$$- \int_{\mathscr{I}^2} \phi'(w)(w-v)g(w)g(v)\,dv\,dw + \frac{1}{2}\int_{\mathscr{I}^2} \phi''(w)\gamma w^2 g(w)g(v)\,dv\,dw$$
$$= - \int_{\mathscr{I}} \phi'(w)(w-m)g(w)\,dw + \frac{\gamma}{2}\int_{\mathscr{I}} \phi''(w)w^2 g(w)\,dw,$$

with $m = \int_{\mathscr{I}} v g(v)\,dv$ being the mean wealth (the mass is set to one for simplicity; otherwise it would appear as well here). After integration by parts, we obtain the right-hand side of (the weak form of) the Fokker-Planck equation

$$\frac{\partial}{\partial t} g(w,t) = \frac{\partial}{\partial w}\Big((w-m)g(w,t)\Big) + \frac{\gamma}{2}\frac{\partial^2}{\partial w^2}\Big(w^2 g(w,t)\Big), \tag{7.21}$$

subject to no-flux boundary conditions (which result from the integration by parts). The same equation has also been obtained by considering the mean-field limit in a trading model described by stochastic differential equations (Bouchaud and Mezard 2000).

## 7.3 Remembering Jun-ichi Inoue

One of us (Enrico Scalas) was expecting to meet Jun-ichi Inoue at the 2015 AMMCS-CAIMS Congress in Waterloo, Ontario, Canada. Together with Bertram Düring (also co-author of this paper), we organised a special session entitled *Wealth distribution and statistical equilibrium in economics* (see: http://www.ammcs-caims2015.wlu.\discretionary-ca/special-sessions/wdsee/). Even if Enrico never collaborated with Jun-ichi on the specific problem discussed in this paper, they co-authored two research papers, one on the nonstationary behaviour of financial markets (Livan et al. 2012) and another one on durations and the distribution of first passage times in the FOREX market (Sazuka et al. 2009). The former was the outcome of a visit of Jun-ichi to the Basque Center for Applied Mathematics in Bilbao from 3 October 2011 to 7 October 2011. Enrico, Jun-ichi and Giacomo Livan met several times in front of blackboards and computers, and the main idea of the paper (nonstationarity of financial data) was suggested by Jun-ichi. The latter is the result of a collaboration with Naoya Sazuka who, among other things, provided the data from Sony Bank FOREX transactions. This paper is connected to Enrico's activity on modelling high-frequency financial data with continuous-time random walks. A third review paper was published on the role of the inspection paradox in finance (Inoue et al. 2010). Before leaving for Canada, Enrico received the sad news of Jun-ichi's death. He had the time to change his presentation in Waterloo to include a short commemoration of Jun-ichi. With Jun-ichi, Enrico lost not only a collaborator, but a friend with an inquisitive mind.

**Acknowledgements** BD acknowledges support by the Leverhulme Trust research project grant "Novel discretisations for higher-order nonlinear PDE" (RPG-2015-69). NG acknowledges support by the EPSRC grant EP/P021409/1.

## References


J. Angle, The surplus theory of social stratification and the size distribution of personal wealth. Soc. Forces **65**, 293–326 (1986); J. Angle, The inequality process and the distribution of income to blacks and whites. J. Math. Sociol. **17**, 77–98 (1992); J. Angle, Deriving the size distribution of personal wealth from "The rich get richer, the poor get poorer". J. Math. Sociol. **18**, 27–46 (1993); J. Angle, How the gamma law if income distribution appears invariant under aggregation. J. Math. Sociol. **31**, 325–358 (1996); J. Angle, The inequality process as a wealth maximizing process. Phys. A **367**, 388–414 (2006)

E. Bennati, Un metodo di simulazione statistica nell'analisi della distribuzione del reddito, Rivista Internazionale di Scienze Economiche e Commerciali **8**, 735–756 (1988); E. Bennati, *La simulazione statistica nell'analisi della distribuzione del reddito. Modelli realistici e metodo di Montecarlo* (ETS Editrice, Pisa, 1988)

J. Bertoin, *Random Fragmentation and Coagulation Processes* (Cambridge University Press, Cambridge, 2006)

J.P. Bouchaud, M. Mezard, Wealth condensation in a simple model of economy. Phys. A **282**, 536–545 (2000)

C. Cercignani, *The Boltzmann Equation and Its Applications*. Springer Series in Applied Mathematical Sciences, vol. 67 (Springer, New York, 1988)

C. Cercignani, R. Illner, M. Pulvirenti, *The Mathematical Theory of Dilute Gases*. Springer Series in Applied Mathematical Sciences, vol. 106 (Springer, New York, 1994)

A.S. Chakrabarti, B.K. Chakrabarti, Statistical theories of income and wealth distribution. Econ Open Acess Open-Assess E-J **4** (2010). http://www.economics-ejournal.org/economics/journalarticles/2010-4

D.G. Champernowne, The graduation of income distributions. Econometrica **20**, 591–615 (1952); D.G. Champernowne, A model of income distribution. Econ. J. **53**, 318–351 (1953)

S. Cordier, L. Pareschi, G. Toscani, On a kinetic model for a simple market economy. J. Stat. Phys. **120**, 253–277 (2005)

L. Devroye, Non-uniform random variate generation (2003). http://www.nrbook.com/devroye/

A. Drăgulescu, V. Yakovenko, Statistical mechanics of money. Eur. Phys. J. B **17**, 723–729 (2000)

B. Düring, D. Matthes, G. Toscani, Kinetic equations modelling wealth redistribution: a comparison of approaches. Phys. Rev. E **78**(5), 056103 (2008)

B. Düring, D. Matthes, G. Toscani, A Boltzmann-type approach to the formation of wealth distribution curves. Riv. Mat. Univ. Parma (8) **1**, 199–261 (2009)

U. Garibaldi, E. Scalas, *Finitary Probabilistic Methods in Econophysics* (Cambridge University Press, Cambridge, 2010)

J. Inoue, N. Sazuka and E. Scalas, On-line trading as a renewal process: waiting time and inspection paradox. Sci. Cult. **76**, 466–470 (2010)

G. Livan, J. Inoue, E. Scalas, On the non-stationarity of financial time series: impact on optimal portfolio selection. J. Stat. Mech.: Theory Exp. **2012**, P07025 (2012)

T. Lux, Emergent statistical wealth distributions in simple monetary exchange models: a critical review, in *Econophysics of Wealth Distributions*, ed. by A. Chatterjee, S. Yarlagadda, B.K. Chakrabarti (Springer, Milan, 2005)

B. Mandelbrot, Stable Paretian random functions and the multiplicative variation of income. Econometrica **29**, 517–543 (1961)

D. Matthes, G. Toscani, On steady distributions of kinetic models of conservative economies. J. Stat. Phys. **130**(6), 1087–1117 (2008)

S.P. Meyn, R.L. Tweedie, *Markov Chains and Stochastic Stability* (Springer, London, 1993)

L. Pareschi, G. Toscani, *Interacting Multiagent Systems: Kinetic Equations and Monte Carlo Methods* (Oxford University Press, Oxford, 2014)

V. Pareto, *Cours d'économie politique, Tome II Livre III. F. Pichon* (Imprimeur-Éditeur, Paris, 1897)

T. Piketty, *Capital in the Twenty-First Century* (Harvard University Press, Cambridge, 2013)

N. Sazuka, J. Inoue, E. Scalas, The distribution of first-passage times and durations in FOREX and future markets. Phys. A: Stat. Mech. Appl. **388**, 2839–2853 (2009)

E. Scalas, U. Garibaldi, S. Donadio, Statistical equilibrium in simple exchange games I: methods of solution and applications to the Bennati-Drăgulescu-Yakovenko (BDY) game. Eur. Phys. J. B **53**, 267–272 (2006); E. Scalas, U. Garibaldi, S. Donadio, Statistical equilibrium in simple exchange games I. Eur. Phys. J. B **60**, 271–272 (2007)

H.A. Simon, On a class of skew distribution functions. Biometrika **42**, 425–440 (1955)

F. Slanina, H. Lavička, Analytical results for the Sznajd model of opinion formation. Eur. Phys. J. B **35**, 279–288 (2003)

K. Sznajd-Weron, J. Sznajd, Opinion evolution in closed community. Int. J. Mod. Phys. C **11**, 1157–1165 (2000)

H.O.A. Wold, P. Whittle, A model explaining the Pareto distribution of wealth. Econometrica **25**, 591–595 (1957)

# Part II
# Complex Network and Sentiments

# Chapter 8
# Document Analysis of Survey on Employment Trends in Japan

**Masao Kubo, Hiroshi Sato, Akihiro Yamaguchi, and Yuji Aruka**

**Abstract** This paper analyses the Survey on Employment Trends to propose methods for discovering the regularity of employment across different industrial sectors. Industrial sectors seem to have cooperative and competitive relationships, for example, through information and logistics. Therefore, some type of regularity in employment is expected across the industrial sectors. We examined the increases and decreases in employment for each industrial sector in Japan using the Survey on Employment Trends and found that most sectors were similar and a simple correlation analysis did not provide a satisfactory solution. Therefore, we developed the idea of applying the latent Dirichlet allocation (LDA) to the employment data; LDA is a method for finding the subjects (or topics) used in documents. This machine learning technique is based on the assumption that words reflect the topic; it helps to find topics from documents without predefining the keywords. We created a hypothetical document interlinking the results of the Survey on Employment Trends between 2000 and 2014, where the latent topic analysis was applied. In this study, we documented the numeric data from the Survey on Employment Trends by thresholding the numeric data to symbolize the data. As a result, we illustrated that there was a significant change in employment in the publication, food, mining, electricity, gas and water industries after 2009.

M. Kubo (✉) • H. Sato
National Defense Academy of Japan, Hashirimizu 1-10-20, Yokosuka, Kanagawa, Japan
e-mail: masaok@nda.ac.jp; hsato@nda.ac.jp

A. Yamaguchi
Fukuoka Institute of Technology, 3-30-1 Wajiro-higashi, Higashi-ku, Fukuoka, Japan
e-mail: aki@fit.ac.jp

Y. Aruka
Faculty of Commerce, Chuo University, Higashinakano Hachioji-shi, Tokyo, Japan
e-mail: aruka@tamacc.chuo-u.ac.jp



Y. Aruka, A. Kirman (eds.), *Economic Foundations for Social Complexity Science*, Evolutionary Economics and Social Complexity Science 9,
DOI 10.1007/978-981-10-5705-2_8

## 8.1 Introduction

The Survey on Employment Trends is statistical data based on a questionnaire survey of businesses and people leaving and entering employment conducted by the Ministry of Health, Labour and Welfare (MHLW). It has been conducted for several decades, and the results have been published on the Internet twice a year since 2000. This survey reported employment information of approximately 45 types of industries (hereinafter called "sectors") categorized on the basis of the Japan Standard Industrial Classification, which excluded agriculture and fisheries. The information includes numeric data on the number of employees, accession rate, separation rate and the text information on age, gender, academic background, reason for joining the company and previous work experience of each new employee (confidentiality issues considered).

In principle, it is assumed that more employees leave companies when employers believe that the current number of employees is in excess; conversely, the number of people entering companies increases when employers believe that there are insufficient employees. Therefore, employer determination significantly affects the increases and decreases in employment; in general, though the sectors vary, some of the common information employers refer to include exchange rates, oil prices (Yoshikawa et al. 2015), changes in economic indicators and political events. Hence, it would not be surprising to see similar changes in employment across different sectors because of similar macroscopic influences (Krugman and Wlells 2009). Also, characteristic changes can be seen in employment across several sectors because of the relationships among them; for example, sectors that receive raw materials are influenced by manufacturers. Similarly, employment in the computer industry is increased by the advances in information technology in business. If we understand these features in advance, we can reduce the number of unemployed people, implement effective employment promotion measures and establish a more resilient society.

Now, it is possible for us to capture such regularities from the huge volumes of statistical data obtained from the Survey on Employment Trends. Studies using the survey include the following: visualization of changes in people entering and leaving companies, visualization of changes in people entering and leaving companies by industry, visualization of changes in people entering and leaving companies by age group, visualization of the reason for leaving companies (e.g. Ministry of Health, Labour and Welfare 2014b), studies related to employment changes (Kamibayashi 2008), studies related to labour transition (Teruyama 2003; Fujioka 2001) and the estimation of labour demand (The Japan Institute for Labour Policy and Training 2014). We can see that many of the studies are related to the visualization of the survey results in an accurate manner or to the calculation of the working population, but feature extraction studies of changes in employment have not been actively conducted. Although there are studies that report the relationship between the accession rate and the separation rate among sectors by using methods like clustering and correlation analysis (Ministry of Health, Labour and Welfare 2014a,b; Kubo et al. 2016), the employment trends in Japan are very similar, as shown later (e.g. Fig. 8.2), and it is not easy to increase the granularity level of the analysis.

However, latent topic analysis (Deerwester 1990; Blei et al. 2003) is a machine learning technique using text, which allows us to find topics in documents without the need to predefine the topic content. Topics refer to extensional expressions of a subject and are represented as the probability of words used frequently in the topic. For example, when the keywords of Topic 1 are *offside*, *shoot*, *World Cup* and *FIFA* and the keywords of Topic 2 are *bat*, *Cy Young Award* and *homerun*, we can assume that the Topic 1 is related to football and Topic 2 is related to baseball. Thus, the principle feature of latent topic analysis is that topics that are not expressed as words can also be found.

Using latent topic analysis, we can easily convert the topic coverage into numbers by determining what topics are covered in a document and to what degree. Traditionally, it was necessary to preform a preliminary check of the keywords for each topic to examine to what degree the topics were covered in the document. However, the latent topic analysis method allows us to find a set of keywords that form a topic and estimate the emergence of the topic simultaneously by assuming that the words belonging to the same topic are often used together. By using the following latent Dirichlet allocation (LDA) (Griffiths and Steyvers 2004), one can allocate multiple topics to a document.

We created a hypothetical document interlinking the results of the Survey on Employment Trends for several years, and we applied the latent topic analysis. If we can apply latent topic analysis to this document, we can then provide the employment analysis field with a method to capture changes in employment across sectors over time in different ways from a rational point of view. However, the Survey on Employment Trends involves many numerical results; therefore, no study has applied this method in the past. We converted numeric data in this study into the written form by thresholding numeric data to symbolize the data. In the experiments, we focused on the accession rate and the separation rate by industry (middle classification), extracted features from several sectors and proved that it was possible to visualize the changes. As a result, we were able to assume that there was a significant change in employment in the publication, food, mining, electricity, gas and water industries after around 2009.

The rest of this paper is structured as follows. Section 8.2 introduces the Survey on Employment Trends and the results of the correlation analysis. Section 8.3 explains LDA; it also describes how to symbolize numeric data. Section 8.4 presents the results of the experiment.

## 8.2 Survey on Employment Trends

This study examined the MHLW Survey on Employment Trends (Ministry of Health, Labour and Welfare 2016). The MHLW conducts the questionnaire surveys twice a year and publishes data on the number of employees, the career changes and the hiring of employees by industry. Table 8.1 lists the data provided by the survey

**Table 8.1** Data examples of the survey on employment trends

| Table No. | |
|---|---|
| 6 | Gender, company size, age group, professional background, employment type and number of hired employees by employment type |
| 7 | Gender, industry (middle classification), professional background, employment type and number of hired employees by employment type |
| 8 | Gender, industry (middle classification), company size (GT/E), professional background, number of hired employees by academic background |
| 9 | Industry (middle classification), gender, number of hired employees by age group |
| 10 | Age group, gender, industry of current employment (large classification), company size of current employment (GT/E), industry of previous employment (large classification), number of hired employees by company size of previous employment (GT/E) |
| 11 | Gender, industry (large classification), company size (GT/E), number of hired employees by prefecture |
| 12 | Industry (middle classification), company size (GT/E), professional background, number of hired employees by channels for job entry |
| 13 | Gender, industry (middle classification), company size (GT/E), age group, number of hired employees by change in wages |
| 14 | Gender, industry (large classification), company size (GT/E), number of hired employees by the use of the Internet and its content before being hired |
| 15 | Gender, industry (large classification), company size (GT/E), employment type, number of hired employees by reason for leaving |
| 16 | Employment type, gender, industry (large classification), company size (GT/E), age group, number of hired employees by period of leaving employment |
| 17 | Gender, job (large classification), age group, number of hired employees by channels for job entry |
| 18 | Gender, current job (large classification), company size of current employment (production), previous job (large classification), number of hired employees by company size (production) of previous job |
| 19 | Gender, region after entering employment, number of hired employees by region before entering employment |
| 20 | Region of current employment, industry (secondary/ tertiary), region of previous employment, number of hired employees by industry of previous employment (primary/ secondary/ tertiary) |
| 21 | Gender, prefecture, professional background, number of hired employees by age group |
| 22 | Professional background, prefecture, number of hired employees from the same prefecture by gender, number of people moving from and out to another prefecture |
| 23 | Prefecture, gender, professional background, number of hired employees by academic background |
| 24 | Gender, industry of current employment (large classification), company size of current employment (GT/E), industry of previous employment (large classification), number of loaned employees by company size of previous employment (GT/E) |

on the employment trends in 2014, which published the results of the sampling surveys of the numbers of hired employees and the numbers of separated employees; it categorized employees based on gender, age, previous jobs and regions. This study used 15 years of annual data between the years 2000 and 2014.

The industry classification is based on the Japan Standard Industrial Classification. The following studies use the industrial sector classification called middle classification. Table 8.2 shows the industrial sector middle classification in 2013.

**Table 8.2** Industry sector examples included in this document

| All industries | Mining and quarrying of stone | Construction |
|---|---|---|
| Manufacturing food | Manufacture of beverages, tobacco and feed | Manufacture of textile products |
| Manufacture of lumber and wood products | Manufacture of furniture and fixtures | Manufacture of pulp, paper and paper products |
| Printing and allied industries | Manufacture of chemical and allied product, manufacture of petroleum and coal products | Manufacture of plastic products |
| Manufacture of rubber products | Manufacture of ceramic, stone and clay products | Manufacture of iron and steel |
| Manufacture of non-ferrous metals and products | Manufacture of metal products | Manufacture of general-purpose machinery |
| Manufacture of production machinery | Manufacture of business-oriented machinery | Electronic parts, devices and electronic circuits |
| Manufacture of electrical machinery, equipment and supplies | Manufacture of information and communication electronics equipment | Manufacture of transportation equipment |
| Miscellaneous manufacturing industries, manufacture of leather tanning, leather products and fur skins | Electricity, gas, heat supply and water | Information and communications |
| Transport and postal services | Wholesale and retail trade | Wholesale |
| Retail trade | Finance and insurance | Real estate and goods rental and leasing |
| Scientific research, professional and technical services | Accommodations , eating and drinking services | Living-related and personal services and amusement services |
| Services for amusement and recreation | Education, learning support | Medical, health care and welfare |
| Medical and other health services | Social insurance, social welfare and care services | Compounded services |
| Services, N.E.C. | Automobile maintenance services, machine, repair services | Miscellaneous business services |

It is a classification that divides the industrial areas, except for agriculture and fisheries, into smaller areas. Although the classification has changed slightly since 2000, classification of this study employs 55 types of industries.

## 8.3 Recent Employment Trends in Japan

Using the data, this study explains the recent employment in Japan and the results of a basic correlation analysis. First, we describe the features of the number of hired and separated employees. The number of hired employees refers to the number of persons who started working in the sector as of this year, whereas the number of separated employees refers to the number of persons who left the sector.

Figure 8.1 projects the ratio of hired employees and separated employees in each sector; this data has been taken from the Survey on Employment Trends in 2014. The horizontal axis shows the sectors, and the vertical axis shows the number of employees in each sector. For example, in sector 30, the number of hired employees is the same as the number of separated employees. This means that the total number of employees is stable, although some employees had been replaced. In sector 35, however, the number of separated employees is 30%, whereas the number of hired employees is 40%. This means there are more employees being hired than being separated, which consequently increases the number of employees in sector 35. Thus, we found that the number of hired employees varies significantly from the separated employees depending on the sector.

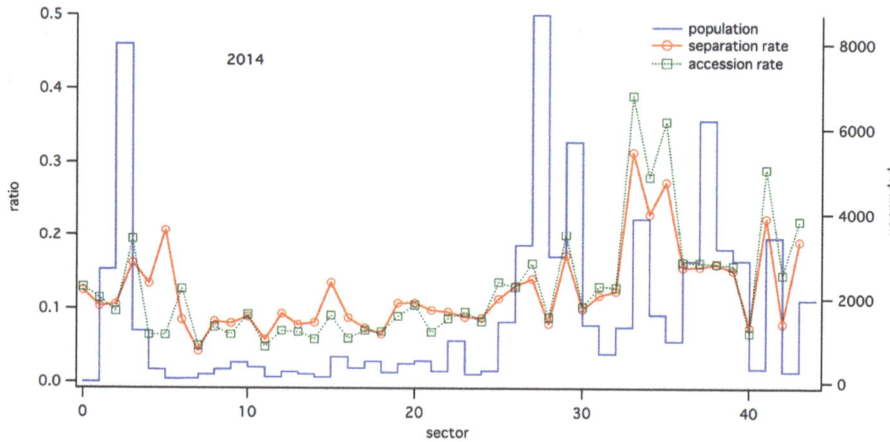


**Fig. 8.1** Results of the survey on employment trends in 2014

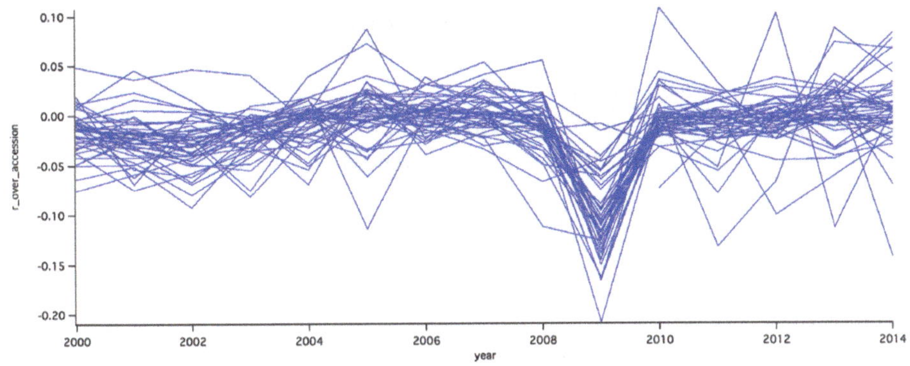


**Fig. 8.2** Increase rates between 2000 and 2014

## *8.3.1 Increase Rate*

In this section, we examined the changes in employment in each sector between 2000 and 2014. The definition of the increase rate $z_{i,t}$ in sector $i$ in the year of $t$ is

$$z_{i,t} = a_{i,t} - s_{i,t} \tag{8.1}$$

In this formula, $a_{i,t}$ represents the accession rate, and $s_{i,t}$ represents the separation rate.

The increase rate of this period was −0.013 on average, and its standard deviation was 0.037. Figure 8.2 illustrates the time series. The horizontal axis shows the years. This study found that changes in the increase rate over time were similar across all sectors in Japan's industries. For example, in the study immediately after 2009 (at the time of the Lehman crisis), the increase rates became negative in almost all sectors, and more people were unable to find jobs. Upon closer examination of the graph, for example, in 2009, it can be seen that the there is a variation in degree depending on each sector from −20% to nearly 0%; however, the increase rates are all negative in all the sectors. Also, in 2001, many of the sectors showed negative values, although a few sectors showed positive values.

This graph shows that the increase rate in 2010 is the same magnitude as that in 2008. Therefore, by 2010, the industry seems to have completely recovered from the impact of the Lehman shock. However, if this event has the same characteristics as a natural disaster, it should be that it will take a long time to recover. Therefore, this complete recovery assumption is incorrect, and a more appropriate analysis is necessary.

### 8.3.2 Correlation Graph of Increase Rates

We created a correlation graph to further examine the similarities between the increase rates. First, we studied the correlation of the increase rates between the sectors. Here, Pearson's product-moment correlation was used:

$$r_{i,j,k} = \frac{\sum_{i=1}^{n}(z_{i,t} - \bar{z_i})(z_{j,t-k} - \bar{z_j})}{\left(\left(\sum_{i=1}^{n}(z_{i,t} - \bar{z_i})^2\right)\left(\sum_{i=1}^{n}(z_{j,t-k} - \bar{z_j})^2\right)\right)^{1/2}} \quad (8.2)$$

In this formula, $k$ represents the time lag, $k \geq 0$.

Figure 8.3 illustrates the correlation of $r_{i,j,k}$ when $k = 0$. The horizontal axis and the vertical axis represent the sectors. The black areas represent $-1$ and the yellow areas represent 1. The graph looks more yellow overall; therefore, the increase rates are more positively correlated. However, more black dots are seen around sectors 30 and 40; this shows a strong negative correlation as well. Both positive and negative correlations are observed in these results; therefore, a better alternative visualization technique is required.

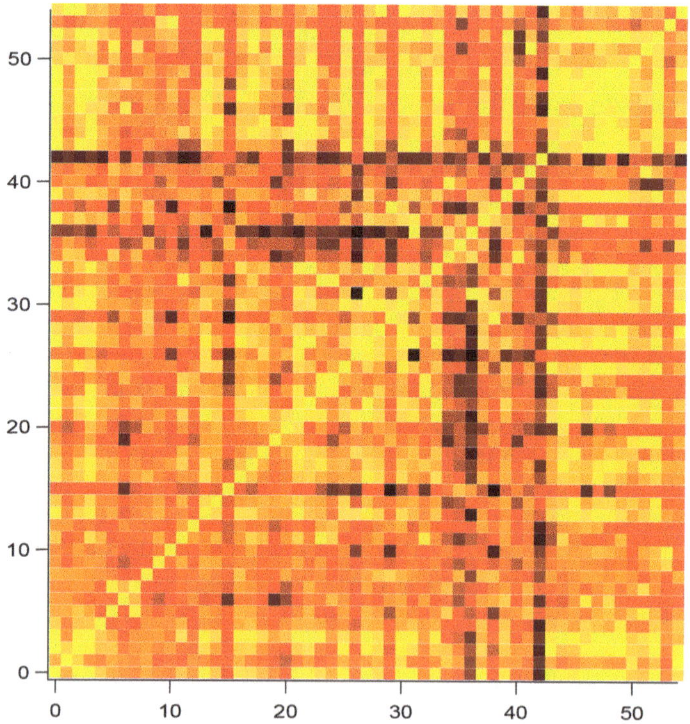


**Fig. 8.3** Correlation matrix of increase rate

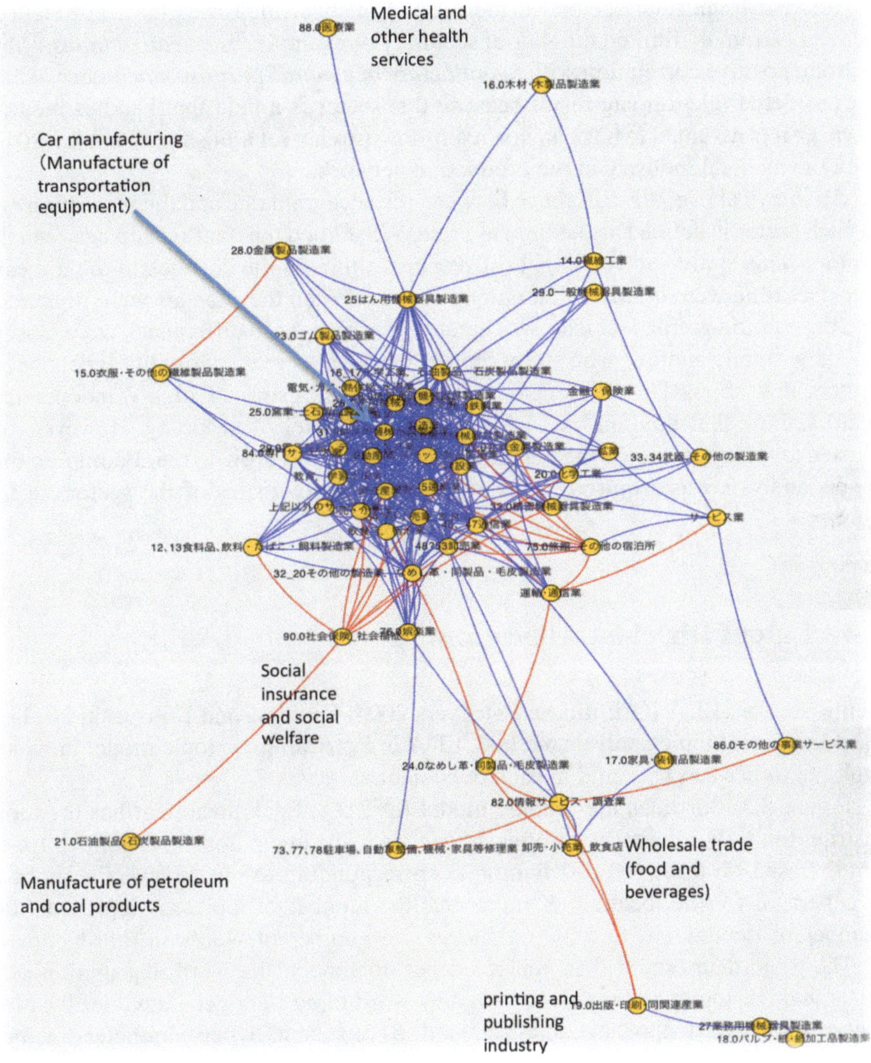


**Fig. 8.4** Graphic representation of autocorrelation of the increase rate in each sector

For a better visualization, only ($|r_{i,j,k=0}| \geq 0.7$) correlations, which show a strong correlation, were linked and shown in a graph (Fig. 8.4). The red links represent a negative correlation, and the blue links represent a positive correlation. The graph was drawn with the higher-order links in the centre. As a result, a cluster of sectors was generated in the centre of the graph, which had higher positive correlations with one another; this cluster included the auto industry. Figure 8.4 shows the relationship between the sectors is too congested in the cluster; therefore, further analysis is required.

We also found that sectors in the oil and medical industries have correlations with an extremely limited number of sectors. For example, the *medical industry* has a strong positive correlation with *manufacture of general-purpose machinery*, which is considered a legitimate result because this sector is a field that handles medical devices; it is assumed to have an upstream-downstream relationship (Carvalho 2014) with the medical industry in the production network.

As shown above, the difference between the accession rate and the separation rate in each sector is defined as the increase rate. We studied the relationship between the sectors. This study showed slight differences, although the changes in the increase rate over time were quite similar among sectors. When the changes were illustrated in a correlation graph, a cluster was created in which these differences were closely linked with one another. The study found that there were a few highly independent sectors that showed a different trend. The characteristics of highly independent sectors show that this analysis assures a certain level of accuracy. However, the cluster to which many of the sectors belonged was so crowded and complex that a new analysis was required to determine the characteristics of the sectors in the cluster.

## 8.4 Latent Dirichlet Allocation

In this section, LDA (Griffiths and Steyvers 2004; Kitajima and Kobayashi 2011) is used as a latent topic analysis method. LDA is a probabilistic topic model in which multiple topics can be found in each document.

Figure 8.5 illustrates the graphic model for LDA. Each document has the topic distribution $\theta$. For the location of each word in a document, topic $z$ is defined based on $\theta$. Based on the word distribution $\phi$ corresponding to the topic $z$, the word $w$ is generated for the location. $K$ represents the number of topics, $D$ represents the number of documents, and $N_d$ represents the number of words in the document $d$. The topic distribution $\theta$ is generated per document, the word distribution $\phi$ is generated per topic, and the word $w$ and its word topic $z$ are generated per location where each word appears. Also, $\alpha$, and $\beta$ represent hyper-parameters, which represent the parameters of the directory distribution followed by the parameters $\theta$ and $\phi$. Out of these variables, the ones actually observed were the words $w$ in

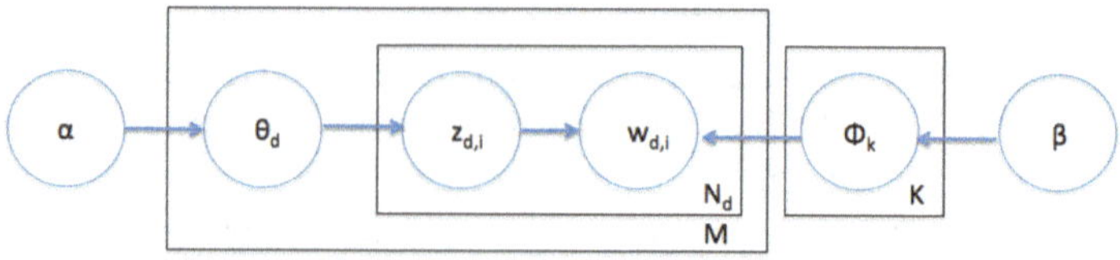


**Fig. 8.5** Graphical representation of latent Dirichlet allocation

the document, and we actually estimated the latent variables using the observed variables. The LDA document creation process is assumed to be as follows:

1. For each topic $k = 1, \ldots, K$
   (a) Based on the directory distribution $Dir(\beta)$, create the word distribution $\phi_k = \{\phi_{k,1}, \ldots, \phi_{k,V}\}$. $V$ represents the number of vocabularies here.
2. For each document $d = 1, \ldots, D$
   (a) Base on the directory distribution $Dir(\alpha)$, create the topic distribution $\theta_d = \{\theta_{d,1}, \ldots, \theta_{d,K}\}$
   (b) For each word $i = 1, \ldots, N_d$ in the document $d$
      (i) Based on the polynomial distribution $Multi(\theta_d)$, create topic $z_{d,i}$.
      (ii) Based on the polynomial distribution $Multi(\phi_{z_{d,i}})$, create word $w_{d,i}$

$\phi_k$ represents the word distribution of the topic $k$, $\theta_d$ represents the topic distribution of the document $d$, $z_{d,i}$, $\theta_d$ represents the latent topic of the $i$ th word in the document $d$, $w_{d,i}$ represents the $i$ th word in the document $d$, $Dir(\cdot)$ represents the directory distribution, and $Multi(\cdot)$ represents polynomial. The likelihoods of the topic set $Z$ and the document set $W$ are described in (8.3). In this expression $P(W|Z, \beta)$ and $P(Z|\alpha)$ can be treated independently and described in (8.4) and (8.5), respectively. Here, $V$ represents the number of vocabularies and $\Gamma(\cdot)$ represents the gamma function.

$$P(Z, W|\alpha, \beta) = P(W|Z, \beta)P(Z|\alpha) \tag{8.3}$$

$$P(W|Z, \beta) = \left(\frac{\Gamma(\beta V)}{\Gamma(\beta)^V}\right)^K \prod_{k=1}^{K} \frac{\prod_{w=1}^{V} \Gamma(N_{k,w} + \beta)}{\Gamma(N_k + \beta V)} \tag{8.4}$$

$$P(Z|\alpha) = \left(\frac{\Gamma(\alpha K)}{\Gamma(\alpha)^K}\right)^D \prod_{d=1}^{D} \frac{\prod_{k=1}^{K} \Gamma(N_{k,d} + \alpha)}{\Gamma(N_d + \alpha V)} \tag{8.5}$$

To estimate the topic set $Z$, we suggest the variational Bayesian method, the collapsed variational Bayesian method and the Gibbs sampling. It has been proved that Gibbs sampling can estimate models with higher accuracy than the variational Bayesian method as long as a sufficient number of iterations are allowed (Teh et al. 2006); therefore, we decided to employ the Gibbs sampling estimation (Griffiths and Steyvers 2004) in this study. The updated expressions of the Gibbs sampling for LDA are as follows:

$$P(z_i|z_{\backslash i}, W) \propto \frac{p(w|z)p(z)}{p(w_{\backslash i}|z_{\backslash i})p(z_{\backslash i})} \tag{8.6}$$

$$= \frac{(n^{v}_{\backslash i,j} + \beta)(n^{d}_{\backslash i,j} + \alpha)}{(n^{\cdot}_{\backslash i,j} + W\beta)(n^{d}_{\backslash i,\cdot} + T\alpha)} \tag{8.7}$$

In the expressions, $z_{\backslash i}$ represents the difference of deducting the topic $z_i$ from topic set $Z$. When the location information $i$ is excluded, $n^{v}_{\backslash i,j}$, $n^{d}_{\backslash i,j}$, $n^{\cdot}_{\backslash i,j}$ and $n^{d}_{\backslash i,\cdot}$ represent the generation frequency of the word $v$ from the topic $j$, the allocation frequency of the topic $j$ in the document $d$, the allocation frequency of the topic $j$ in the entire corpus and the generation frequency of the words in the document $d$, respectively, when their location $i$ information is excluded.

The estimated distributions of the topic distribution $\theta$ of each document and the word distribution $\phi$ of each topic are calculated for the sample obtained by the Gibbs sampling. When the topic $k$ is chosen, $\hat{\theta}^k_d$ (the estimate of the probability to generate the topic $k$ in the document $d$) and $\hat{\phi}^w_k$ (the estimate of the probability to generate the word $w$) are obtained by the following expressions:

$$\hat{\theta}^k_d = \frac{N_{d,k} + \alpha}{N_d + \alpha K} \tag{8.8}$$

$$\hat{\phi}^w_k = \frac{N_{k,w} + \beta}{N_k + \beta V} \tag{8.9}$$

## 8.5 Latent Topic Analysis of Increase Rate Using LDA

Here, LDA is applied to the increase rates. Increase rates cannot be used for LDA because they are numeric values. Therefore, we divided the increase rates into five levels and symbolized them (see in Table 8.3). For example, when the increase rate $z_i$ in the sector $i$ is 0.2, 0.2 is symbolized a "substantial increase". Combining the 5 levels and the 55 sectors, we prepared 275 words (i.e. 55 × 5). However, in the data of the 15 years used here, only 173 words of them are used (namely, $|V| = 173$). The actual examples of the conversion are illustrated in Table 8.4. For example, for the increase rate in the construction industry of −0.1161, sector number 3 and "substantial decrease" were combined. This data was converted and stored as a Word 10.

### 8.5.1 Analytical Results

We then analysed the increase rates by changing the number of the context $K$. Figure 8.6a–d represent the probability $\theta$ of the topic when $K = \{2, 3, 4, 5\}$. The horizontal axis represents the years.

**Table 8.3** Symbolized rule of $z_{i,t}$

| $z_i < -0.1$ | $-0.1 \le z_i < -0.01$ | $-0.01 \le z_i < 0.01$ | $0.01 \le z_i < -0.1$ | $0.1 \le z_i$ |
|---|---|---|---|---|
| Substantial decrease | Decrease | Stable | Increase | Substantial increase |

**Table 8.4** An example of the vocabulary of words in the symbolized $z_{i,t}$ document

| Sector | $z_{i,t}$ | Sector's id | Symbol | Word |
|---|---|---|---|---|
| Construction | −0.1161 | 3 | Substantial decrease | Word10 |
| Manufacture of lumber and wood products | −0.0416 | 8 | Decrease | Word23 |
| Manufacture of lumber and wood products | 0.0017 | 8 | Stable | Word24 |
| Manufacture of lumber and wood products | 0.0869 | 8 | Increase | Word25 |
| Manufacture of business-oriented machinery | 0.1020 | 54 | Substantial increase | Word168 |

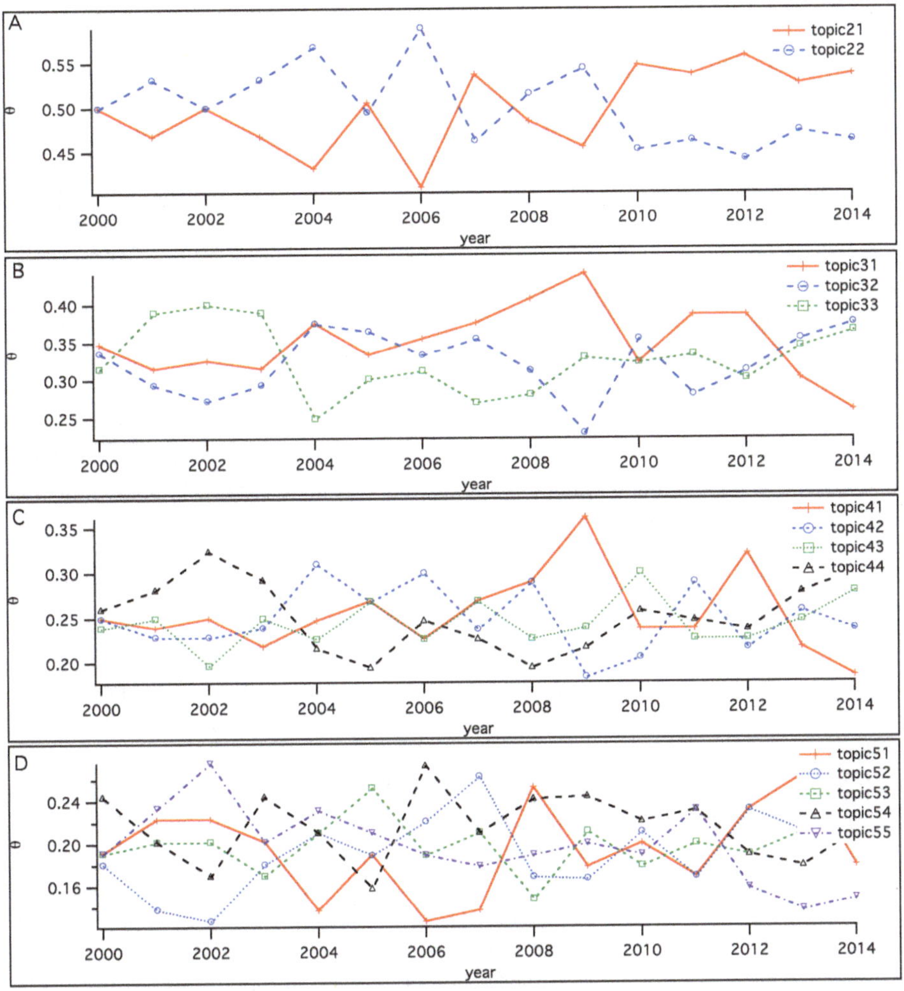


**Fig. 8.6** LDA application results

When $K = \{2, 3, 4\}$, there was a significant change in topics between 2009 (immediately after the Lehman crisis) and 2010. When analysing using $K = 5$, it is difficult to read the impact of the Lehman crisis. The fact that the crisis had a significant impact on the world economy is acknowledged by everyone. In the analysis, when $K = 5$, the presupposed number of topics was excessive for this document. Now, we further analysed the increase rates when $K = \{2, 3, 4\}$. When $K = 2$, the probabilities $\theta$ of the two topics were exactly the same as 2000. However, the probability of Topic 22 exceeded that of Topic 21 after that. This shows that Topic 22 was the mainstream topic in the world during this period. However, the probabilities of the two topics (topic 21 and 22) replaced each other and remained so between 2009 and 2010. The analysis when $K = 2$ proved that the crisis had a significant impact on the world.

These changes in the two topics were very drastic, and it is difficult to read the social changes during this 15-year period. If the Lehman crisis did such considerable damage to society as the great earthquake, long-term efforts would still be required to recover from the crisis, and gradual changes would be observed. If such indexes could be identified, social conditions could be finely quantified behind dramatic social changes.

In the analysis when $K = 3$, the probabilities of the three topics were almost the same in 2000, although those of Topic 31 were a little higher. From 2000 to 2003, the probability of Topic 33 was high, whereas the probabilities of Topic 31 and 32 were relatively low. However, the probability of Topic 33 dropped drastically in 2004, whereas that of Topic 31 slowly increased until 2009. The probability also dropped in 2010 and gradually declined afterward. Currently both Topic 32 and 33 have high probability, which is something both have never experienced before.

In the analysis, when $K = 4$, the probability $\theta$ of Topic 41 slowly increased overall between 2000 and 2009 and dropped drastically in 2010; then it increased again in 2012, like that of Topic 31. To compare social changes before and after the Lehman crisis, Topic 31 and 41 are good indexes.

### 8.5.2 Employment Trends from Topic 31

We now analysed Topic 31, which showed slower changes than Topic 41 after the Lehman crisis. Figure 8.7 shows the probability $\phi$ of each word in the three topics in the $K = 3$ analysis. Tables 8.5, 8.6 and 8.7 show the five most frequent words in each topic. The frequent words in Topic 31 are Words 33, 14, 4, 84 and 155, which show that the employment in publication, food, mining, electricity, gas and water industries will decline. We found that the employment in these industries gradually decreased from 2000 and the Lehman crisis in 2008; employment in these industries gradually increased after the Lehman crisis. Also, food, electricity, gas and water fell under the non-random term category in household consumption (Aruka et al. 2014). If the changes in employment in these industries were compared to domestic life, they could be compared to the fact that the consumption of indispensable items for households had at least slowed since 2000 and recovered after the Lehman crisis.

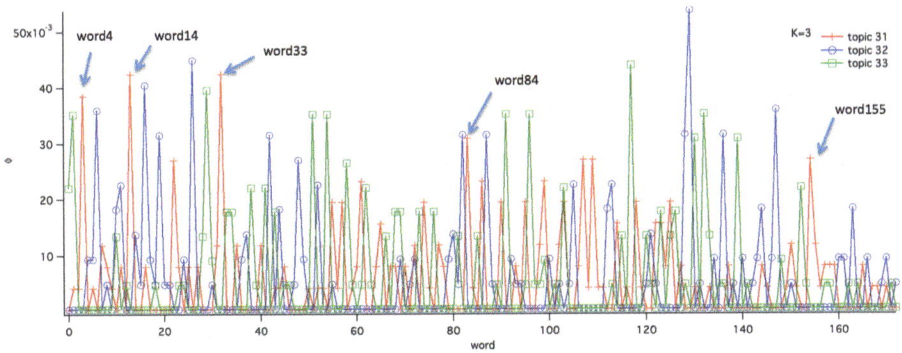


**Fig. 8.7** Attribute of each topic when $K = 3$

**Table 8.5** Topic 31

| Rank | Word | Sector | Symbol | $\phi$ |
|---|---|---|---|---|
| 1 | 33 | Publishers, except newspapers, and printing and allied industries | Decrease | 0.0423 |
| 1 | 14 | Food, beverages and tobacco related | Decrease | 0.0423 |
| 3 | 4 | Mining | Decrease | 0.0385 |
| 3 | 84 | Electricity and gas supply and water | Decrease | 0.0385 |
| 5 | 155 | Wholesale and retail trade | Stable | 0.0270 |

**Table 8.6** Topic 32

| Rank | Word | Sector | Symbol | $\phi$ |
|---|---|---|---|---|
| 1 | 130 | Medical and other health services | Increase | 0.0537 |
| 2 | 27 | Manufacture of furniture and fixtures | Decrease | 0.0448 |
| 2 | 17 | Manufacture of textile products | Decrease | 0.0403 |
| 4 | 148 | Education, learning support | Stable | 0.0359 |
| 4 | 7 | Construction | Decrease | 0.0359 |

**Table 8.7** Topic 33

| Rank | Word | Sector | Symbol | $\phi$ |
|---|---|---|---|---|
| 1 | 118 | Services for amusement and recreation | Decrease | 0.0438 |
| 2 | 30 | Manufacture of pulp, paper and paper products | Decrease | 0.0395 |
| 3 | 133 | Services not included | Stable | 0.0351 |
| 3 | 97 | Wholesale (48–53) | Stable | 0.0351 |
| 3 | 92 | Communications (46,47) | Decrease | 0.0351 |

## 8.6 Conclusions

To clarify recent employment trends in Japan, this study suggested a method for estimating the employment trends across sectors based on an employment trend survey using LDA. We applied latent topic analysis to the estimation by focusing on the differences between the accession rate and the separation rate for each industry (middle classification) and by symbolizing numeric values to five symbols (substantial decrease, decrease, stable, increase and substantial increase). As a result, we successfully visualized the changes in employment for the 15-year period before and after the Lehman crisis.

## References

Y. Aruka, Y. Kichikawa, H. Iyetomi, A macroscopic order of consumer demand due to heterogenous consumer behaviors on Japanese household demand tested by the random matrix theory (Chapter 10), in *Econophysics of Agent-Based Models*, ed. by F. Abergel et al. New economic windows, vol 187 (Springer International Publishing, Switzerland, 2014). doi:10.1007/978-3-319-00023-7_10

D.M. Blei, A.Y. Ng, M.I. Jordan, Latent Dirichlet allocation. J. Mach. Learn. Res. **3**, 993–1022 (2003)

V.M. Carvalho, From micro to macro via production networks. J. Econ. Perspect. **28**(4), 23–48 (2014)

S. Deerwester, S.T. Dumais, G.W. Furnas, T.K. Landauer, R. Harshman, Indexing by latent semantic analysis, J. Am. Soc. Inf. Sci. **41**(6), 391–407 (1990)

M. Fujioka, Estimation of occupational mobility using census data. Econ. Rev. Shizuoka Univ. **5**(4), 51–96, 2001–02 (2001). http://doi.org/10.14945/00004986

T.L. Griffiths, M. Steyvers, Finding scientific topics. Proc. Natl. Acad. Sci. **101**, 5228–5235 (2004)

R. Kamibayashi, The decomposition of job flows in Japan: 1991–2005 (2008). http://hdl.handle.net/10086/15136. Cited 2 Nov 2016

R. Kitajima, I. Kobayashi, Latent topic estimation based on events in a document, in *The 25th Annual Conference of the Japanese Society for Articial Intelligence, 3H2-OS3-4* (2011)

M. Kubo, H. Sato, A. Yamaguchi, Y. Aruka, A result of employment analysis on open data of Japan, in *Proceedings of Joint Agent Workshops and Symposium* (2016), pp. 256–260

P. Krugman, R. Wlells, Macroeconomics (Worth Publishers, New York, 2009)

Ministry of Health, Labour and Welfare, Survey on employment trends (2014a). http://www.mhlw.go.jp/english/database/db-l/employment_trends.html. Cited 2 Nov 2016

Ministry of Health, Labour and Welfare, 2014 digest of the summary report of survey on employment trends (2014b). http://www.mhlw.go.jp/english/database/db-l/employment_trends_2014_digest.html. Cited 2 Nov 2016

Ministry of Health, Labour and Welfare, Survey on employment trends (2016). http://www.mhlw.go.jp/toukei/list/9-23-1.html. Cited 2 Nov 2016

Y.W. Teh, D. Newman, M. Welling, A collapsed variational Bayesian inference algorithm for latent Dirichlet allocation, in *Advances in Neural Information Processing Systems Conference*, vol 19 (2006), pp. 1353–1360

H. Teruyama, Job reallocation and labor reallocation in Japan: evidence from establishment-level data, ESRI discussion paper series No. 31 (2003). http://www.esri.go.jp/jp/archive/e_dis/e_dis031/e_dis031.pdf. Cited 2 Nov 2016

The Japan Institute for Labour Policy and Training, Mechanism of employment portfolio formation-empirical study through qualitative analysis. JILPT Res. Rep. **166** (2014), http:// http://www.jil.go.jp/english/reports/jilpt_research/2014/no.166.html. Cited 2 Nov 2016

H. Yoshikawa, H. Aoyama, Y. Fujiwara, H. Iyetomi, Deflation and money (2015). http://voxeu.org/ article/deflation-and-money. Cited 2 Nov 2016

# Chapter 9
# Extraction of Bi-graph Structures Among Multilingual Financial Words Using Text-Mining Methods

**Enda Liu, Tomoki Ito, Kiyoshi Izumi, Kota Tsubouchi, and Tatsuo Yamashita**

**Abstract** Vector representation of words such as word2vec is an efficient method used in text mining. However, few papers have focused on multilingual studies. In this chapter, we present a comparative study on English and Japanese text data, investigating possible relationships between the two vector models in two languages. We first extract two word2vec models by using news resources spanning ten years and then cluster them on the basis of their cosine similarities for both Japanese and English. Second, we extract the words related to finance and create a dictionary in two languages based on the models obtained. Finally, we compare cross-lingual clusters with the help of the dictionary and attempt to establish relationships between English clusters and Japanese clusters.

## 9.1 Introduction

Financial text mining is a significant component of data mining and financial analysis. The many studies in this domain based on machine learning algorithms and natural language processing have led to intelligent models for stock and market price prediction with the help of text data mining and analysis using Twitter data and news reports (Vu et al. 2012). However, not all news articles and reports are written in a language with which we are familiar, and this means we may overlook some critical foreign-language messages that may affect the stock and market price slightly or even dramatically during the construction of the predicting model. Although there are some popular solutions for multilingual translation, they are neither designed nor specialized for the financial domain. For example, Wan

E. Liu • T. Ito • K. Izumi (✉)
Department of Systems Innovation, School of Engineering, The University of Tokyo, 7-3-1 Hongo, Bunkyo, 113-8656 Tokyo, Japan
e-mail: izumi@sys.t.u-tokyo.ac.jp

K. Tsubouchi • T. Yamashita
Yahoo Japan Corporation, Tokyo, Japan



Y. Aruka, A. Kirman (eds.), *Economic Foundations for Social Complexity Science*, Evolutionary Economics and Social Complexity Science 9,
DOI 10.1007/978-981-10-5705-2_9

(2009) developed a co-training method for cross-lingual sentimental classification in financial text mining, where sentimental analysis is considered an indispensable component of prediction models in the financial field. However, this method is limited by the accuracy of commercial machine translation services,[1] which are not designed for the financial domain, suggesting that the embedded dictionaries cannot comprehend the financial content precisely. This highlights the need for a system that can establish relationships among multilingual text data for financial text mining so that both developers and users are able to handle foreign information promptly and accurately.

In this chapter given multilingual financial text data in Japanese and English, after preprocessing, we extract and compare the bi-graph structure among multilingual financial words using text mining methods including word2vec distributed representation as well as clustering based on the cosine distances. We attempt to create mapping relationships between the English and Japanese text based on previous modeling and clustering results. We also analyze and discuss the two different results obtained when conducting the same processing procedures for two sets of multilingual text data derived from a stock message board and Thomson Reuters news.

In the first section of this chapter, we elaborate upon related theories and on the framework utilized in this research. In the second section, we discuss the results of a mapping experiment among cross-lingual clusters conducted from three aspects along with comments and explanations. We conclude the chapter with a brief summary.

## 9.2 Theories and Framework

This section introduces the framework of our multilingual bi-graph comparison system and also covers explanations of the related theories applied. Generally, our framework includes five main steps: data retrieving and preprocessing, word2vec modeling, dictionary extraction (vector assignment), clustering with spherical $k$-means, and finally mapping for cross-lingual clusters (the mapping results are also partially discussed in Sect. 9.3). Figure 9.1 illustrates the basic processing flow.

### 9.2.1 *Data Preparation and Preprocessing*

The data preparation and preprocessing consist of four parts: data retrieving, cleaning, tagging, and lemmatisation of the original raw text data.

[1]Google Translate Service: https://translate.google.com/

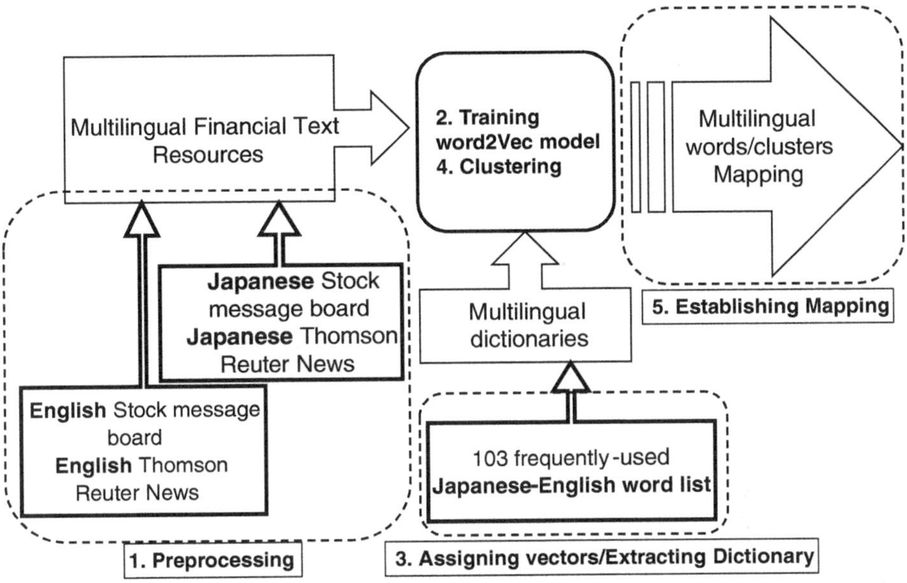


**Fig. 9.1** Basic processing flow and framework for multilingual clustering and mapping

#### 9.2.1.1 Data Retrieving

We examine two data sets in the finance domain. One consists of text data from a stock message board in 2010. Specifically, the English data are from StockTwits,[2] a system that automatically collects English information about stocks from the Internet, and the Japanese data are from a stock message board on the Finance channel of Yahoo! Japan.[3] Although all data are related to the topic of finance, they are extracted from different sources located in different countries and therefore cannot be considered as either paralleled translated pairs or discussions regarding the same events. Figure 9.2 shows examples of StockTwits news and messages from the stock message board, where we can clearly see there are significant differences regardless of whether items belong to the same financial topic.

The other data set is retrieved from Thomson Reuters news from 2010. Thomson Reuters,[4] a worldwide news agency located in London, England, provides thousands of news report in various languages with a especial focus on the economics and finance domains, suggesting a high possibility to extract paralleled multilingual articles that are translated based on the same original English draft withthe same

[2]http://stocktwits.com/

[3]http://finance.yahoo.co.jp/

[4]http://www.reuters.com/

**English samples of StockTwits**

- options trade in nordstrom today $jwn.
- bloomberg: fed discusses limited bond sales to withdraw stimulus --fed's bal sheet not part of the $1.44t new '09 debt.
- bloom--treasuries head for worst performance among g-7 on supply woes--but long bond started '09 at absurd 136 level.
- rt @ftfirehose uk considers full-body scanning at airports for this to work, everybody has to scan.

**Japanese samples of Yahoo! Japan stock message board**

- (株)XXXXグループ　株価がＭＡ線より上だし　順張りで飛び乗る人も多かろう<br />先週末には長い下髭も出たことだし上へ向かうと考えて順当だろう<br /><br />ＮＹも高い、日本円も１０４円台と円安<br />新内閣の発足もあるし諸環境も問題なし　此処か否かはともかくＧＰＩＦも乗り出してくるだろうし、全体の出来高も相当増えるのではないか
- (株)YYYY　Re: 大口取引成立の風景　正解!!
- (株)ZZZZ　s○tて言うやつ現れた(^_^)いつも激アツ銘柄に大暴落と言ってるやつや！

**Fig. 9.2** Samples of raw data from financial stock message board and StockTwits

content. Furthermore, since our purpose is to build a system optimized for the finance domain, we filter out all unrelated articles and focus only on finance, economics, or markets.

As we investigated Japanese news articles, we found that some of them were marked by the phrase *see English original version through:* followed by a unique code that seems to be the number of the original raw news article before translation. Hence, although we retrieved a total of 6,000 pairs of Japanese-English articles with similar contents, we found that, after investigating the pairs in more detail, the Japanese articles were not perfectly translated from the original English reports sentence by sentence. Rather, most of them were translated and then edited, and in the end, only the general meaning of the English articles was expressed in Japanese, indicating that only the consistency of content could be guaranteed. Even so, compared with the first data set (messages from the finance message board), it has a relatively higher reliability.

#### 9.2.1.2 Data Cleaning

Both of the raw data sets obtained contain some unsolvable elements, such as special characters, http and e-mail addresses, typos, and face marks, all of which need to be eliminated with the help of regular expressions in order to de-noise the raw data. Figure 9.2 shows examples of English and Japanese raw text data containing numerous items of useless and noisy information.

**Results of Preprocessing of English Text**

- option trade nordstrom today jwn
- bloomberg feed discuss limited bond sale withdraw stimulus fed 's bal sheet not part t new debt
- bloom treasury head bad performance g-7 supply woe long bond start absurd level
- rt uk consider full-body scanning airport work everybody have scan

**Results of Preprocessing of Japanese Text**

- 株 XXXXグループ 株価 MA線 より 上 順張り 飛び乗る 人 多 先週末 長い 下髭 出る こと 上 向かう 考える 順当 NY 高い 日本円 104 円台 と円安 新内閣 発足 ある 諸 環境 問題 此処 否 ともかく GPIF 乗り出す くる 全体 出来高 相当 増える
- 株 YYYY Re 大口取引成立の風景 正解
- 株 ZZZZ 言う やつ 現れる いつも 激アツ 銘柄 大暴落 言うつや

**Fig. 9.3** Samples of text data after preprocessing in English and in Japanese

#### 9.2.1.3 Tagging and Lemmatization

After the data cleaning, tagging and lemmatization are conducted on both sets of text data, since we need to perform the vector representation of words by means of word2vec, an implementation of skip-gram, and a continuous bag-of-word vectorization algorithm (Tomas et al. 2013a), which requires us to eliminate possible morphologies in order to derive reliable models. A tagger, also known as a part-of-speech tagger, assigns every element or token appearing in a sentence a label such as noun, verb, adjective, etc. For the English text data, we implement the Stanford NLP (Toutanova et al. 2003) tool as tagger and NLTK (Taghva et al. 2005) as a lemmatizer, which is in charge of the transformation of plural nouns, comparative adjectives, past tense verbs, and adverbs to their base forms. Similarly, we utilize MeCab (Kudo et al. 2004) with a neologism dictionary[5] for analyzing Japanese resources during tokenization, tagging, and then lemmatizing. We remove unnecessary and meaningless semantic elements including determiners, such as the word *the*, punctuation marks, conjunctions, and foreign words in order to derive more accurate word2vec models. Figure 9.3 shows the results after preprocessing.

#### 9.2.1.4 Phrase Detector

In some languages, such as English, the meaning of a word is different when it appears on its own compared to when it appears in a phrase, such as in the phrase *New York Times*. On the other hand, there are some proper nouns containing words

[5]Neologism dictionary implementation on Mecab-ipadic: https://github.com/neologd/mecab-ipadic-neologd

$$\boxed{\text{Japan}} = \begin{pmatrix} -0.232 \\ -1.054 \\ 0.158 \\ \vdots \\ \vdots \\ 0.403 \\ 0.389 \\ -0.063 \end{pmatrix} \quad \boxed{\text{Economic}} = \begin{pmatrix} 0.0432 \\ -0.104 \\ 0.458 \\ \vdots \\ \vdots \\ -0.0034 \\ 0.109 \\ -0.713 \end{pmatrix} \quad \boxed{\text{growth}} = \begin{pmatrix} -1.054 \\ 0.109 \\ 0.389 \\ \vdots \\ \vdots \\ -0.232 \\ 0.403 \\ -0.713 \end{pmatrix}$$

**Fig. 9.4** Example of word2vec modeling results for English words

that do not correspond to their internal meaning. We therefore need to apply a phrase detector to consider them as a single *phrase unit*, such as making the phrase *new york times* be a single unit *new_york_time*. We implement a data-driven approach (Tomas et al. 2013b) commonly used before feeding words for training the word2vec model.

### *9.2.2 Word2vec Modeling*

Google's word2vec is a prevalent tool based on deep learning that provides an efficient implementation of the continuous bag-of-words and skip-gram architectures for computing vector representations of words (Tomas et al. 2013a). In this study, with the help of the Gensim platform,[6] we apply the word2vec scheme to train word vector models for the preprocessed English and Japanese text data (including Thomson Reuters news in English, Thomson Reuters news in Japanese, messages from Yahoo Stock board in Japanese, and messages from StockTwits in English) with the dimensionality of 200, which is considered a reasonable size for training. Figure 9.4 shows an example of the modeling results for English words.

### *9.2.3 Dictionary Extraction*

We choose a list containing 103 crucial financial keywords and phrases frequently used in the financial domain (i.e., considered to be critical for the prediction of stock price), in both the English translations and original Japanese form.

Before moving to the next step, our 103 keyword dictionary is not normalized and consists of various tenses and plural dorms. We therefore first implement the tagging and lemmatizing again, similar to the preprocessing section discussed above, and then remove any useless elements such as articles (e.g., *the*) and prepositions (e.g., *of*). Then we recombine the phrases into a single unit.

In the case of words that are not included in our trained word2vec model, we remove them directly. In the case of phrases that are not detected by the phrase detector, we first retrieve the vector representation for each word appearing in the

[6]https://radimrehurek.com/gensim/

phrase and then conduct a summation on these vectors and consider the result as the vector representation for the whole phrase. This is proposed on the basis of recent findings (Mikolov et al. 2013) that word2vec can represent many linguistic regularities, suggesting that vector addition and subtraction are still able to represent the relative meaning of the phrase.

### 9.2.4 *Spherical K-Means Clustering*

On the basis of the vector representation generated via the word2vec model discussed above, we then conduct spherical $k$-means clustering for both the Japanese and English word lists with centroids $k$ as 10, 15, 20, and 25 accordingly. Concerning word and document vector clustering, it has been reported that the $k$-means algorithm combined with cosine similarity, also known as spherical $k$-means, can produce satisfactory clustering results (Dhillon et al. 2002). We initialize with $k$-means++ seeding (Arthur and Vassilvitskii 2007) and overwrite the algorithm (Banerjee et al. 2005) with the help of the scikit-learn platform.[7]

A sample of clustering results for cross-lingual financial news reports of Thomson Reuters in both English and Japanese is shown in Table 9.1 with the number of cluster centroid $k = 10$. Here, we define the notations $EN_i^m$ and $JP_j^n$ as $i, j, m, n \in (1,2,3,\ldots,k)$ referring to English cluster names and Japanese cluster names, respectively, where $m$ and $n$ indicate the number of centroids we choose during clustering and $i$ and $j$ show the cluster name in terms of number. Normally, as the default size of cluster centroids is 10, the superscript $m = 10$ and $n = 10$ would be omitted.

These results indicate that for each cluster, the spherical $k$-means approach can effectively cluster words with similar meanings into the same cluster.

**Table 9.1** Example of English word clustering using Thomson Reuters news

| Cluster name (English) | English word | Corresponding Japanese word | Cluster name (Japanese) |
|---|---|---|---|
| $EN_5$ | Improvement | Kaizen (*Improvement*) | $JP_5$ |
| $EN_5$ | Jump | Koutou (*Jump*) | $JP_1$ |
| $EN_5$ | Fall | Teika (*Fall*) | $JP_1$ |
| $EN_5$ | Decline | Genshou (*Decline*) | $JP_9$ |
| $EN_9$ | Decrease of profit | Zoushuu (*Decrease*) | $JP_2$ |
| $EN_9$ | Increase of income | Zoushuu (*Increase*) | $JP_2$ |
| $EN_7$ | Concern | Kennen (*Concern*) | $JP_{10}$ |
| $EN_7$ | Risk | Risuku (*Risk*) | $JP_{10}$ |
| $EN_7$ | Aggravation | Akka (*Aggravation*) | $JP_{10}$ |

[7] http://scikit-learn.org/stable/index.html

## 9.3 Mapping Experiments

Since our goal is to compare and extract the bi-graph structure of cross-lingual text data, in the following subsections, we investigate and exploit the clustering results both in Japanese and in English through two approaches.

Please note that in Sects. 9.3.1 and 9.3.2, we conduct experiments using the Thomson Reuters news, and in Sect. 9.3.3, we compare the results using other text sources including StockTwits as well as the stock message board.

### *9.3.1 Mapping for Cross-Lingual Clusters*

In addition to the properties mentioned in the previous sections, the clustering results in Table 9.1 show that some Japanese words whose English translations have been clustered into the same cluster, for example, $EN_7$, are also categorized into the cluster $JP_{10}$, indicating that there might be potential relationships between English clusters and Japanese clusters.

For the convenience of later discussion, we define the concept *common words* for clusters $EN_i$ and $JP_j$ as any translation pair where the English translation belongs to $EN_i$ while the Japanese translation belongs to $JP_j$, denoted as $C_{(i,j)}$. According to Table 9.1, for example, the Japanese-English pair *Decline* and *Genshou* forms *common words* for clusters $EN_5$ and $JP_9$.

By simple statistic methods, listing every *common word* for all possible combinations of English clusters and Japanese clusters, we can now derive a mapping table for cross-lingual clusters as shown in the Table 9.2, where the size of each cluster is attached following the cluster name embraced by brackets. Note that three fourths of the words from cluster $EN_{10}$ are Japanese pairs are from $JP_4$, indicating the relationships among cross-lingual clusters. Figure 9.5 illustrates such relationships among cross-lingual clusters.

**Table 9.2** Comparison of English clusters and Japanese clusters

| Common words | $JP_1$ [20] | $JP_2$ [8] | $JP_3$ [6] | $JP_4$ [10] | $JP_5$ [6] | $JP_6$ [8] | $JP_7$ [10] | $JP_8$ [16] | $JP_9$ [8] | $JP_{10}$ [12] |
|---|---|---|---|---|---|---|---|---|---|---|
| $EN_1$[12] | 0 | 1 | 1 | 0 | 1 | 4 | 2 | 2 | 0 | 1 |
| $EN_2$[16] | 0 | 3 | 1 | 2 | 1 | 0 | 2 | 7 | 0 | 0 |
| $EN_3$[9] | 1 | 0 | 0 | 1 | 3 | 1 | 0 | 2 | 0 | 1 |
| $EN_4$[7] | 0 | 0 | 0 | 0 | 0 | 1 | 1 | 1 | 1 | 3 |
| $EN_5$[12] | 7 | 0 | 1 | 0 | 0 | 0 | 0 | 1 | 1 | 2 |
| $EN_6$[15] | 0 | 2 | 0 | 0 | 1 | 1 | 3 | 1 | 5 | 2 |
| $EN_7$[9] | 1 | 1 | 1 | 1 | 0 | 0 | 1 | 1 | 0 | 3 |
| $EN_8$[6] | 0 | 1 | 0 | 3 | 1 | 0 | 0 | 1 | 0 | 0 |
| $EN_9$[4] | 1 | 0 | 2 | 0 | 0 | 0 | 1 | 0 | 0 | 0 |
| $EN_{10}$[4] | 0 | 0 | 0 | 3 | 0 | 0 | 0 | 1 | 0 | 0 |

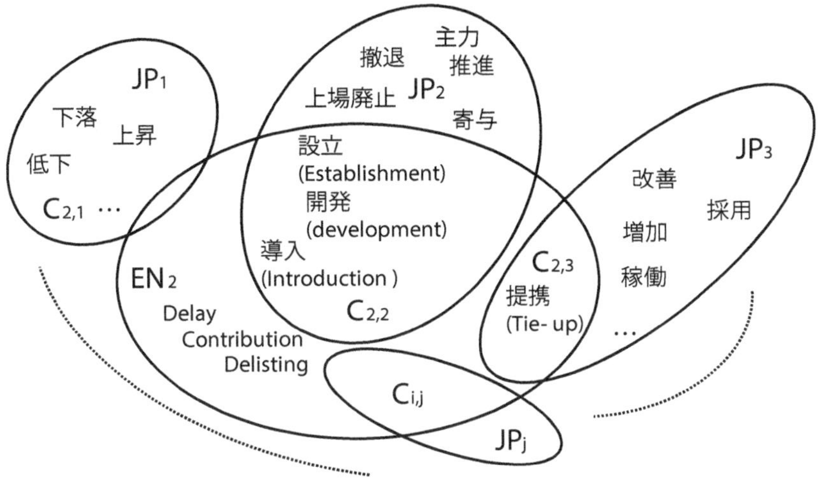


**Fig. 9.5** Illustration of cross-lingual cluster structures

**Table 9.3** Cluster pairs with maximum cross-lingual cluster similarity

| | $EN_1$ | $EN_2$ | $EN_3$ | $EN_4$ | $EN_5$ | $EN_6$ | $EN_7$ | $EN_8$ | $EN_9$ | $EN_{10}$ |
|---|---|---|---|---|---|---|---|---|---|---|
| Most similar cluster | $JP_6$ | $JP_8$ | $JP_5$ | $JP_{10}$ | $JP_1$ | $JP_9$ | $JP_{10}$ | $JP_4$ | $JP_3$ | $JP_4$ |
| Corresponded similarity score | 0.4 | 0.44 | 0.4 | 0.32 | 0.64 | 0.43 | 0.29 | 0.38 | 0.4 | 0.43 |

In order to identify the *cross-lingual cluster similarity* as the similarity among multilingual clusters quantitatively, we define the *similarity between any two clusters* as

$$\mathrm{sim}(\mathrm{EN}_i, \mathrm{JP}_j) = \frac{2 \times \mathrm{size}[C_{(i,j)}]}{\mathrm{size}[EN_i] + \mathrm{size}[\mathrm{JP}_j]}, \tag{9.1}$$

where $C_{(i,j)}$ (as mentioned previously) denotes the common words of $\mathrm{EN}_i$ and $\mathrm{JP}_j$, while the notation size[$A$] refers to the size of a set $A$. In this equation, the similarity among clusters could reach 1 when $\mathrm{size}[C_{(i,j)}] = \mathrm{size}[\mathrm{EN}_i] = \mathrm{size}[\mathrm{JP}_j]$, whereas it becomes 0 when there are no common words, that is, $\mathrm{size}[C_{(i,j)}] = 0$.

For example, as shown in Table 9.2, $\mathrm{JP}_1$ contains 10 Japanese words in all, while $\mathrm{EN}_5$ contains 12 English words in all, and they have 7 common words, which means $\mathrm{size}[C_{5,1}] = 7$. Hence, we can calculate their similarity as $\mathrm{sim}(\mathrm{EN}_i, \mathrm{JP}_j) = (2 \times 7)/(12 + 10) = 0.64$.

Then, if we choose the maximum cross-lingual similarity according to Table 9.2, a mapping relation can be obtained as shown in Table 9.3, where each Japanese cluster has its most similar English pairs mapped.

### 9.3.2 Mapping for Cross-Lingual Clusters Extended

In this subsection, we go further to exploit more accurate mapping relationships among cross-lingual clusters. Considering full combinations of clusters from set $EN_i$ and $JP_j$, where $i, j \in (1, 2, \ldots, k)$ and $k = 10$ for the present case, we then obtain two larger sets, denoted as $ENF_q$ and $JPF_p$, where $q, p \in (1, 2, \ldots, Q)$. Here, Q is the number of the full combinations, normally calculated as

$$Q = \sum_{n=1}^{k} \frac{k!}{n!(k-n)!}, \tag{9.2}$$

and $k$ is the size of cluster centroids, which is 10 by default. For illustration, when $k = 10$, the new larger sets after full combination for Japanese clusters should include:

- $JP_1, JP_2, \ldots, JP_{10}$
- $(JP_1+JP_2)$, $(JP_1+JP_2)$, ... , $(JP_9+JP_{10})$
- $(JP_1+JP_2+JP_3)$, $(JP_1+JP_2+JP_4)$, ... , $(JP_8+JP_9+JP_{10})$
- ...
- $(JP_1+JP_2+\ldots+JP_{10})$

To make them easy to distinguish, name the new combinations of base clusters as *groups*.

Then, redefine the similarity equation (Eq. 9.1) to be a more general one for cross-lingual *groups* instead of clusters:

$$\text{sim}(EN_i, JPF_p) = \frac{2 \times \text{size}[C_{(i,p)}]}{\text{size}[EN_i] + \text{size}[JPF_p]} \tag{9.3}$$

Here, we still use the notation $EN_i$ instead of $ENF_q$ because we are expected to build mapping relations for each English cluster.

For example, as shown in Fig. 9.5, for $EN_2$ consisting of 16 English words, considering the combination of $JP_1$, $JP_2$, and $JP_3$ as a new group that contains $10 + 8 + 6 = 24$ Japanese words in all, the size of their common words would be $0 + 3 + 1 = 4$. Then, the similarity between $EN_2$ and the new group JPF could be calculated as $\text{sim}(EN_2, JPF_p) = (2 \times 4)/(24 + 16) = 0.2$.

Now we extend the mapping bases from clusters to *groups* and again calculate the similarity score for all Japanese groups listed previously with respect to each English cluster $EN_i$, where $i \in (1,2,\ldots,k)$ $(k = 10)$, and then filter out the groups with maximum group similarity scores, as shown in Table 9.4.

In order to quantify improvements for the new mapping method, simply define the *average similarity* of cross-lingual mapping as the mean value of $k$ $(k = 10)$ maximum group similarity scores with respect to English clusters. The improvements of our new model are listed in Table 9.5.

**Table 9.4** Japanese group with maximum cross-lingual similarity with respect to English clusters

| English clusters | Most similar cluster | Similarity scores |
|---|---|---|
| $EN_1$ | $JP_6$ | 0.4 |
| $EN_2$ | $(JP_1+JP_2)$ | 0.5 |
| $EN_3$ | $JP_5$ | 0.4 |
| $EN_4$ | $JP_{10}$ | 0.32 |
| $EN_5$ | $JP_1$ | 0.64 |
| $EN_6$ | $(JP_9+JP_7+JP_2)$ | 0.49 |
| $EN_7$ | $(JP_{10}+JP_3)$ | 0.30 |
| $EN_8$ | $JP_4$ | 0.38 |
| $EN_9$ | $JP_3$ | 0.4 |
| $EN_{10}$ | $JP_4$ | 0.43 |

**Table 9.5** Average similarity for basic and extended mapping methods

| | Simple mapping | Mapping with cluster extension |
|---|---|---|
| Mean similarity scores | 0.413 | 0.426 |

### *9.3.3 Comparison of News Reports and Stock Board Messages*

So far, our experiments with respect to the financial text resources based on Japanese-English news and reports from Thomson Reuter have produced reasonable mapping results. As comparison, we also perform the proposed mapping methods over the multilingual stock message board from the Yahoo! Finance channel as well as StockTwits. Unlike the multilingual news and reports, due to the different markets they are targeting, where one is mainly the Japanese financial market and the other is the global financial market excluding Japan, their content is widely divergent in spite of the same aspect to which they belong, which is expected to result in a worse mapping model.

Table 9.6 shows the mapping results for extended clustering using the text resources from the stock message boards and StockTwits. These results are much worse than those using the Thomson Reuters news. This indicates that, for our proposed system, the higher the similarity between multilingual text recourse in terms of content, the more accurate the mapping will be.

On the other hand, we can also conclude that the accuracy of the cross-lingual skip-gram-based vector models depends highly on the similarity of multilingual text resources (Table 9.7).

## 9.4 Conclusion

In this work, with the help of a financial dictionary created using multilingual financial text data by means of word2vec, spherical clustering, and other techniques, we compare the structure between English and Japanese text models and build a

**Table 9.6** Japanese group with maximum cross-lingual similarity with respect to English clusters

| English clusters | Most similar cluster | Similarity scores |
|---|---|---|
| $EN_1$ | $(JP_1+JP_5+JP_7+JP_8+JP_9)$ | 0.42 |
| $EN_2$ | $(JP_1+JP_4+JP_8)$ | 0.34 |
| $EN_3$ | $(JP_2+JP_3+JP_7)$ | 0.34 |
| $EN_4$ | $(JP_7+JP_{10})$ | 0.17 |
| $EN_5$ | $JP_6$ | 0.57 |
| $EN_6$ | $(JP_1+JP_5)$ | 0.25 |
| $EN_7$ | $JP_7$ | 0.21 |
| $EN_8$ | $(JP_3+JP_4+JP_5)$ | 0.26 |
| $EN_9$ | $(JP_4+JP_7)$ | 0.33 |
| $EN_{10}$ | $JP_{10}$ | 0.36 |

**Table 9.7** Average similarity for different text bases

| | Reuters news | Stock message board |
|---|---|---|
| Mean similarity scores | 0.426 | 0.325 |

mapping relationship for cross-lingual clusters. Our experiments demonstrate that one of the necessary requirements to build a good mapping model is the similarity of the original multilingual text data, where higher similarity results in better mapping.

**Acknowledgements** We are grateful to StockTwits Inc. and Yahoo Japan Corporation for providing the textual data. This study is partially supported by JSPS KAKENHI Grant Numbers JP26282089 and JP15H02745.

## References


D. Arthur, S. Vassilvitskii, k-means++: the advantages of careful seeding, in *Proceedings of the Eighteenth Annual ACM-SIAM Symposium on Discrete Algorithms* (Society for Industrial and Applied Mathematics, 2007)

A. Banerjee et al., Clustering on the unit hypersphere using von Mises-Fisher distributions. J. Mach. Learn. Res. **6**, 1345–1382 (2005)

I.S. Dhillon, Y. Guan, J. Kogan, Refining clusters in high-dimensional text data, in *Proceedings of the Workshop on Clustering High Dimensional Data and Its Applications at the Second SIAM International Conference on Data Mining* (2002)

T. Kudo, K. Yamamoto, Y. Matsumoto, Applying conditional random fields to Japanese morphological analysis, in *EMNLP*, vol. 4 (2004)

T. Mikolov, W.-T. Yih, G. Zweig, Linguistic regularities in continuous space word representations, in *HLT-NAACL*, vol. 13 (2013)

K. Taghva, R. Elkhoury, J.S. Coombs, Arabic stemming without a root dictionary, in *ITCC*, vol 1 (2005)

M. Tomas, K. Chen, G. Corrado, J. Dean, Efficient estimation of word representations in vector space, in *Proceedings of Workshop at ICLR* (2013a)

M. Tomas, I. Sutskever, K. Chen, G. Corrado, J. Dean, Distributed representations of words and phrases and their compositionality, in *Proceedings of NIPS* (2013b)

K. Toutanova et al., Feature-rich part-of-speech tagging with a cyclic dependency network, in *Proceedings of the 2003 Conference of the North American Chapter of the Association for Computational Linguistics on Human Language Technology*, vol. 1 (Association for Computational Linguistics, 2003)

T.-T. Vu et al., An experiment in integrating sentiment features for tech stock prediction in Twitter (2012), pp. 23–38

X. Wan, Co-training for cross-lingual sentiment classification, in *Proceedings of the Joint Conference of the 47th Annual Meeting of the ACL and the 4th International Joint Conference on Natural Language Processing of the AFNLP*, vol. 1–vol. 1 (Association for Computational Linguistics, 2009)

# Chapter 10
# Transfer Entropy Analysis of Information Flow in a Stock Market

**Kiyoshi Izumi, Hiroshi Suzuki, and Fujio Toriumi**

**Abstract** Using the transfer entropy method and order book data, we analysed the dynamics of the relationship between index futures and individual stocks when an external shock affected a financial market. There were three main findings. First, the information flows between assets were enhanced during the impact of external shocks such as the Great East Japan Earthquake. Second, order information became the source of information flow for high-frequency relationship during the external shocks. Finally, index futures tended to have a significant effect on the price changes of the other stocks during the external shocks.

## 10.1 Introduction

With the recent advances of information and communications technology in the financial market, high-frequency trading (HFT) has rapidly become widespread. HFT is a computerised trading that places a large number of orders at very fast speeds on the order of milliseconds, microseconds and nanoseconds. In the US stock markets, over 50% of trades are estimated to be submitted by HFT, and over 30% depend on HFT in the Europe stock markets (World Federation of Exchanges 2013). In the Tokyo Stock Exchange, more than 25% of trades accounted for HFT in 2013 (Hosaka 2014). These rates are estimated to keep growing.

There is a compelling need for new methods to analyse large-scale, high-frequency financial data because of the rapid changes of the trading environment. Especially in stock markets, many financial professionals are seeking strongly analytical technologies to extract the high-frequency relationship between multiple stocks. A huge amount of simultaneous orders for multiple stocks at high speed has become popular due to the spread of HFT. These simultaneous orders are said to enlarge the correlation between the prices of multiple stocks just before large fluctuations in the financial markets (Harmon et al. 2011). Therefore, it is very

K. Izumi (✉) • H. Suzuki • F. Toriumi
Department of Systems Innovation, School of Engineering, University of Tokyo, 7-3-1 Hongo, Bunkyo, Tokyo, 113–8656, Japan
e-mail: izumi@sys.t.u-tokyo.ac.jp



Y. Aruka, A. Kirman (eds.), *Economic Foundations for Social Complexity Science*, Evolutionary Economics and Social Complexity Science 9,
DOI 10.1007/978-981-10-5705-2_10

important to analyse the high-frequency relationship between multiple stocks from large-scale and high-frequency data for risk management in investment.

In this work, we propose an analytical method that extracts the high-frequency relationship between assets by using transfer entropy. To test the proposed method, we used the high-frequency order book data in the Tokyo Stock Exchange (time series data of the transactions and quotes of each stock).

## 10.2 Related Works

Many of the current studies that analyse the relationship between stocks focus on correlation between stock prices. For example, Kullman et al. (2002) calculated the correlation between stock prices with the time difference of each stock and tried to reveal the direction of the effect between stock prices. Although such correlation can be used to roughly analyse the existence of the relationship between data, it cannot deal with more complicated relationship such as nonlinearity between time series data. Since the relation between the stocks within a short time is particularly complicated and becomes nonlinear, the correlation cannot be used to examine the high-frequency relationship between stocks.

Transfer entropy (Schreiber 2000) is a nonlinear analytic method that considers the direction of the information flow between time series data. It can measure the complicated relation between the observational data at one time and that at the next time. Marschinski and Kantz showed that there was propagation of price change from the Dow Jones average to a DAX index using transfer entropy (Marschinski and Kantz 2002). Kwon and others increased the types of index to analyse to 25 and visualised the relationship between stocks as an information flow network (Kwon and Yang 2008). However, these studies analysed the relation between indexes, not between individual stocks. The interval of the time series data that they analysed is longer than one day, and their purpose is not the analysis of high-frequency relationship.

As mentioned above, transfer entropy was introduced in order to measure the flow of information between two discrete stationary processes (Schreiber 2000). Consider the two discrete stationary processes $I$ and $J$ shown in Fig. 10.1.

When the last $l$ samples from process $I$ and the last $k$ samples from process $J$ are used, transfer entropy from $J$ to $I$ is:

$$T_{J\to I} = \sum p\left(i_{n+1}, \boldsymbol{i}_n^{(k)}, \boldsymbol{j}_n^{(l)}\right) \log \frac{p\left(i_{n+1}|\boldsymbol{i}_n^{(k)}, \boldsymbol{j}_n^{(l)}\right)}{q\left(i_{n+1}|\boldsymbol{i}_n^{(k)}\right)},$$

where $i_n$ ($j_n$) indicates the discrete state at time $n$ in processes $I$ ($J$). $\boldsymbol{i}_n^{(k)} = (i_n, i_{n-1}, \ldots, i_{n-k+1})$ is a vector including the states at the last $k$ steps from time $n$ of process $I$. Similarly, $\boldsymbol{j}_n^{(l)} = (j_n, j_{n-1}, \ldots, j_{n-l+1})$. The joint probability $p\left(i_{n+1}, \boldsymbol{i}_n^{(k)}, \boldsymbol{j}_n^{(l)}\right)$ is a probability that $i_{n+1}$, $\boldsymbol{i}_n^{(k)}$ and $\boldsymbol{j}_n^{(l)}$ take specific values. The

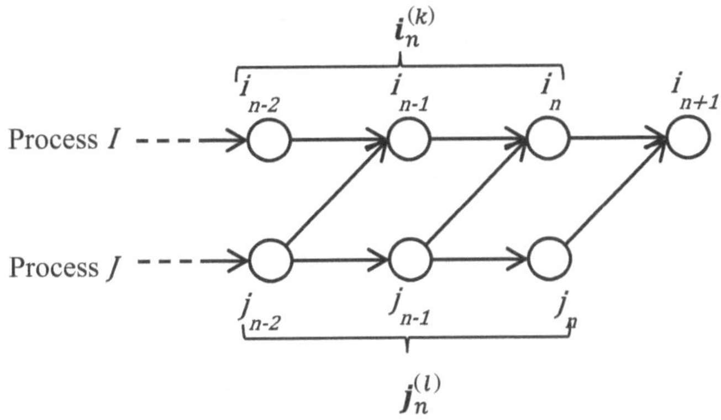


**Fig. 10.1** Framework of transfer entropy

conditional probabilities $p\left(i_{n+1}|\boldsymbol{i}_n^{(k)},\boldsymbol{j}_n^{(l)}\right)$ and $q\left(i_{n+1}|\boldsymbol{i}_n^{(k)}\right)$ are the probability that $i_{n+1}$ becomes a specific value when $\boldsymbol{i}_n^{(k)}$ and $\boldsymbol{j}_n^{(l)}$ or only $\boldsymbol{i}_n^{(k)}$, is determined, respectively. The transfer entropy with reversed direction $T_{I\to J}$ is calculable by replacing the $i$ and $j$ of the joint probability and conditional probability of $T_{J\to I}$. Transfer entropy can measure the direction of information propagation between two time series because its calculation is asymmetrical about $i_n$ and $j_n$.

This study analyses the relationship between individual stocks using order book information and transfer entropy. The relationship between stocks here is the information propagation showing of which stock the order book and price of which stock affect an order book and a price. We compare the stock relationship before and after an external shock occurs. From the analysis results, we aim to provide useful information for the risk management of investors.

## 10.3 Analysis Method of Information Propagation Among Stocks

### *10.3.1 Analytical Periods and Used Data*

We analysed the data of the following three periods as examples of periods experiencing large external shocks.

1. Four weeks around the Great East Japan Earthquake on March 11, 2011 (from February 28, 2011 to March 27, 2011)
2. Four weeks around Big Special Quote (SQ) Day on March 8, 2013 (February 25, 2013 to March 24, 2013)
3. Four weeks around the Bernanke shock and subsequent Asian market decline on May 23, 2013 (May 6, 2013 to June 2, 2013)

**Table 10.1** Sectors and stocks

| Sectors | Stocks |
|---|---|
| Index futures | Nikkei 225 future |
| Chemicals and pharmaceuticals | Teijin Ltd.; Toray Industries, Inc.; Kuraray Co., Ltd.; Asahi Kasei Corp. |
| Pharmaceuticals | Takeda Pharmaceutical Company, Ltd.; Astellas Pharma Inc.; Eisai Co., Ltd.; Daiichi Sankyo Co., Ltd. |
| Automotive | Nissan Motor Co., Ltd.; Toyota Motor Corp.; Mitsubishi Motors Corp.; Mazda Motor Corp.; Honda Motor Co., Ltd. |
| Banking | Mitsubishi UFJ Financial Group, Inc.; Sumitomo Mitsui Financial Group, Inc.; Mizuho Financial Group, Inc. |
| Real estate | Mitsui Fudosan Co., Ltd.; Mitsubishi Estate Co., Ltd.; Sumitomo Realty & Development Co., Ltd. |
| Electric power | Tokyo Electric Power Co., Inc.; Kansai Electric Power Co., Inc.; Chubu Electric Power Co., Inc. |
| Others | Fanuc Crop.; Fast Retailing Co., Ltd.; Softbank Corp. |

The 26 stocks shown in Table 10.1 were used for the analysis. Several stocks with large trading volume and aggregate market value were chosen from each sector. For the Nikkei 225 futures, data on quote prices and volume from the best bid and ask order to the tenth best bid and ask order in the Osaka Securities Exchange were used. For individual stocks, data on quote prices and volume from the best bid and ask order to the eighth best bid and ask order in the Tokyo Stock Exchange were used. The best bid (best ask) is the highest buying order (lowest selling order). The $n$th best bid (ask) is the $n$th highest buying order ($n$th lowest selling order). We also used the execution prices of the stocks in Table 10.1.

### 10.3.2 Analysis Method

First, the quote and execution data were divided into separate weeks, and the data during each week were treated as one item of time series data. Next, time series data were divided into various periods with intervals $\Delta t$, and the variables in Table 10.2 were calculated from the order book data for every period. Variables with different parameter values were distinguished as different variables. These variables were placed into two groups: execution (E1–E6) and order book (O1–O8) information. We used 21 time intervals $\Delta t$: 1 sec., 2 sec., 3 sec., 4 sec., 5 sec., 6 sec., 7 sec., 8 sec., 9 sec., 10 sec., 15 sec., 20 sec., 25 sec., 30 sec., 45 sec., 1 min., 1.5 min., 2 min., 3 min., 4 min. and 5 min. The values of each variable were classified into three bins $\{1, 0, -1\}$ according to their size, and the amount of data in all three bins was equalised. Using these discrete data, transfer entropy $T_{J \to I}$ was calculated for each week. The variable $I$, which is affected by the other variables, is E6 (absolute return) or E1 (trading volume) of a certain stock. The variable $J$, a source of information

**Table 10.2** Variables used in the calculation of transfer entropy

| | | Variables | Parameters |
|---|---|---|---|
| Execution information | E1 | Trading volume | – |
| | E2 | Number of executions | – |
| | E3 | Average volume | – |
| | E4 | Cumulative price changes | – |
| | E5 | Price return | – |
| | E6 | Absolute return | – |
| Order book information | O1 | Spread | – |
| | O2 | Order imbalance | – |
| | O3 | Order imbalance revised | – |
| | O4 | Order imbalance BD ($k$) | $k = 2, 3, 4, 5$ |
| | O5 | Cost-to-trade | $t = 0.01, 0.001, 0.0001$ |
| | O6 | Dispersion | $n = 2, 3, 4, 5$ |
| | O7 | Bi-dimensional liquidity measure (BLM) | $s = 0.5, 1.0, 2.0, 2.5, 3.0$ |
| | O8 | Slope | $N_A$ ,$N_B$=2, 3, 4, 5 |

Details of the variables are provided in Appendix

propagation, is one of the variables {E1–E6 and O1–O8} of another stock. We calculated transfer entropy $T_{J\rightarrow I}$ with all combinations $\{\Delta t \times I \times J\}$ for each week and compared them before and after the external shocks.

## 10.4 Case 1: Great East Japan Earthquake

In Case 1, we examined the 4 weeks around the Great East Japan Earthquake on March 11, 2011. Table 10.3 shows the average values of transfer entropy $T_{J\rightarrow I}$ in all stock pairs when time interval $\Delta t$ is 3 s and the affected variable $I$ is absolute return (E6). Transfer entropies of almost all variables just after the earthquake (March 14–March 18) were much larger than the other weeks. With the shorter time intervals, the transfer entropies with number of executions (E2) were the largest before the earthquake. The transfer entropies with spread (O1) had the largest values just after the earthquake. Transfer entropies showed the same trend when $\Delta t$ is shorter than 15 s, but there was no difference among the weeks when $\Delta t > 1$ min.

For the period around the earthquake, we visualised the information flow between individual stocks using transfer entropies with spread (O1), which had the largest values with the shorter interval. In the networks in Fig. 10.2, a link exists from stock $j$ to stock $i$ when transfer entropy $T_{J\rightarrow I}$ is larger than $3.0 \times 10^{-4}$. The affecting variable $J$ is spread O1 of stock $j$, the affected variable $I$ is absolute return E6 of stock $i$, and the time interval $\Delta t$ is 3 s. These findings show that the information flow between stocks just after the earthquake (Fig. 10.2b) was denser than at the other periods (Fig. 10.2a, c). There was no link during the period from February 28 to March 4, 2011.

**Table 10.3** Averages of transfer entropy $T_{J \to I}$ ($\Delta t = 3$ sec. and $I$ = E6) in all stock pairs ($\times 10^{-5}$)

| Affecting variables $J$ | Parameters | Feb. 28–Mar. 4 | Mar. 7–Mar. 11 | Mar. 14–Mar. 18 | Mar. 21–Mar. 25 |
|---|---|---|---|---|---|
| E1 | – | 6.81 | 10.39 | 16.65 | 11.84 |
| E2 | – | **8.46** | **11.43** | 24.58 | 13.75 |
| E3 | – | 6.81 | 10.29 | 16.65 | 11.19 |
| E4 | – | 6.00 | 8.82 | 23.29 | 10.67 |
| E5 | – | 2.96 | 4.00 | 9.21 | 4.75 |
| E6 | – | 5.25 | 7.86 | 20.77 | 9.41 |
| O1 | – | 5.86 | 9.42 | **41.38** | **15.32** |
| O2 | – | 2.54 | 2.91 | 27.97 | 5.39 |
| O3 | – | 2.40 | 2.53 | 6.31 | 4.08 |
| O4 | $k = 2$ | 3.17 | 3.16 | 25.90 | 6.40 |
| O4 | $k = 3$ | 4.06 | 4.42 | 25.77 | 8.70 |
| O4 | $k = 4$ | 4.74 | 4.86 | 25.70 | 10.59 |
| O4 | $k = 5$ | 5.19 | 5.24 | 24.54 | 11.44 |
| O5 | $t = 0.01$ | 6.53 | 7.42 | 31.20 | 10.91 |
| O5 | $t = 0.001$ | 5.39 | 8.75 | 36.17 | 12.49 |
| O5 | $t = 0.0001$ | 5.72 | 8.90 | 37.84 | 14.29 |
| O6 | $n = 2$ | 3.75 | 3.62 | 14.20 | 6.02 |
| O6 | $n = 3$ | 4.49 | 4.76 | 17.28 | 7.45 |
| O6 | $n = 4$ | 4.99 | 4.69 | 18.39 | 8.06 |
| O6 | $n = 5$ | 4.73 | 4.82 | 20.26 | 11.53 |
| O7 | $s = 0.5$ | 7.93 | 10.50 | 32.64 | 14.01 |
| O7 | $s = 1.0$ | 7.93 | 10.50 | 32.64 | 14.01 |
| O7 | $s = 1.5$ | 7.90 | 10.37 | 32.71 | 13.92 |
| O7 | $s = 2.0$ | 7.90 | 10.37 | 32.71 | 13.92 |
| O7 | $s = 2.5$ | 7.81 | 10.24 | 33.04 | 13.67 |
| O7 | $s = 3.0$ | 7.81 | 10.24 | 33.04 | 13.67 |
| O8 | $N_A, N_B = 2$ | 8.36 | 9.40 | 36.77 | 12.10 |
| O8 | $N_A, N_B = 3$ | 8.32 | 9.45 | 36.93 | 12.11 |
| O8 | $N_A, N_B = 4$ | 8.31 | 9.45 | 37.01 | 12.11 |
| O8 | $N_A, N_B = 5$ | 8.32 | 9.43 | 37.07 | 12.09 |

Bold-faced numbers show the largest values for each week

Next, we examined the information flows between the sections in Table 10.1. The influence quantity $\overline{\mathrm{TE}}_{s_1 \to s_2}$ of the information from section $s_1$ to section $s_2$ is calculated by the average of the transfer entropies of all stock pairs, as:

$$\overline{\mathrm{TE}}_{s_1 \to s_2} = \frac{1}{N_{s_1} N_{s_2}} \sum_{i \in s_1, j \in s_2} T_{i \to j},$$

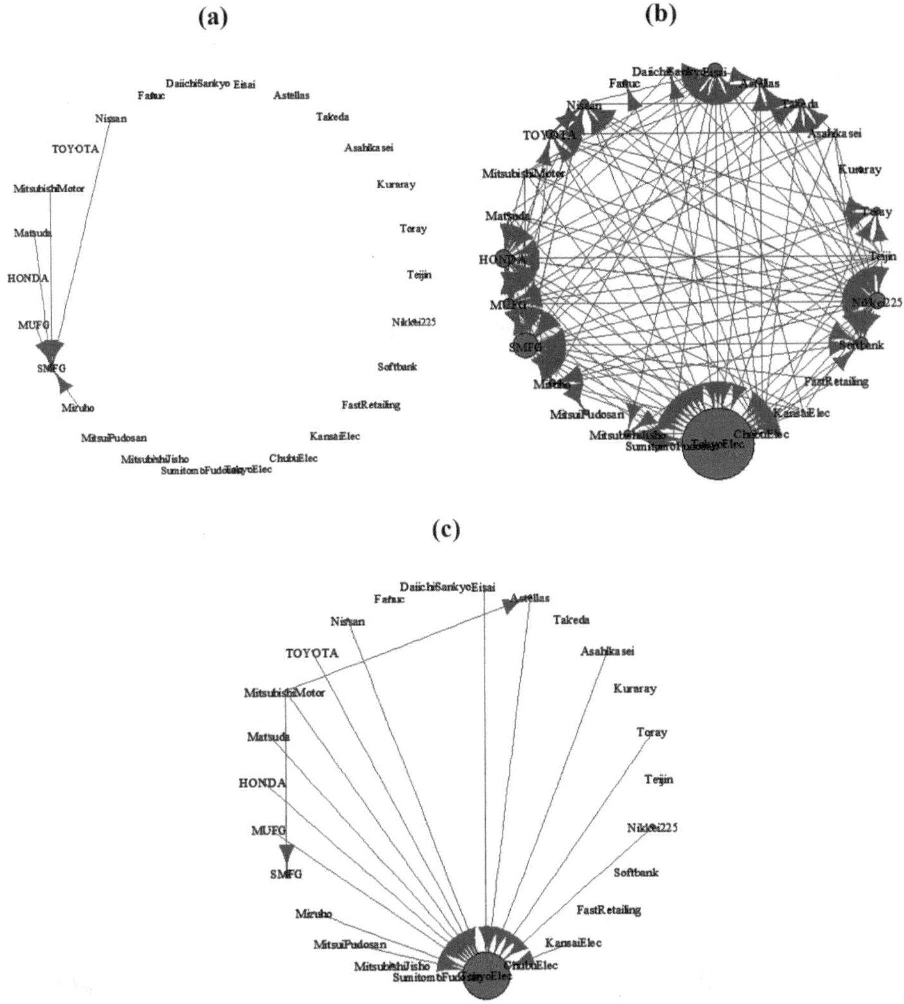


**Fig. 10.2** Information flow networks between stocks. A link exists from stock *j* to stock *i* when transfer entropy $T_{J \to I}$ ($J$ is O1, $I$ is E6 and $\Delta t$ is 1 sec) is larger than $3.0 \times 10^{-4}$. The size of nodes shows the total amount of transfer entropies incoming to each stock. (**a**) March 7–March 11, 2011. (**b**) March 14–March 18, 2011. (**c**) March 21–March 25, 2011

where $N_s$ denotes the number of the stocks included in section $s$. $T_{i \to j}$ is the transfer entropy from spread (O1) of stock $i$ 1 s ago to absolute return (E6) of stock $j$. The outgoing information flow from sector $s$, the degree of leading to other stocks, is represented by $\overline{\mathrm{TE}}_{s \to s^{-1}}$, where $s^{-1}$ denotes all stocks that are not included in section $s$. The information flow incoming to sector $s$, the degree of lagging behind other stocks, is represented by $\overline{\mathrm{TE}}_{s^{-1} \to s}$. $\Delta \mathrm{TE}_s$ is an indicator that shows the overall direction of information flow from and to section $s$:

$$\Delta \mathrm{TE}_s = \frac{\overline{\mathrm{TE}}_{s \to s-1} - \mathrm{TE}_{s-1 \to s}}{\overline{\mathrm{TE}}_{\mathrm{ALL}}},$$

where $\overline{TE}_{\mathrm{ALL}}$ is an average of $\overline{\mathrm{TE}}_{s \to s-1}$ and $\overline{\mathrm{TE}}_{s-1 \to s}$ for all sectors. A positive value of $\Delta\mathrm{TE}_s$ shows a larger influence of the spread of stocks in section $s$ on the price movements of the other stocks, while a negative value of $\Delta\mathrm{TE}_s$ indicates that the price movements of the stocks in section $s$ are affected by the spreads of other stocks.

In order to test the efficiency of transfer entropy, we also calculated a similar indicator using correlation functions. The influence quantity using correlation $\overline{\mathrm{COR}}_{s_1 \to s_2}$ from section $s_1$ to section $s_2$ is calculated as:

$$\overline{\mathrm{COR}}_{s_1 \to s_2} = \frac{1}{N_{s_1} N_{s_2}} \sum_{i \in s_1, j \in s_2} \left| \mathrm{COR}\left(i, j_{+\Delta t}\right) \right|,$$

where $|\mathrm{COR}(i, j_{+\Delta t})|$ is an absolute value of correlation between stock $i$'s price and stock $j$'s price with a time lag $\Delta t = 1$ s. Like $\Delta\mathrm{TE}_s$ for transfer entropy, $\Delta\mathrm{COR}_s$ is an indicator that shows the overall direction of information using correlation:

$$\Delta \mathrm{COR}_s = \frac{\overline{\mathrm{COR}}_{s \to s-1} - \overline{\mathrm{COR}}_{s-1 \to s}}{\overline{\mathrm{COR}}_{\mathrm{ALL}}}$$

Figure 10.3 shows indexes both by transfer entropy and correlation coefficient for the week just after the earthquake in 2011. Overall, $\Delta\mathrm{TE}_s$, the index by transfer entropy, shows the lead and lagging differences of every sector and the information

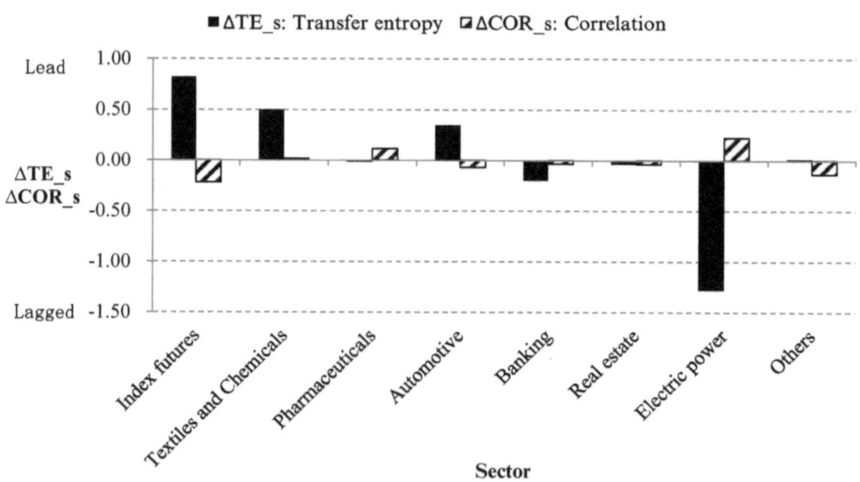


**Fig. 10.3** Indexes of information flow using transfer entropy and correlation just after the Great East Japan Earthquake (March 14–March 18, 2011)

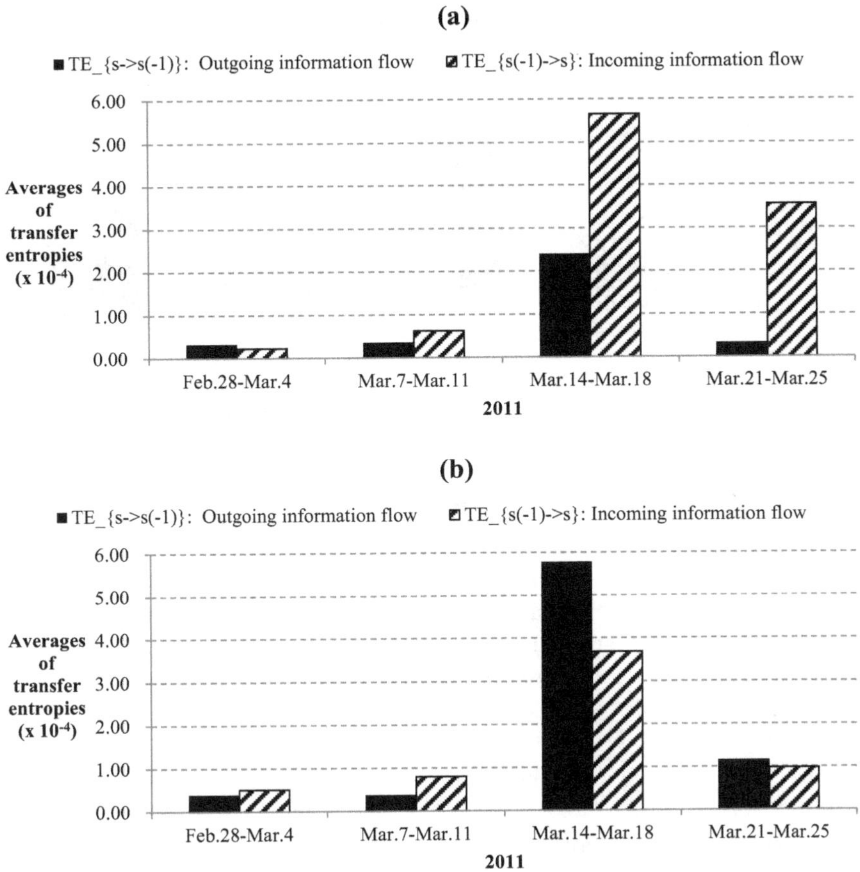


**Fig. 10.4** Averages of transfer entropies of outgoing and incoming information flows around the Great East Japan Earthquake in 2011. (**a**) Electric power sector. (**b**) Index futures

flow between the sectors in a short interval. In contrast, with $\Delta COR_s$, the index with correlation, no large difference was seen among sections. This demonstrates that transfer entropy is advantageous for the analysis of relationship between stocks with high-frequency data.

Figure 10.4 shows the average values of the outgoing and incoming transfer entropies, $\overline{TE}_{s \to s-1}$ and $\overline{TE}_{s-1 \to s}$, of the electric power sector and index futures for each week. The results of transfer entropy analysis show that the spread of index futures had a large effect on the other assets (Fig. 10.4a) and that the stock prices in the electric power sector were significantly affected by the spread of other assets (Fig. 10.4b). These results were prominently visible in the week just after the quake (March 14–March 18, 2011).

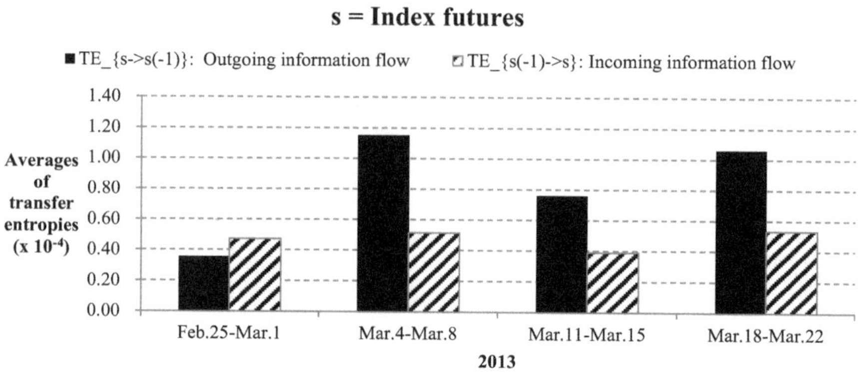


**Fig. 10.5** Averages of transfer entropies of outgoing and incoming information flows around the Big SQ Day (March 8, 2013)

## 10.5 Case 2: Big Special Quote (SQ) Day

Big Special Quote (SQ) Days in Japan's stock market are days at which the equity index futures expire. Big SQ Days are typically characterised by significant spikes in trading volumes and large fluctuation of prices. The same as in Section 3.4, we calculated $T_{J \to I}$ transfer entropy from a variable of one stock in Table 10.3 to E6, absolute return of another stock with the time lag, $\Delta t$. Similar to Case 1, the transfer entropies with spread (O1) had the largest values during weeks close to the Big SQ Day, when $\Delta t$ is shorter than 30 s. Thus, we used $J$ = O1 (spread) and $\Delta t$ = 1 s for the calculation of $T_{J \to I}$. Figure 10.5 shows the average values of outgoing and incoming transfer entropies, $\overline{\text{TE}}_{s \to s-1}$ and $\overline{\text{TE}}_{s-1 \to s}$, for each week for the index futures. The results show that the spread of index futures had a significant effect on the other assets.

## 10.6 Case 3: Bernanke Shock and Asian Market Decline

The prices of Japanese stocks widely plunged on May 23, 2013, after the statement by Bernanke (then-Chair of the Federal Reserve Bank) suggesting quantitative easing on May 22 and the aggravation of the Chinese economic indicator. We calculated $T_{J \to I}$ transfer entropy from a variable of one stock in Table 10.3 to E6, absolute return of another stock with the time lag, $\Delta t$. When $\Delta t$ is shorter than 30 s, transfer entropies of order information had the larger values, but when $\Delta t$ is longer than 30 s, that of the execution information became larger during the external shock. We used $J$ = E4 (cumulative price changes) and $\Delta t$ = 1 min for the calculation of $T_{J \to I}$. Figure 10.6 shows the average values of outgoing and incoming transfer entropies, $\overline{\text{TE}}_{s \to s-1}$ and $\overline{\text{TE}}_{s-1 \to s}$, for each week for the index futures. The results show that the cumulative price changes of index futures had a significant effect on the other assets.

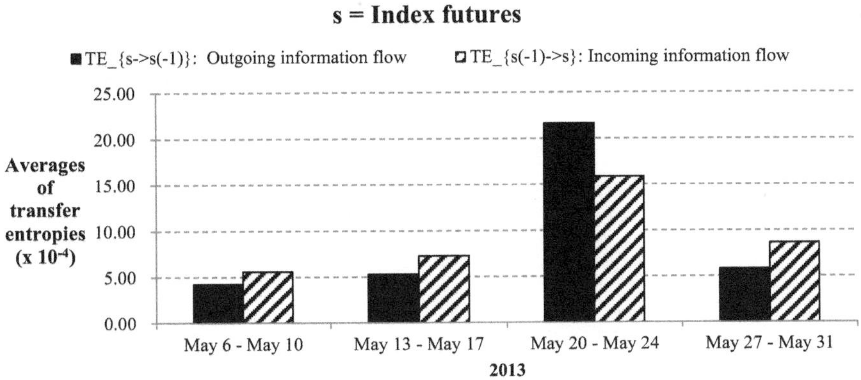


**Fig. 10.6** Averages of transfer entropies of outgoing and incoming information flows around Bernanke shock and Asian market decline (May 23, 2013)

## 10.7 Comparative Analysis Among Three Periods

Table 10.4 shows the variables $J$ with the largest transfer entropies $T_{J \to I}$ to the other stocks' absolute returns $I$ = E6 for the periods before and after the external shocks. External information like E2 (number of executions) and E4 (cumulative price changes) became the origin of information flows before the external shocks. In contrast, the order information like O1 (spread), O4 (order imbalance) and O8 (slope) became the origin of information flows during the external shocks when the time lag $\Delta t$ was shorter than 30 s.

In cases where the affected variable $J$ was E1 (trading volume), the results had the same trend as in Table 10.1. The variables based on order book information were effective for analyses of the short-term relationship between stocks during external shocks. In contrast, when the external shock had no influence on a market, variables based on external information were effective in all the intervals. This is because many more orders than usual are placed on a financial market during an external shock. Since orders unconnected to an execution also increase, an order book has more information for short-term price changes.

## 10.8 Conclusion

This study revealed the dynamics of the relationship between stocks during an external shock using transfer entropy and order book information.

1. Information flows between stocks were much larger during the external shocks than usual.
2. Order information became the source of information flow for high-frequency relationship during the external shocks.

**Table 10.4** Variables with the largest transfer entropies to other stocks' absolute returns

| | 1. Earthquake | | 2. Big SQ Day | | 3. Bernanke shock and Asian market decline | |
|---|---|---|---|---|---|---|
| $\Delta t$ sec. | Mar. 7– Mar. 11 2011 | Mar. 14– Mar. 18 2011 | Feb. 25– Mar. 1 2013 | Mar. 4– Mar. 8 2013 | May 13– May 17 2013 | May 20– May 24 2013 |
| 1 | ***O7*** | ***O1*** | ***O7*** | ***O1*** | ***O4*** | ***O8*** |
| 2 | E2 | | E2 | | | |
| 3 | | | | | | |
| 4 | | | | | | |
| 5 | | ***O8*** | | | **O4**, E2 | |
| 6 | | ***O1*** | | | E2 | |
| 7 | ***O1*** | | | | | |
| 8 | E2 | ***O8*** | | | | ***O4*** |
| 9 | | | | | | ***O8*** |
| 10 | | | | | | |
| 15 | | | | | | |
| 20 | | E4 | | | E4 | |
| 25 | | | | | | |
| 30 | | | | E2 | | E4 |
| 45 | | | E4 | E4 | | |
| 60 | | E6 | | | | |
| 90 | | E4 | | | | |
| 120 | E4 | | | | | E6 |
| 180 | E2 | | | | | |
| 240 | E4 | | | | | |
| 300 | E2 | | | | E2 | |

Order information has the largest values in bold italic

3. Index futures tended to have a significant effect on the price changes of other stocks during the external shocks.

We also compared transfer entropy with correlation coefficient and demonstrated the advantages of transfer entropy for the analysis of high-frequency relationship between stocks.

**Acknowledgements** This research is partially supported by JSPS KAKENHI Grant Numbers JP14504030 and JP15547331. This research is partially supported by MEXT as "Exploratory Challenges on Post-K computer (Study on multilayered multiscale space-time simulations for social and economic phenomena)".

## A.1 Appendix: Calculation of Variables

The details of the calculation of the variables in Table 10.3 are based on previous papers (Pascual and Veredas 2010; Jain et al. 2011). These are well-known variables that many traders use.

### *A.1.1 Execution Information*

- E1 (trading volume): The total quantity of shares traded during a specified period, $\Delta t$.
- E2 (number of executions): The number of completions of buy or sell orders during a specified period, $\Delta t$.
- E3 (average volume): Average value of trading volumes per an execution during a specified period, $\Delta t$.
- E4 (cumulative price changes): Sum of absolute values of price changes during a specified period, $\Delta t$.
- E5 (price return): Difference of log values of prices $p_t$ during a specified period, $\Delta t$.

$$\text{Return} = \log \frac{p_{t+\Delta t}}{p_t}$$

- E6 (absolute return): An absolute value of the price return.

### *A.1.2 Order Information*

- O1 (spread): The difference between the best ask (the lowest price of selling quotes) and the best bid price (the highest price of buying quotes) of an asset. In this study, the difference is normalised by the average of the best ask and the best bid price.

$$\text{Spread} = \frac{\text{Best Ask} - \text{Best Bid}}{(\text{Best Ask} + \text{Best Bid})\,/2}$$

- O2 (order imbalance): The difference between an order volume at the best ask price and that at the best bid price.
- O3 (order imbalance revised): The order imbalance normalised by the average of an order volume at the best ask price and that at the best bid price.

- O4 (order imbalance BD ($k$)): The difference between $\mathrm{BD}^a(k)$ and $\mathrm{BD}^b(k)$. $\mathrm{BD}^a(k)$ ($\mathrm{BD}^b(k)$) is a cumulative order volume from the best ask (bid) price to the $k$th best ask (bid) price.
- O5 (cost-to-trade): An indicator of the depth of the order book (Kang and Yeo 2008). It measures the size of price deviation caused by a large volume of virtual market orders.

$$\mathrm{Cost-to-trade} = \frac{\sum_{\tau=1}^{n} I_\tau^A \left(p_\tau^A - \mathrm{midquote}\right) + \sum_{\tau=1}^{n} I_\tau^B \left(\mathrm{midquote} - p_\tau^B\right)}{T \times \mathrm{midquote}}$$

$T$ is the total volume of the virtual market orders, $p_\tau^A$ ($p_\tau^B$) is the $\tau$th best bid (ask) price, and *midquote* is the average of the best bid and ask prices. $I_\tau^A$ and $I_\tau^B$ refer to the number of shares, respectively, bought or sold by the virtual market orders at each price point.

$$I_\tau^A = \begin{cases} w_j^A & \text{if } T > \sum_{j=1}^{\tau} w_j^A \\ \left(T - \sum_{j=1}^{\tau-1} w_j^A\right) & \text{if } T > \sum_{j=1}^{\tau-1} w_j^A \text{ and } T < \sum_{j=1}^{\tau} w_j^A \\ 0 & \text{otherwise} \end{cases}$$

$$I_\tau^B = \begin{cases} w_j^B & \text{if } T > \sum_{j=1}^{\tau} w_j^B \\ \left(T - \sum_{j=1}^{\tau-1} w_j^B\right) & \text{if } T > \sum_{j=1}^{\tau-1} w_j^B \text{ and } T < \sum_{j=1}^{\tau} w_j^B \\ 0 & \text{otherwise} \end{cases}$$

$w_j^A$ ($w_j^B$) is an order volume at the $j$th best ask (bid) price. The larger the cost-to-trade, the larger the trading volumes at prices far from the fair value (midquote) when a large volume of market orders are placed. Thus, the trading cost becomes larger.

- O6 (dispersion): An indicator of the tightness of the order book (Kang and Yeo 2008).

$$\mathrm{Dispersion} = \frac{1}{2}\left(\frac{\sum_{\tau=1}^{n} w_\tau^A Dst_\tau^A}{\sum_{\tau=1}^{n} w_\tau^A} + \frac{\sum_{\tau=1}^{n} w_\tau^B Dst_\tau^B}{\sum_{\tau=1}^{n} w_\tau^B}\right)$$

$w_\tau^A$ ($w_\tau^B$) is an order volume at the $\tau$th best ask (bid) price. $Dst_\tau^A$ ($Dst_\tau^B$) is the price interval between the $\tau$th best bid (ask) and the ($\tau + 1$)th best bid (ask). The larger (smaller) the dispersion, the sparser (denser) the order book and the smaller (larger) the liquidity.

- O7 (bi-dimensional liquidity measure, BLM): The size of the price change caused by a virtual market order (Irvine et al. 2000; Pascual and Veredas 2010).

$$\mathrm{BLM}(s) = \frac{PI^A(s) - PI^B(s)}{\mathrm{midquote}}$$

$\mathrm{PI}^A(s)$ ($\mathrm{PI}^B(s)$) is the price impact of a fictitious buyer-initiated (seller-initiated) trade with size equal to $s$ times the normal market size (the median trade size per stock and month). $\mathrm{PI}^A(s)$ is the difference between the weighted average ask quote at which the fictitious buyer-initiated market order would be executed minus the quote midpoint, *midquote*. $\mathrm{PI}^B(s)$ is the difference between the quote midpoint and the weighted average bid quote at which the hypothetical seller-initiated market order would be executed.

- O8 (slope): The shape of the cumulative distribution of the order volumes from the best quote.

$$\mathrm{Slope} = \frac{\mathrm{SE} + \mathrm{DE}}{2}$$

SE (DE) is an ask-side (bid-side) slope.

$$\mathrm{SE} = \frac{1}{N_A}\left\{\frac{v_1^A}{p_1^A\Big/\mathrm{midquote} - 1} + \sum_{\tau=1}^{N_A-1}\frac{v_{\tau+1}^A\Big/v_\tau^A - 1}{p_{\tau+1}^A\Big/p_\tau^A - 1}\right\}$$

$$\mathrm{DE} = \frac{1}{N_B}\left\{\frac{v_1^B}{\left|p_1^B\Big/\mathrm{midquote} - 1\right|} + \sum_{\tau=1}^{N_B-1}\frac{v_{\tau+1}^B\Big/v_\tau^B - 1}{p_{\tau+1}^B\Big/p_\tau^B - 1}\right\}$$

$p_\tau^A$ ($p_\tau^B$) is the $\tau$th best bid (ask) price, and $v_\tau^A$ ($v_\tau^B$) is the natural logarithm of the cumulative sum of the order volumes from the best ask (bid) to the $\tau$th best bid (ask).

## References


G. Hosaka, Analysis of High-frequency Trading at Tokyo Stock Exchange, JPX Working Paper, vol. 4 (2014) http://www.jpx.co.jp/english/corporate/research-study/working-paper/

D. Harmon, M.A.M. Aguiar, D.D. Chinellato, D. Braha, I.R. Epstein, Y. Bar-Yam, Predicting economic market crises using measures of collective panic. arXiv:1102.2620v1. (2011)

P.J. Irvine, G.J. Benston, E. Kandel, *Liquidity beyond the inside spread: Measuring and using information in the limit order book*, working paper, Emory University (2000)

P. Jain, P. Jain, T.H. McInish, The predictive power of limit order book for future volatility, trade price, and speed of trading, SSRN (2011)

W. Kang, W.Y. Yeo, *Liquidity beyond the best quote: A Study of the NYSE limit order book*, WFA 2008 conference paper. (2008)

L. Kullman, J. Kertész, K. Kaski, Time-dependent cross-correlations between different stock returns: A directed network of influence. Phys. Rev. E **66**, 026125 (2002)

O. Kwon, J.-S. Yang, Information flow between stock indices. Europhys. Lett. **82**(6), 68003 (2008)

R. Marschinski, H. Kantz, Analysing the information flow between financial time series: An improved estimator for transfer entropy. Eur. Phys. J. B. **30**, 275–281 (2002)

R. Pascual, D. Veredas, Does the open limit order book matter in explaining informational volatility? J. Financ. Econ. **8**(1), 57–87 (2010)

T. Schreiber, Measuring information transfer. Phys. Rev. Lett. **85**(2), 461–464 (2000)

World Federation of Exchanges, *Understanding High Frequency Trading*, Focus No.243-May 2013 (2013), pp. 3–6

# Part III
# Empirical Laws in Financial Market

# Chapter 11
# Sectoral Co-movements in the Indian Stock Market: A Mesoscopic Network Analysis

**Kiran Sharma, Shreyansh Shah, Anindya S. Chakrabarti, and Anirban Chakraborti**

**Abstract** In this article, we review several techniques to extract information from large-scale stock market data. We discuss recurrence analysis of time series, decomposition of aggregate correlation matrices to study co-movements in financial data, stock level partial correlations with market indices, multidimensional scaling, and minimum spanning tree. We apply these techniques to daily return time series from the Indian stock market. The analysis allows us to construct networks based on correlation matrices of individual stocks on one hand, and on the other, we discuss dynamics of market indices. Thus, both microlevel and macrolevel dynamics can be analyzed using such tools. We use the multidimensional scaling methods to visualize the sectoral structure of the stock market and analyze the co-movements among the sectoral stocks. Finally, we construct a mesoscopic network based on sectoral indices. Minimum spanning tree technique is seen to be extremely useful in order to group technologically related sectors, and the mapping corresponds to actual production relationship to a reasonable extent.

## 11.1 Introduction

In this paper, we present a coherent analysis of the Indian stock market employing several techniques recently proposed in the econophysics literature. Stock market is a fascinating example of a rapidly evolving multi-agent interacting system that

K. Sharma (✉) • A. Chakraborti
School of Computational and Integrative Sciences, Jawaharlal Nehru University, 110067, New Delhi, India
e-mail: kiransharma1187@gmail.com; anirban@jnu.ac.in

S. Shah
Indian Institute of Technology, Banaras Hindu University, 221005, Varanasi, India
e-mail: shreyansh.shah.mec13@iitbhu.ac.in

A.S. Chakrabarti
Economics Area, Indian Institute of Management, 380015, Ahmedabad/Vastrapur, Gujarat, India
e-mail: anindyac@iima.ac.in



Y. Aruka, A. Kirman (eds.), *Economic Foundations for Social Complexity Science*, Evolutionary Economics and Social Complexity Science 9,
DOI 10.1007/978-981-10-5705-2_11

generates an enormous amount of very well-defined and well-documented data. Due to the sheer volume of data, it becomes possible to construct large-scale correlation matrices across stocks that contain information about the aggregate market. Thus, the loss of information due to aggregation can be minimized to a great extent. Several useful techniques to analyze such large-scale data have been proposed, and there are multiple resources reviewing them. Interested readers can refer to Mantegna and Stanley (1999) and Bouchaud and Potters (2009) for excellent and quite extensive textbook expositions.

We present a series of analysis on the Bombay stock exchange, using both macroscale and microscale data. Even though there are separate attempts in a few other papers that presented analysis on similar datasets, this probably is the first attempt to systematically analyze Indian stock market data in a comprehensive manner. At the beginning of the discussion on every technique, we point out the papers that proposed the techniques and subsequent analyses, if any, on Indian or any other emerging market data.

India, being an emerging market, is an interesting example. Several papers Pan and Sinha (2007) and Bastos and Caiddo (2011) have pointed out that there are systematic differences between the dynamic behaviors of developed economies and emerging economies. Earlier hypothesis was that, as the financial market develops in a country, the market dynamics changes monotonically or at least in a clearly discernible way. Although it sounds intuitive, there is no clear demonstration of changing dynamical structure of the markets along with the process of development (Kuyyamudi et al. 2015). Instead, what we find is that, in a cross-sectional sense, such differences in the market behavior exist across countries.

In the following, we focus only on the Indian stock market. To summarize the findings, we see that the recurrence analysis is not very useful for the present dataset. Correlation decomposition techniques do not show any strong group correlation structure, which is consistent with the literature (Pan and Sinha 2007). Different clustering algorithms have been applied to understand sectoral concentration. Finally, we end with a section on sectoral correlation networks. A nontrivial finding is that technologically related sectors show very similar kind of fluctuations in the stock market. We quantify the relationship using network theoretic tools.

## 11.2 Nonlinear Dynamics: Recurrence Plot Analysis

For a very long time, it had been conjectured that the stock market indices may have certain features of highly nonlinear dynamical system. It originated from certain speculations that economic systems, in general, may show chaotic behavior (see, e.g., Baumol and Benhabib 1989). Brock and Sayers (1988) considered an idea that the aggregate macro dynamics of an economy may show chaotic behavior. By and large, such theories are no longer considered to be useful descriptions of economic dynamics. However, in recent times there have been some attempts to analyze the stock index behavior, by using recurrence analysis based on phase space reconstruction.

In general, the technique's usefulness comes from the fact that it is nonparametric, does not make any assumptions about the data, and can work with nonstationary data. In particular, the technique is useful for detecting sudden large change in a time series. A stock market crash has often been thought as a phase transition indicating a large abrupt change in the behavior (Sornette 2004). However, the technique is useful for recovering patterns in potentially highly nonlinear but recursive systems, an assumption that is not satisfied by the stock market. We follow the mode of analysis presented in details in Guhathakurta et al. (2010) and Bastos and Caiddo (2011).

Here, we describe the construction of recurrence plots. It is based on the idea of recurrence within a phase space, and the plot exhibits times, when a nonlinear system revisits the same phase space during the process of evolution. Consider a time series $\{x(i)\}_{i=1}^{N}$ representing an index of a stock market. We know, from Takens' theorem (Takens 1981), that it is possible to extract information about the phase space from the time series (see also Bastos and Caiddo 2011). We start by embedding $\{x\}$ into an $m$ dimensional space given by

$$y(i) = [x(i), x(i + \delta), x(i + 2\delta), \ldots, x(i + (m - 1)\delta)] \tag{11.1}$$

where $\delta$ is the time delay. Together these two parameters constitute the set of embedding parameters. Thus $y(i)$ is a point in the $m$ dimensional Euclidean space, representing the evolution of the system in the reconstructed phase space. We collect all such $y(i)$'s and present element-by-element difference with Euclidean norm to create a two-dimensional plot. Such a plot exhibits if there is any recurrence as explained below.

Let us define a matrix $R$ such that its $i,j$-th elements ($i,j = 1, \ldots, n$, with $n = N - (m - 1)\delta$) are expressed as

$$R_{ij}(\varepsilon) = \begin{cases} 0 \text{ if } |y(i) - y(j)| > \epsilon \\ 1 \text{ if } |y(i) - y(j)| \leq \epsilon \end{cases}$$

where $\|.\|$ is the Euclidean norm and $\epsilon$ is the threshold applied which is a positive real number. Recurrence plots are exactly symmetric along the diagonal.

**Inference based on structures:** In recurrence plots, we see multiple patterns including dots, as well as diagonal, vertical and horizontal lines, and all possible combinations of them.

- Isolated points exist if states are rare, or persistence is low, or if they represent high fluctuations.
- Existence of a diagonal line $R_{i+m,j+m} = 1$ (for $m = 1, \ldots, l$ where $l$ is the length of the diagonal line) indicates presence of recurrence, i.e., a segment of the time

series revisits the same area in the phase space at a lag. If there are lines parallel to the line of identity, it represents the parallel evolution of trajectories.

- Existence of a vertical/horizontal line $R_{i,j+m} = 1$ (for $m = 1, \ldots, v$ where $v$ is the length of the line) indicates a stage during evolution where the system gets trapped for some time and does not evolve fast. This can be an intermittent behavior.

Now we conduct recurrence quantification analysis (RQA) by studying the structure of the plots numerically. Such an analysis is based essentially on densities of isolated points, diagonal lines, as well as vertical lines. We borrow the discussion presented below from Bastos and Caiddo (2011). The measures, which we have considered, are as follows:

- RR: Recurrence rate.
- DET: Fraction of points in the plot forming diagonal lines. This indicates determinism and hence predictability.
- $\langle L \rangle$: Average lengths of the diagonal lines.
- LMAX: Length of the longest diagonal line (except the line of identity). Its inverse is associated with the divergence of the trajectory in phase space.
- ENTR: Shannon entropy defined over the distribution of lengths of diagonal lines indicates diversity of the diagonal lines.
- LAM: Fraction of points forming vertical lines indicates existence of laminar states in the system.
- TT: Average length of the vertical lines. This value estimates the trapping time.

We have computed the RQA measures for the BSE index, under a range of embedding dimension. In each case, we have set the delay equal to 1.

In Figs. 11.1 and 11.2, we present recurrence analysis on logarithmic return series ($r_\tau = \ln P_\tau - \ln P_{\tau-1}$) constructed from BSE index data. As it is apparent, there is no clearly discernible pattern in the data. Next, we follow the standard approach and use the level of the price data upon normalization by the maximum value of the time series ($\tilde{P}_\tau = (P_\tau - P_{\min})/P_{\max}$). Table 11.1 contains the RQA measures. Most of the prior literatures on stock market data consider high values of embedding parameter. It is evident that, in general, recurrence rates are very low and determinism is very high. However, this approach has an inherent problem that it is not particularly good at differentiating non-recursive series from recursive series. In general, we found that when we construct similar measures for standard recursive series, it is not clear from such RQA measures that they can be easily separated from a stochastic series. Thus, it does not really shed much light on the problem, as the primary focus is to figure out determinism or lack thereof. Bastos and Caiddo (2011) discusses a possible application that these measures still retain some usefulness for cross-country analysis. Since in this case we are focusing on one country only, it is not very helpful. So, we consider a fully stochastic framework in the rest of the analysis.

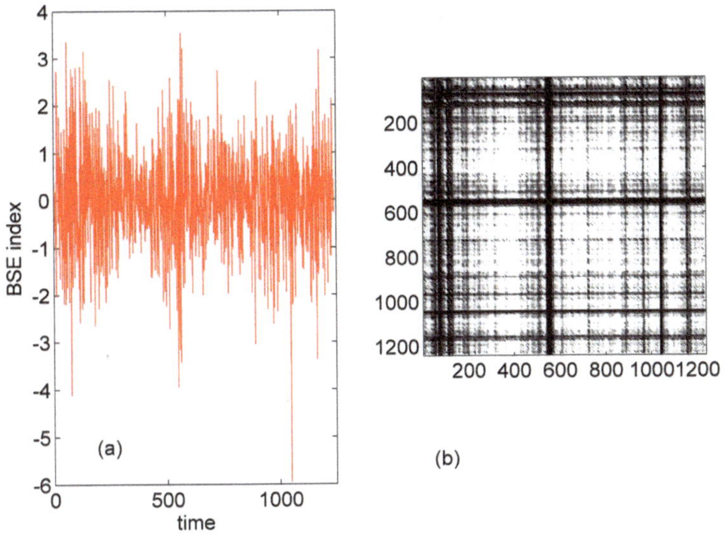


**Fig. 11.1** *Left panel*: Normalized daily return series constructed from BSE index data for five years (June 6, 2011, to June 6, 2016). *Right panel*: Recurrence plot constructed from the same data with an embedding dimension equals to 11 and time delay 1

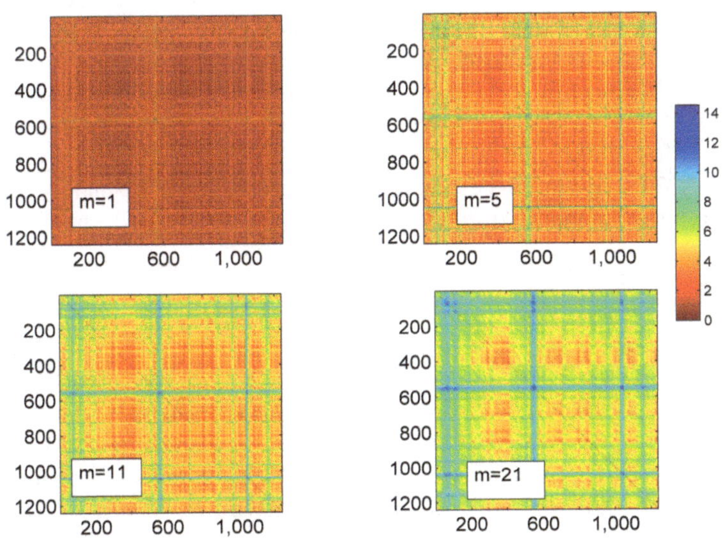


**Fig. 11.2** Distance plots constructed from the BSE index for different values of the embedding dimensions ($m = 1, 5, 11, 21$)

**Table 11.1** Measures based on recurrence analysis of normalized BSE data. Generated by the CRP toolbox (Marwan et al. 2002, 2007). Threshold for calculating neighbors set at the default value 0.1

| Quantity | $m = 1$ | $m = 2$ | $m = 5$ | $m = 11$ |
|---|---|---|---|---|
| RR | 0.0758 | 0.0442 | 0.0075 | $3.1049 \times 10^{-4}$ |
| DET | 0.9029 | 0.8817 | 0.8516 | 0.9211 |
| $\langle L \rangle$ | 4.3854 | 3.9302 | 3.8841 | 4.6667 |
| LMAX | 146 | 88 | 58 | 18 |
| ENTR | 2.0319 | 1.8575 | 1.8095 | 1.8527 |
| LAM | 0.9479 | 0.9074 | 0.7234 | 0.2763 |
| TT | 5.7416 | 4.4451 | 3.2874 | 2.2703 |

## 11.3 Empirical Study of the Correlation Structure of the Indian Stock Market

In this section, we analyze the empirical cross-correlation matrices constructed from the stock market data.

### *11.3.1 Data Specification, Notations, and Definitions*

In order to study correlations and co-movements in the stock price time series, the popular Pearson correlation coefficient was commonly used. However, with the electronic markets producing data at different frequencies (low to high), it is now known that several factors, viz., the statistical uncertainty associated with the finite-size time series, heterogeneity of stocks, heterogeneity of the average inter-transaction times, and asynchronicity of the transactions, may affect the applicability/reliability of this estimator. In this article, we have mainly focused on the daily returns computed from closure prices, for which the Pearson coefficient works well.

#### 11.3.1.1 Dataset

We have used the freely downloadable daily adjusted closure prices from Yahoo finance for $N = 199$ companies in the Bombay stock exchange (BSE) SENSEX (https://in.finance.yahoo.com/q/hp?s=%5EBSESN), for 5 years, over a period spanning from June 6, 2011, to June 6, 2016. Also, we have downloaded 199 stock prices of companies chosen randomly from the BSE and 13 sectoral indices of the BSE, for the period May 27, 2011, to May 27, 2016. The lists are given in the Tables 11.2 and 11.3, in Appendices A and B, respectively.

### 11.3.2 Correlation Matrices

We construct the correlation matrix from individual stock returns in the following way.

#### 11.3.2.1 Pearson Correlation Coefficient

In order to study the equal time cross correlations between $N$ stocks, we first denote the adjusted closure price of stock $i$ in day $\tau$ by $P_i(\tau)$ and determine the logarithmic return of stock $i$ as $r_i(\tau) = \ln P_i(\tau) - \ln P_i(\tau - 1)$. For the window of $T$ consecutive trading days, these returns form the return vector $r_i$. We use the equal time Pearson correlation coefficients between stocks $i$ and $j$ defined as

$$C_{ij} = \frac{\langle r_i r_j \rangle - \langle r_i \rangle \langle r_j \rangle}{\sqrt{[\langle r_i^2 \rangle - \langle r_i \rangle^2][\langle r_j^2 \rangle - \langle r_j \rangle^2]}}, \tag{11.2}$$

where $\langle \ldots \rangle$ indicates an average over the window of $T$ successive trading days in the return series. Naturally, such correlation coefficients satisfy the usual condition of $-1 \leq C_{ij} \leq 1$, and we can create an $N \times N$ correlation matrix $C$ by collecting all values (Chakraborti 2006; Tilak et al. 2012). By construction, the matrix is symmetric, and it serves as the basis of the rest of the present article.

### 11.3.3 Decomposition Analysis

For the present section, we are following the sequence of methods discussed by Pan and Sinha (2007) which is one of the first few papers that applied this technique. Suppose we have $N$ return time series of length $T$ that are pairwise uncorrelated. The correlation matrix generated by collecting all pairwise correlations for $N$ of such series is called the Wishart matrix. In the limits $N \to \infty$ and $T \to \infty$, such that the ratio $Q \equiv T/N > 1$, the eigenvalue distribution of this matrix has a specific distributional form,

$$f(\lambda) = (Q/2\pi) \frac{\sqrt{(\lambda_{\max} - \lambda)(\lambda - \lambda_{\min})}}{\lambda}, \tag{11.3}$$

for $\lambda_{\min} \leq \lambda \leq \lambda_{\max}$ and 0 otherwise. This distribution is clearly bounded by $\lambda_{\max,\min} = [1 \pm (1/\sqrt{Q})]^2$. In the BSE data, we considered $Q = 5$. Thus, the Wishart matrix should have the following bounds: $\lambda_{\min} = 0.3056$ and $\lambda_{\max} = 2.0944$. The distribution of eigenvalues unexplained by the Wishart matrix sheds light on the interaction structures and the coevolution process of the stocks in the market.

The largest eigenvalue corresponds to the market mode which captures the aggregate dynamics of the market that is common across all stocks. The eigenvectors associated with the next few eigenvalues (we took the next five dominant eigenvalues) describe the sectoral dynamics. The rest of the eigenvectors correspond to the random mode. From such a segregation, it is possible to reconstruct the contributions of different modes to the aggregate correlation matrix.

Following the literature to filter the data to remove market mode and the random noise, we first decompose the aggregate correlation matrix as

$$C = \sum_{i=0}^{N-1} \lambda_i a_i a_i^T, \tag{11.4}$$

where $\lambda_i$ are the eigenvalues of the correlation matrix $C$. An easy way to handle the reconstruction of the correlation matrix is to sort the eigenvalues in descending order. Then we rearrange the eigenvectors $a_i$ in corresponding ranks. This allows us to decompose the matrix into three separate components, viz., market, group, and random:

$$C = C^M + C^G + C^R, \tag{11.5}$$

$$= \lambda_0 a_0 a_0^T + \sum_{i=1}^{N_G} \lambda_i a_i a_i^T + \sum_{i=N_G+1}^{N-1} \lambda_i a_i a_i^T, \tag{11.6}$$

where $N_G$ is taken to be 5, i.e., it corresponds to the five largest eigenvalues except the first one. It is worth noting that the exact value of $N_G$ is not crucial for the result as long as it is kept within the same ballpark. The decomposition is shown in Fig. 11.3.

An important finding is that the group mode almost coincides with the random mode, whereas the market mode is segregated by a large margin from the rest. Thus, the sectoral dynamics are almost absent, whereas the market mode is very strong. This is in line with the prior literature (see, e.g., Pan and Sinha 2007).

Following standard procedure (see, e.g., Pan and Sinha 2007), we also calculate the inverse participation ratio (IPR) to extract information about contribution of different stocks to the eigenvalues. IPR is defined for the $k$-th eigenvector as the sum of fourth power of all individual components of the corresponding eigenvector, $I_k \equiv \sum_{i=1}^{N} [a_{ki}]^4$, where $a_{ki}$ are the components of eigenvector $k$. The result is presented in Fig. 11.4. Intuitively, if a single stock dominates in terms of contribution to any particular eigenvector, then the IPR would go to 1. For example, consider a limiting case of $a_{k1} = 1$ and $a_{ki} = 0$ for $i \neq 1$. On the other hand, if all elements were equal to $1/\sqrt{N}$, then we would get IPR $= 1/N$. Thus by considering IPR, we can understand if there is significant contribution coming from specific stocks or a more diversified bundle of stocks.

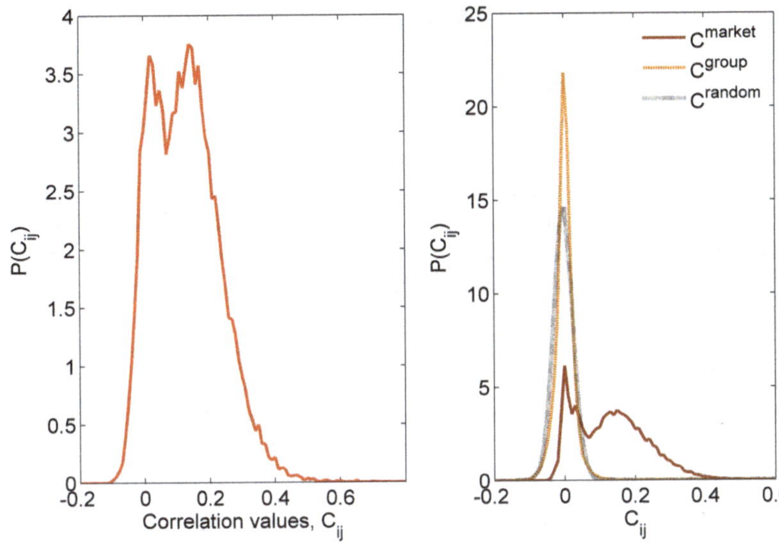


**Fig. 11.3** *Left panel*: Probability density function of the cross-correlation coefficients of 199 BSE stocks. *Right panel*: Decomposition of the correlation matrix into market mode, group mode, and random mode

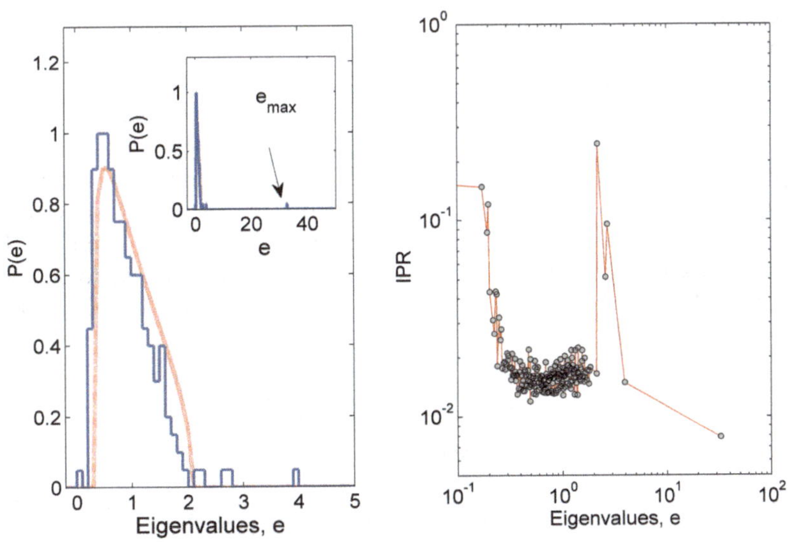


**Fig. 11.4** Eigenvalue decomposition of the correlation matrix. *Left panel*: Probability density function of eigenvalues. Inset shows the full distribution. *Right panel*: Inverse participation ratio with respect to the corresponding eigenvalues

### 11.3.4 Partial Correlation Analysis

Partial correlation is a newly introduced tool to investigate the effects of third stock on the correlation between pairs of stocks. Kenett et al. (2015) introduced this analysis for multiple stock markets. In the present paper, we apply their technique to the Bombay stock exchange data. To describe its usefulness, consider three stocks, $i, j$, and $k$, with significant correlation between all three pairs of the stocks. Suppose we think that the high value of $C_{ij}$ is the result of their own correlations with $k$, i.e., part of $C_{ij}$ might be spurious correlation arising from a third variable effect (in this case $k$), we should remove such effects to figure out the actual correlation across $i$ and $j$. Then, we can recalculate $C_{ij}$, after controlling for the effect of $k$. The resultant correlation value is called the partial correlation. The difference between the raw correlation value for a pair of stocks and the corresponding partial correlation tells us how much third variable effect was there.

For this purpose, we again use the same daily log return $r_i(t)$. However, we need to adjust for one more factor. From the preceding analysis, we already know that there is a significant market mode. Therefore, that will act as a common driving factor. Hence, the market mode should also be controlled in order to extract the actual correlation values for the exact same reason. In this case, the market mode is given by a market index. Note the difference from the earlier analysis. For constructing the market mode from the eigenvalue analysis, the market mode arises endogenously from the panel data itself, whereas in this case, we take the market mode to be given by an exogenous index time series. Hence, these two types of analysis complement each other.

Following the notation of Kenett et al. (2015), let $x$ and $y$ be two time series and let $M$ be the BSE index for the same time frame. The partial correlation, $(x, y|M)$, is defined as the standard Pearson correlation coefficient (described above) between $x$ and $y$ after controlling for $M$. More technically, this is the correlation between the residuals of $x$ and $y$ which are unexplained by the market index represented by $M$. So first, we need the residuals of the two time series. A simple way to do it would be to regress both on M. Then we can work with the resulting variables. Formally, the correlation is given as

$$C_{x,y|M} = \frac{(C_{x,y} - C_{x,M}.C_{y,M})}{\sqrt{[1 - C^2_{x,M}].[1 - C^2_{y,M}]}} \tag{11.7}$$

In the same way, when the same two stocks $x$ and $y$ are affected by a common stock $z$, we can control for that effect as well. Given a third stock $z$, the partial correlation between $x$ and $y$ after controlling for both the market factor and the third stock $z$ is given by the following formula:

$$C_{x,y|M,z} = \frac{C_{x,y|M} - C_{x,z|M}.C_{y,z|M}}{\sqrt{[1 - C^2_{x,z|M}].[1 - C^2_{y,z|M}]}} \tag{11.8}$$

If it is found that the third stock has an important effect on pairs of stocks, then it is useful to define the "influence quantity" (see Kenett et al. 2015)

$$d(x, y|z) = C_{x,y|M} - C_{x,y|M,z}. \tag{11.9}$$

Magnitude of this quantity will reflect how much influence the third stock has on a certain pair of stocks. A natural extension of this idea is to consider the average influence $d(x|z)$ of stock $z$ on the correlations between a given stock $x$ and all other stocks except $x$ itself and $z$. Kenett et al. (2015) defined this index as the following:

$$d(x|z) = \langle d(x, y|z) \rangle_{y \neq x} \, . \tag{11.10}$$

This quantity captures the average influence from stock $z$ to stock $x$ through the third variable effect after controlling for the market index. We present all results of our analysis in Fig. 11.5. Panel (a) shows the correlation coefficients of all stocks after controlling for the market index. Since the bulk of it is below the 45° line, we conclude that the market index has a positive effect on pairwise correlations. This is consistent with the results from the eigenvalue analysis and also with Kenett et al. (2015). Similarly, in panel (b), we show the data for the same correlation coefficients, after controlling for all possible third variable effects. In panel (d),

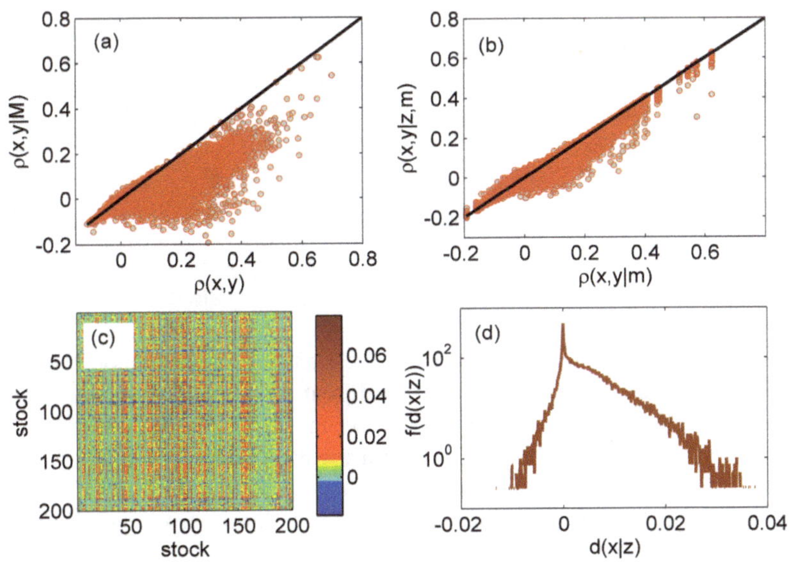


**Fig. 11.5** Correlation matrix after controlling for market mode (BSE index). Panel (**a**): Partial correlation after controlling for the market mode as a function of raw correlation coefficients. Panel (**b**): Same after controlling for the third variable effect. Panel (**c**): Influence of all stocks as the third variable (on $x$-axis) on all other stocks (on $y$-axis). Panel (**d**): Probability density function of average influence quantity

we present the probability density function of the *influence quantity*. Again bulk of the distribution is in the positive quadrant implying positive effect on average.

## 11.4 Network Analysis

In this section, we present network analysis based on the empirical correlation matrix.

### *11.4.1 Distance Metric*

To obtain "distances," the following transformation

$$d_{ij} = \sqrt{2(1 - C_{ij})} \tag{11.11}$$

is used, which clearly satisfies $2 \geq d_{ij} \geq 0$. Collecting all distances, one can form an $N \times N$ distance matrix $D$, such that all elements of the matrix are "ultrametric" (Rammal et al. 1986). The concept of ultrametricity appears in multiple papers. Interested readers can refer to the detailed discussions by Mantegna (1999), Onnela et al. (2003a,b), and Chakraborti (2006) among others. There are multiple possible ultrametric spaces. We opt for the subdominant ultrametric, as it is simple to work with, and its associated topological properties. The choice of the nonlinear function is again arbitrary, as long as all the conditions of ultrametricity are met.

### *11.4.2 Multidimensional Scaling (MDS)*

Multidimensional scaling is a method to analyze large-scale data that displays the structure of similarity in terms of distances, given by Eq. (11.11), as a geometrical picture or map, where each stock corresponds to a set of coordinates in a multidimensional space. MDS arranges different stocks in this space according to the strength of the pairwise distances between stocks, – two similar stocks are represented by two sets of coordinates that are close to each other, and two stocks behaving differently are placed far apart (see Borg and Groenen 2015) in the space. We construct a distance matrix consisting of $N \times N$ entries from the $N$ time series available, defined using Eq. (11.11):

$$\begin{bmatrix} d_{11} & d_{12} & \dots & d_{1N} \\ d_{21} & d_{22} & \dots & d_{21} \\ \vdots & \vdots & \ddots & \vdots \\ d_{N1} & d_{N2} & \dots & d_{NN} \end{bmatrix}$$

Given $D$, the aim of MDS is to generate $N$ vectors $x_1, \dots, x_N \in \Re^D$, such that

$$\|x_i - x_j\| \approx d_{ij} \qquad \forall i, j \in N, \tag{11.12}$$

where $\|.\|$ represents vector norm. We can use the Euclidean distance metric as is done in the classical MDS. Effectively, through MDS we try to find a mathematical embedding of the $N$ objects into $\Re^D$ by preserving distances. In general, we choose the embedding dimension $D$ to be 2, so that we are able to plot the vectors $x_i$ in the form of a map, representing $N$ stocks. It may be noted that $x_i$ are not necessarily unique under the assumption of the Euclidean metric, as we can arbitrarily translate and rotate them, as long as such transformations leave the distances $\|x_i - x_j\|$ unaffected. Generally, MDS can be obtained through an optimization problem, where $(x_1, \dots, x_N)$ is the solution of the problem of minimization of a cost function, such as

$$\min_{x_1, \dots, x_N} \sum_{i<j} (\|x_i - x_j\| - d_{ij})^2. \tag{11.13}$$

In order to capture the sectoral behavior of the market visually, we have generated the MDS plot of 199 stocks as described before, for the time window of 250 trading days between May 2015 and May 2016. As before, using the correlation matrix as input, we computed the distance matrix using the transformations (given by Eq. (11.11)). The distance matrix was then used as an input to the inbuilt MDS function in MATLAB (http://in.mathworks.com/help/stats/cmdscale.html). The outputs of the MDS were the sets of coordinates, which were plotted as the MDS map as shown in Fig. 11.6.

The coordinates are plotted in a manner such that the centroid of the map coincides with the origin $(0, 0)$. It is interesting to follow the positions of certain sectors, (i) sugar, (ii) textiles, and (iii) pharmaceuticals, which will be discussed in details in Sect. 11.5.

### *11.4.3 Dendrogram*

Dendrogram is basically a tree diagram. This is often used to depict the arrangement of multiple nodes through hierarchical clustering. We have used the inbuilt function in MATLAB (http://in.mathworks.com/help/stats/dendrogram.html) to generate the

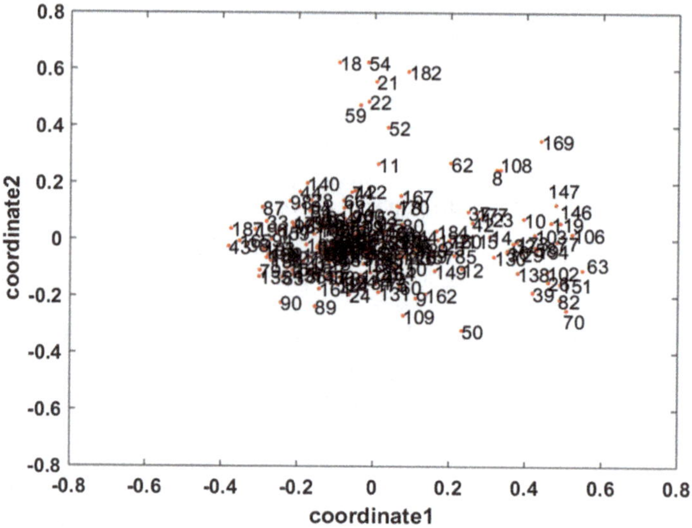


**Fig. 11.6** Multidimensional scaling of the sample data for the time window May 2015–May 2016. There is a cluster of stocks with identifiers 18, 54, 21, etc. at the top, all of which belong to the sugar industry (Appendix A)

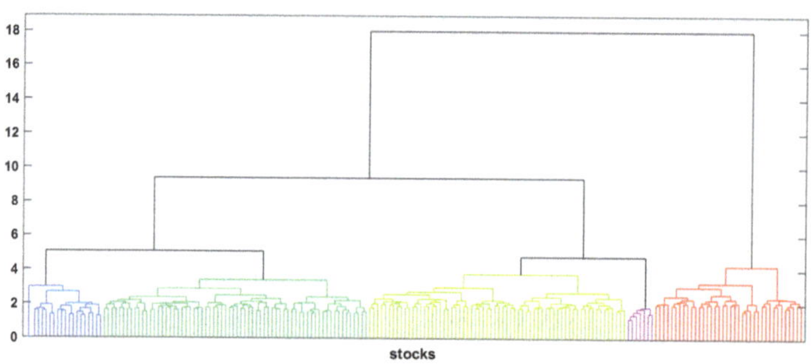


**Fig. 11.7** Dendrogram of 199 stocks

hierarchical binary cluster tree (dendrogram) of $N$ stocks connected by many U-shaped lines (as shown in Fig. 11.7), such that the height of each U represents the distance (given by Eq. (11.11)) between the two data points being connected. Thus, the vertical axis of the tree captures the similarity between different clusters, whereas the horizontal axis represents the identity of the objects and clusters. Each joining (fusion) of two clusters is represented on the graph by the splitting of a vertical line into two vertical lines. The vertical position of the split, shown by the short horizontal bar, gives the distance (similarity) between the two clusters. We set the property "linkage type" as "Wards minimum variance," which requires the distance method to be Euclidean that results in group formation such as the pooled within-group sum of squares which would be minimized. In other words, at every iteration, two clusters in the tree are connected such that it results in the least possible increment in the relevant quantity, i.e., pooled within-group sum of squares. Figure 11.7 shows the dendrogram of all the 199 stocks clustered in five different colors (by using "color threshold" property in MATLAB). The magenta color represents the cluster of "sugar industries."

### *11.4.4 Minimum Spanning Tree*

A minimum spanning tree is a spanning tree of a connected, undirected graph such that all the $N$ vertices are connected together with the minimal total weighting for its $N-1$ edges (total distance is minimum). The distance matrix defined by Eq. (11.11) was used as an input to the inbuilt MST function in MATLAB (http://in.mathworks.com/help/bioinfo/ref/graphminspantree.html). See MATLAB documentation for all details. Here we state Kruskal and Prim algorithms for the sake of completeness of the present article. Description of the two algorithms (source: see http://in.mathworks.com/help/bioinfo/ref/graphminspantree.html):

- Kruskal – *This algorithm extends the minimum spanning tree by one edge at every discrete time interval by finding an edge, which links two separate trees in a spreading forest of growing minimum spanning trees.*
- Prim – *This algorithm extends the minimum spanning tree by one edge at every discrete time interval by adding a minimal edge, which links a node in the growing minimum spanning tree with one other remaining node.*

Figure 11.8 shows the MST for all the 199 stocks. MATLAB algorithms set the root node as the first node in the largest connected component, which in our case is node 43.

## 11.5 Sectoral Co-movements: Mesoscopic Network

After quantifying the general cross-correlation structure of the market, we probe deeper into the sectoral co-movements. There are multiple ways to analyze the data. First, we can impose a threshold on the group cross-correlation matrix and

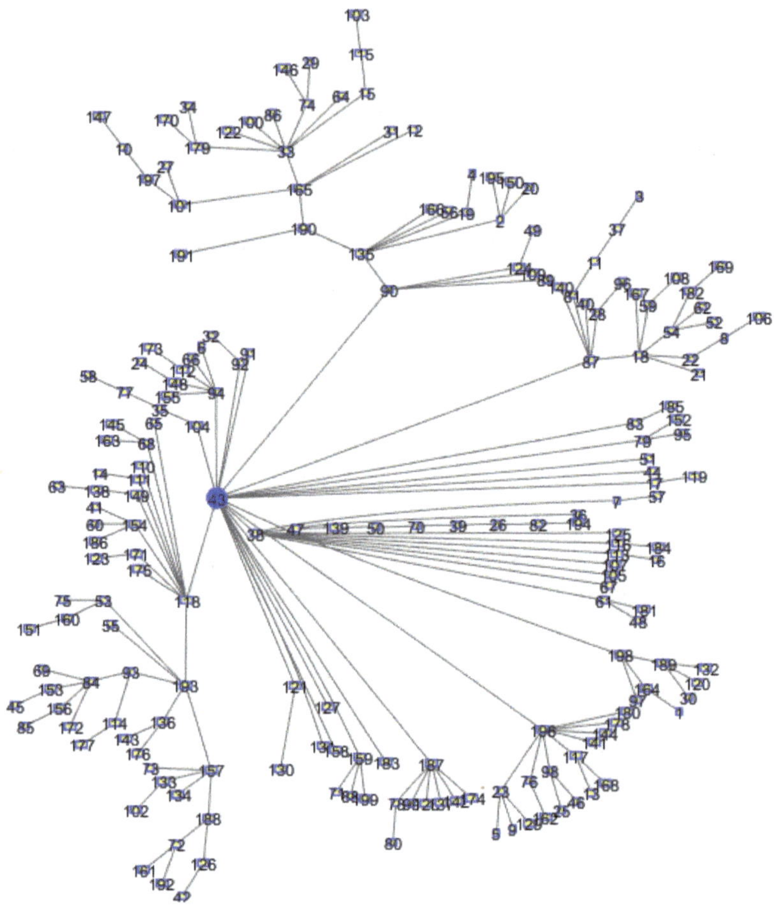

**Fig. 11.8** Minimum spanning tree (MST) of the sample data. For such microlevel data, there is no clear pattern. However, the pattern becomes much clearer with sectoral data (Sect. 11.5)

construct a network of stocks which move closely. This is the approach that is followed in Pan and Sinha (2007), for example. This approach has some problems. First, the threshold has to be exogenous, hence basically arbitrary. Second, even with such networks, it is difficult to identify clusters that match with actual industry classifications. An alternative way is to follow the industry classifications first, and then try to see if they form clusters.

To study the sectoral behavior in the market, we have selected stocks from the list of BSE from the industries: (i) sugar, (ii) textiles, and (iii) pharmaceuticals. Following the same methodologies, as described in the previous Sects. 11.4.2, 11.4.3, and 11.4.4, we have generated the plots given in Figs. 11.9, 11.10, and 11.11. By looking at the diagram, it becomes clear that the method is partially successful to segregate the market into clusters, but not fully. Therefore, we construct a new

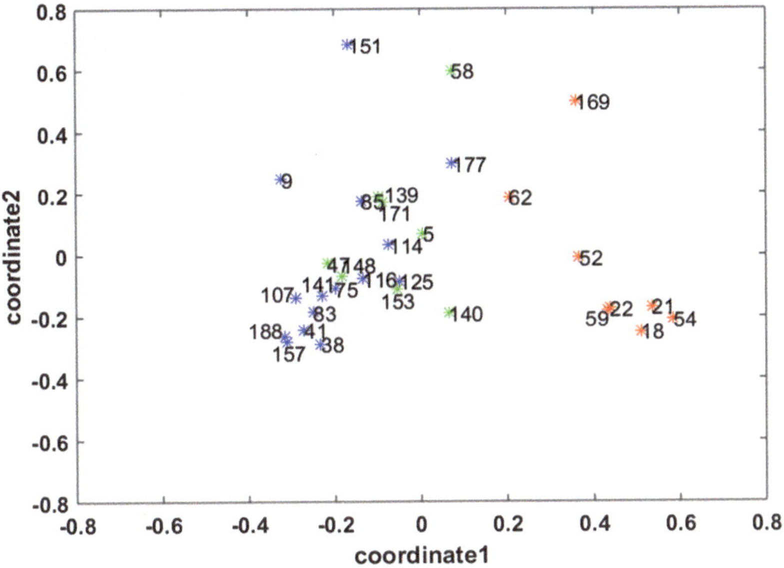


**Fig. 11.9** Plot of MDS for stocks within (i) sugar (*red*), (ii) textile (*blue*), and (iii) pharmaceuticals (*green*) industries. Stock details are given in Appendix B

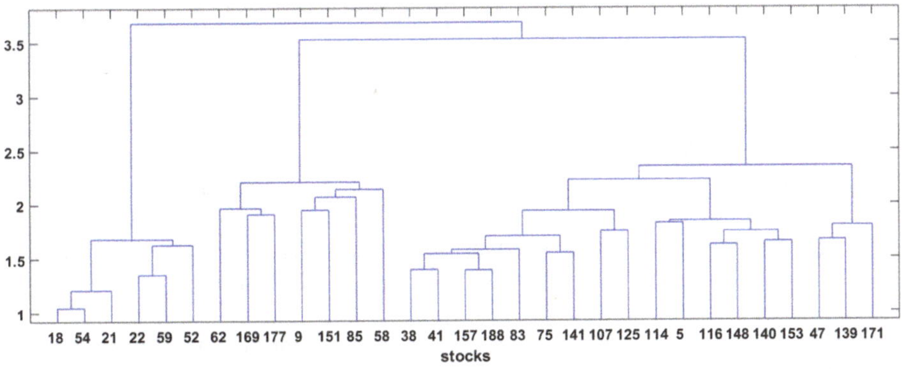


**Fig. 11.10** Plot of dendrogram for stocks within (i) sugar, (ii) textile, and (iii) pharmaceuticals industries. Stock details are given in Appendix B

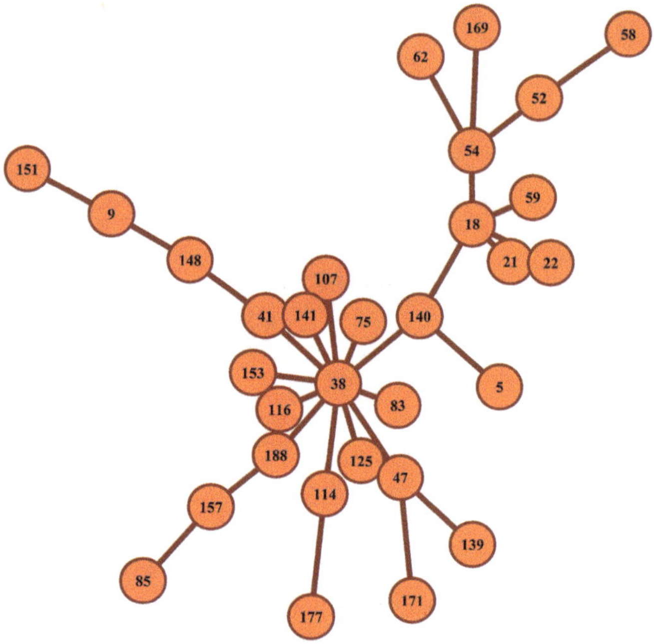


**Fig. 11.11** Plot of minimum spanning tree (MST) for stocks within (i) sugar, (ii) textile, and (iii) pharmaceuticals industries. Stock details are given in Appendix B

network. Rather than working with actual stock returns, we work with sectoral index returns. This marks a prominent departure from the usual mode of analysis. Typically, most studies focus on either an aggregate macrolevel market index like S&P 500 or consider collective dynamics of microlevel individual stock returns. Here, we consider a *mesoscopic* network to characterize correlations.

Empirically, we used the 13 sectoral indices from the BSE (list given in Appendix B) for the time window May 2015–May 2016. The resulting multidimensional scaling results have been plotted in Fig. 11.12, dendrogram in Fig. 11.13, and minimum spanning tree in Fig. 11.14.

The MDS algorithm cannot segregate the markets into clusters in a way that corresponds to the industry classifications. Dendrogram produces better results than that. Finally, the minimum spanning tree corresponds to a fairly intuitive market structure. Note that the only information used was sectoral returns' correlations. The MST shows that the banks and realty sectors are most closely related to the finance sector. Energy sector is most closely associated with oil and gas sector and so on. Thus we see that the sectoral MST approximates the industrial relations in a fairly correct manner.

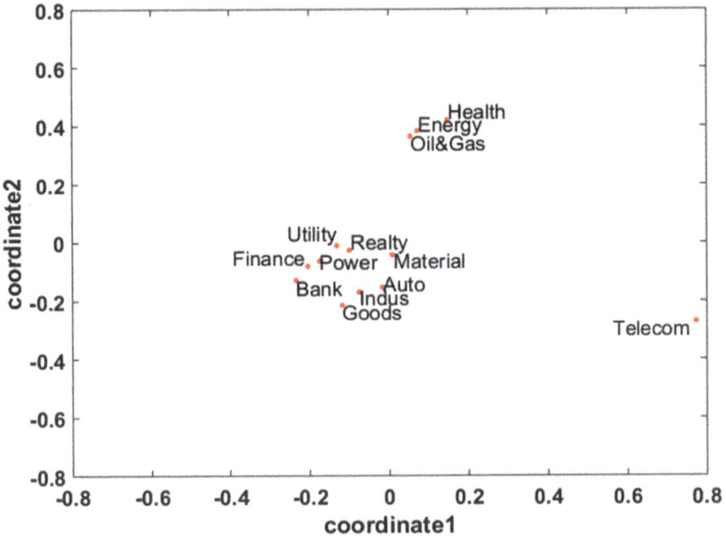


**Fig. 11.12** Plot of clustering with multidimensional scaling (MDS) algorithm is shown for the BSE indices. As it is clear from the figure, it is not very useful for segregating sectors, at least with the present sample. See Appendix A for sector details

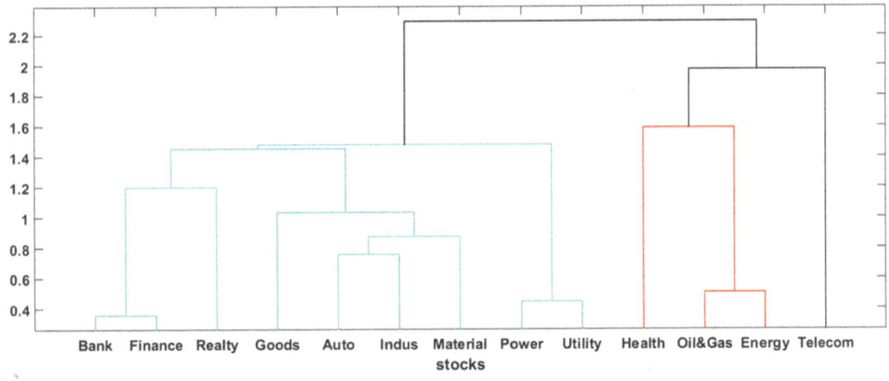


**Fig. 11.13** Plot of dendrogram for the BSE indices. This algorithm clusters related sectors. But MST (Fig. 11.14) presents a clearer picture. See Appendix A for sector details

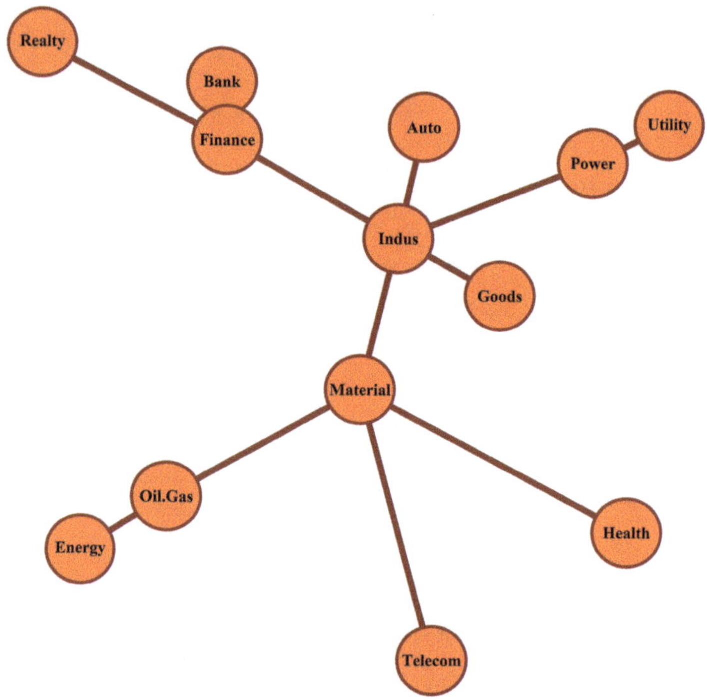


**Fig. 11.14** Plot of the minimum spanning tree (MST) of BSE indices. Technologically related sectors are closer to each other, e.g., bank and realty are related to finance only. Thus return fluctuations of technologically related sectors co-move significantly more than other sectors. See Appendix A for sector details

## 11.6 Summary

In this article, we have applied multiple techniques to analyze daily data from Bombay stock exchange. Our analyses cover a large spectrum of tools proposed in the econophysics literature in the last two decades. Using eigen decomposition method, we show that the market cross-correlation structure shows a very prominent market mode. Consistent with the literature, we show that the group mode is not very strong for emerging countries and, in fact, is very difficult to differentiate from the random mode. Then we carry out partial correlation analysis, which is a newly proposed method, on the Indian data. This helps us to explicitly characterize and quantify the average "third variable" effect in the cross correlations.

Finally, we turn to network analysis to study the core-periphery structure. We use multidimensional scaling and dendrograms to identify clusters. In general, we do not find any significant pattern between such clusters and the industrial classifications. However, a much more intuitive picture emerges when we construct a mesoscopic network with the sectoral indices. We see that minimum spanning tree across the indices clearly segregates nodes according to their industrial classification, just by using the return cross correlations.

# A Appendix

**Table 11.2** List of all sectoral indices. The first column has the serial number, the second column has the abbreviation, the third column has the full name of the sector, and the fourth column has the category of the sector as given in the BSE

| S.No. | ID | Name | Category |
|---|---|---|---|
| 1 | SI1900 | S&P BSE AUTO | AUTO |
| 2 | SIBANK | S&P BSE BANKEX | BANK |
| 3 | SPBSBMIP | S&P BSE BASIC MATERIALS | MATERIAL |
| 4 | SI0200 | S&P BSE CAPITAL GOODS | GOODS |
| 5 | SPBSEIP | S&P BSE ENERGY | ENERGY |
| 6 | SPBSFIIP | S&P BSE FINANCE | FINANCE |
| 7 | SPBSIDIP | S&P BSE INDUSTRIALS | INDUS |
| 8 | SI1400 | S&P BSE OIL & GAS | OIL & GAS |
| 9 | SIPOWE | S&P BSE POWER | POWER |
| 10 | SIREAL | S&P BSE REALTY | REALTY |
| 11 | SPBSTLIP | S&P BSE TELECOM | TELECOM |
| 12 | SPBSUTIP | S&P BSE UTILITIES | UTILITY |
| 13 | SI0800 | S&P BSE HEALTHCARE | HEALTH |

# B Appendix

**Table 11.3** List of all stocks considered for the analysis. The first column has the serial number, the second column has the abbreviation, the third column has the full name of the stock, and the fourth column specifies the sector as given in the BSE

| S.No. | ID | Name | Category |
|---|---|---|---|
| 1 | ABB | ABB India Limited | Heavy electrical equipment |
| 2 | ABIRLANUVO | Aditya Birla Nuvo Ltd. | Diversified |
| 3 | AEGISLOG | Aegis Logistics Ltd. | Oil marketing and distribution |
| 4 | AMARAJABAT | Amara Raja Batteries Ltd. | Auto parts and equipment |
| 5 | AMBALALSA | Ambalal Sarabhai Enterprises Ltd. | Pharmaceuticals |
| 6 | ANDHRAPET | ANDHRA PETROCHEMICALS LTD. | Commodity chemicals |
| 7 | ANSALAPI | ANSAL PROPERTIES and INFRASTRUCTURE LTD. | Realty |
| 8 | APPLEFIN | APPLE FINANCE LTD. | Finance (including NBFCs) |
| 9 | ARVIND | ARVIND LTD. | Textiles |
| 10 | ASIANHOTNR | ASIAN HOTELS (NORTH) LIMITED | Hotels |

(continued)

**Table 11.3** (continued)

| S.No. | ID | Name | Category |
|---|---|---|---|
| 11 | ASSAMCO | ASSAM COMPANY (INDIA) LIMITED | Tea and coffee |
| 12 | ATFL | AGRO TECH FOODS LTD. | Other agricultural products |
| 13 | ATUL | ATUL LTD. | Agrochemicals |
| 14 | ATVPR | ATV PROJECTS INDIA LTD. | Construction and engineering |
| 15 | AUTOLITIND | AUTOLITE (INDIA) LTD. | Auto parts and equipment |
| 16 | AUTORIDFIN | AUTORIDERS FINANCE LTD. | Finance (including NBFCs) |
| 17 | BAJAJELEC | BAJAJ ELECTRICALS LTD. | Household appliances |
| 18 | BAJAJHIND | BAJAJ HINDUSTHAN SUGAR LIMITED | Sugar |
| 19 | BAJFINANCE | BAJAJ FINANCE LIMITED | Finance (including NBFCs) |
| 20 | BALLARPUR | BALLARPUR INDUSTRIES LTD. | Paper and paper products |
| 21 | BALRAMCHIN | BALRAMPUR CHINI MILLS LTD. | Sugar |
| 22 | BANARISUG | BANNARI AMMAN SUGARS LTD. | Sugar |
| 23 | BANCOINDIA | BANCO PRODUCTS (INDIA) LTD. | Auto parts and equipment |
| 24 | BASF | BASF INDIA LTD. | Specialty chemicals |
| 25 | BATAINDIA | BATA INDIA LTD. | Footwear |
| 26 | BEL | BHARAT ELECTRONICS LTD. | Defense |
| 27 | BEML | BEML LTD. | Commercial vehicles |
| 28 | BEPL | BHANSALI ENGINEERING POLYMERS LTD. | Specialty chemicals |
| 29 | BHAGGAS | BHAGAWATI GAS LIMITED | Industrial gases |
| 30 | BHEL | BHARAT HEAVY ELECTRICALS LTD. | Heavy electrical equipment |
| 31 | BHUSANSTL | BHUSHAN STEEL LTD. | Iron and steel/interm.products |
| 32 | BIHSPONG | BIHAR SPONGE IRON LTD. | Iron and steel/interm.products |
| 33 | BINANIIND | BINANI INDUSTRIES LTD. | Holding companies |
| 34 | BIRLACORPN | BIRLA CORPORATION LIMITED | Flagship company |
| 35 | BIRLAERIC | BIRLA ERICSSON OPTICAL LTD. | Other elect.equip./prod. |
| 36 | BLUESTARCO | BLUE STAR LTD. | Consumer electronics |
| 37 | BNKCAP | BNK CAPITAL MARKETS LTD. | Other financial services |
| 38 | BOMDYEING | BOMBAY DYEING and MFG.CO.LTD. | Textiles |
| 39 | BPL | BPL LTD. | Consumer electronics |
| 40 | CAMPHOR | CAMPHOR and ALLIED PRODUCTS LTD. | Commodity chemicals |
| 41 | CENTENKA | CENTURY ENKA LTD. | Textiles |
| 42 | CENTEXT | CENTURY EXTRUSIONS LTD. | Aluminum |

(continued)

**Table 11.3** (continued)

| S.No. | ID | Name | Category |
|---|---|---|---|
| 43 | CENTURYTEX | CENTURY TEXTILES and INDUSTRIES LTD. | Cement and cement products |
| 44 | CESC | CESC LTD. | Electric utilities |
| 45 | CHAMBLFERT | CHAMBAL FERTILISERS and CHEMICALS LTD. | Fertilizers |
| 46 | CHENNPETRO | CHENNAI PETROLEUM CORPORATION LTD. | Refineries/petro-products |
| 47 | CIPLA | CIPLA LTD. | Pharmaceuticals |
| 48 | CMIFPE | CMI FPE LTD. | Industrial machinery |
| 49 | CRISIL | CRISIL LTD. | Other financial services |
| 50 | CROMPGREAV | CROMPTON GREAVES LTD. | Heavy electrical equipment |
| 51 | DABUR | DABUR INDIA LTD. | Personal products |
| 52 | DALMIASUG | DALMIA BHARAT SUGAR AND INDUSTRIES LTD | Sugar |
| 53 | DCW | DCW LTD. | Petrochemicals |
| 54 | DHAMPURSUG | DHAMPUR SUGAR MILLS LTD. | Sugar |
| 55 | DIAMINESQ | DIAMINES and CHEMICALS LTD. | Commodity chemicals |
| 56 | DICIND | DIC INDIA LTD. | Specialty chemicals |
| 57 | DISAQ | DISA INDIA LTD. | Industrial machinery |
| 58 | DRREDDY | DR.REDDY'S LABORATORIES LTD. | Pharmaceuticals |
| 59 | EIDPARRY | E.I.D.-PARRY (INDIA) LTD. | Sugar |
| 60 | ELANTAS | ELANTAS BECK INDIA LTD. | Commodity chemicals |
| 61 | ELECTCAST | ELECTROSTEEL CASTINGS LTD. | Construction and engineering |
| 62 | EMPEESUG | EMPEE SUGARS and CHEMICALS LTD. | Sugar |
| 63 | ENVAIREL | ENVAIR ELECTRODYNE LTD. | Industrial machinery |
| 64 | ESABINDIA | ESAB INDIA LTD. | Other Industrial goods |
| 65 | ESSELPRO | ESSEL PROPACK LTD. | Containers and packaging |
| 66 | ESTER | ESTER INDUSTRIES LTD. | Commodity chemicals |
| 67 | EXIDEIND | EXIDE INDUSTRIES LTD. | Auto parts and equipment |
| 68 | FEDDERLOYD | FEDDERS LLOYD CORPORATION LTD. | Other elect.equip./prod. |
| 69 | FERROALL | FERRO ALLOYS CORPORATION LTD. | Iron and steel/interm.products |
| 70 | FGP | FGP LTD. | Finance (including NBFCs) |
| 71 | FINCABLES | FINOLEX CABLES LTD. | Other elect.equip./prod. |
| 72 | FORCEMOT | FORCE MOTORS LTD. | Cars and utility vehicles |
| 73 | FOSECOIND | FOSECO INDIA LTD. | Commodity chemicals |
| 74 | GANESHBE | GANESH BENZOPLAST LTD. | Commodity chemicals |
| 75 | GARDENSILK | GARDEN SILK MILLS LTD. | Textiles |
| 76 | GHCL | GHCL LTD. | Commodity chemicals |

(continued)

**Table 11.3** (continued)

| S.No. | ID | Name | Category |
|---|---|---|---|
| 77 | GLFL | GUJARAT LEASE FINANCING LTD. | Finance (including NBFCs) |
| 78 | GODFRYPHLP | GODFREY PHILLIPS INDIA LTD. | Cigarettes-tobacco products |
| 79 | GODREJIND | GODREJ INDUSTRIES LTD. | Commodity chemicals |
| 80 | GOLDENTOBC | Golden Tobacco Ltd. | Cigarettes-tobacco products |
| 81 | GOODRICKE | Goodricke Group Ltd. | Tea and coffee |
| 82 | GOODYEAR | Goodyear India Ltd. | Auto tires and rubber products |
| 83 | GRASIM | GRASIM INDUSTRIES LTD. | Textiles |
| 84 | GTL | GTL LTD. | Other telecom services |
| 85 | GTNINDS | GTN INDUSTRIES LTD. | Textiles |
| 86 | GUJFLUORO | GUJARAT FLUOROCHEMICALS LTD. | Industrial gases |
| 87 | HCC | HINDUSTAN CONSTRUCTION CO.LTD. | Construction and engineering |
| 88 | HCIL | HIMADRI CHEMICALS and INDUSTRIES LTD. | Commodity chemicals |
| 89 | HDFC | HOUSING DEVELOPMENT FINANCE CORP.LTD. | Housing finance |
| 90 | HDFCBANK | HDFC BANK LTD | Banks |
| 91 | HEIDELBERG | HEIDELBERGCEMENT INDIA LTD. | Cement and cement products |
| 92 | HEROMOTOCO | HERO MOTOCORP LTD. | 2/3 wheelers |
| 93 | HFCL | HIMACHAL FUTURISTIC COMMUNICATIONS LTD. | Telecom cables |
| 94 | HINDOILEXP | HINDUSTAN OIL EXPLORATION CO.LTD. | Exploration and production |
| 95 | HINDPETRO | HINDUSTAN PETROLEUM CORPORATION LTD. | Refineries/petro-products |
| 96 | HINDUJAVEN | HINDUJA VENTURES LTD. | Broadcasting and cable TV |
| 97 | HINDZINC | HINDUSTAN ZINC LTD. | Zinc |
| 98 | HMT | HMT LTD. | Commercial vehicles |
| 99 | HOTELEELA | HOTEL LEELAVENTURE LTD. | Hotels |
| 100 | HSIL | HSIL LTD. | Containers and packaging |
| 101 | IDBI | IDBI BANK LTD. | Banks |
| 102 | IFCI | IFCI LTD. | Financial institutions |
| 103 | IFSL | INTEGRATED FINANCIAL SERVICES LTD. | Finance (including NBFCs) |
| 104 | IGPL | I G PETROCHEMICALS LTD. | Commodity chemicals |
| 105 | INDIAGLYCO | INDIA GLYCOLS LTD. | Commodity chemicals |
| 106 | INDLEASE | INDIA LEASE DEVELOPMENT LTD. | Finance (including NBFCs) |
| 107 | INDORAMA | INDO RAMA SYNTHETICS (INDIA) LTD. | Textiles |

(continued)

**Table 11.3** (continued)

| S.No. | ID | Name | Category |
|---|---|---|---|
| 108 | INDSUCR | INDIAN SUCROSE LTD. | Beverage store |
| 109 | INFY | INFOSYS LTD. | IT Consulting and software |
| 110 | INGERRAND | INGERSOLL-RAND (INDIA) LTD. | Industrial machinery |
| 111 | INSILCO | INSILCO LTD. | Other industrial goods |
| 112 | IONEXCHANG | ION EXCHANGE (INDIA) LTD. | Industrial machinery |
| 113 | ITHL | INTERNATIONAL TRAVEL HOUSE LTD. | Travel support services |
| 114 | JASCH | JASCH INDUSTRIES LTD. | Textiles |
| 115 | JAYKAY | JAYKAY ENTERPRISES LTD. | Finance (including NBFCs) |
| 116 | JCTLTD | JCT LTD. | Textiles |
| 117 | JINDALPOLY | JINDAL POLY FILMS LTD. | Commodity chemicals |
| 118 | JISLJALEQS | JAIN IRRIGATION SYSTEMS LTD. | Plastic products |
| 119 | JKLAKSHMI | JK LAKSHMI CEMENT LTD. | Cement and cement products |
| 120 | JSWSTEEL | JSW STEEL LTD. | Iron and steel/interm.products |
| 121 | KAJARIACER | KAJARIA CERAMICS LTD. | Furniture-furnishing-paints |
| 122 | KAKATCEM | KAKATIYA CEMENT SUGAR and INDUSTRIES LTD. | Cement and cement products |
| 123 | KANELIND | KANEL INDUSTRIES LTD. | Comm.Trading and distribution |
| 124 | KANSAINER | KANSAI NEROLAC PAINTS LTD. | Furniture-furnishing-paints |
| 125 | KGDENIM | KG DENIM LTD. | Textiles |
| 126 | KINETICENG | KINETIC ENGINEERING LTD. | 2/3 wheelers |
| 127 | KIRLFER | KIRLOSKAR FERROUS INDUSTRIES LTD. | Iron and Steel/Interm.Products |
| 128 | KIRLOSBROS | KIRLOSKAR BROTHERS LTD. | Industrial machinery |
| 129 | KIRLOSIND | KIRLOSKAR INDUSTRIES LTD. | Industrial machinery |
| 130 | KOTAKBANK | KOTAK MAHINDRA BANK LTD. | Banks |
| 131 | KSBPUMPS | KSB PUMPS LTD. | Industrial machinery |
| 132 | KSL | KALYANI STEELS LTD. | Iron and steel/interm.products |
| 133 | LAXMIMACH | LAKSHMI MACHINE WORKS LTD. | Industrial machinery |
| 134 | LGBBROSLTD | L.G.BALAKRISHNAN and BROS.LTD. | Auto parts and equipment |
| 135 | LICHSGFIN | LIC HOUSING FINANCE LTD. | Housing finance |
| 136 | LML | LML LTD. | 2/3 Wheelers |
| 137 | LOKHSG | LOK HOUSING and CONSTRUCTIONS LTD. | Realty |
| 138 | LORDSCHLO | LORDS CHLORO ALKALI LTD. | Commodity chemicals |
| 139 | LUPIN | LUPIN LTD. | Pharmaceuticals |
| 140 | LYKALABS | LYKA LABS LTD. | Pharmaceuticals |

(continued)

**Table 11.3** (continued)

| S.No. | ID | Name | Category |
|---|---|---|---|
| 141 | MAFATIND | MAFATLAL INDUSTRIES LTD. | Textiles |
| 142 | MAHSCOOTER | MAHARASHTRA SCOOTERS LTD. | 2/3 wheelers |
| 143 | MAHSEAMLES | MAHARASHTRA SEAMLESS LTD. | Construction and engineering |
| 144 | MAJESAUT | MAJESTIC AUTO LTD. | 2/3 wheelers |
| 145 | MANALIPETC | MANALI PETROCHEMICAL LTD. | Petrochemicals |
| 146 | MARGOFIN | MARGO FINANCE LTD. | Finance (including NBFCs) |
| 147 | MAVIIND | MAVI INDUSTRIES LTD. | Plastic products |
| 148 | MERCK | MERCK LTD. | Pharmaceuticals |
| 149 | METROGLOBL | METROGLOBAL LTD. | Paper and paper products |
| 150 | MFSL | MAX FINANCIAL SERVICES LTD. | Life insurance |
| 151 | MIDINDIA | MID INDIA INDUSTRIES LTD. | Textiles |
| 152 | MIRCELECTR | MIRC ELECTRONICS LTD. | Consumer electronics |
| 153 | MOREPENLAB | MOREPEN LABORATORIES LTD. | Pharmaceuticals |
| 154 | MRF | MRF LTD. | Auto tires and rubber products |
| 155 | MRPL | MANGALORE REFINERY & PETROCHEMICALS LTD. | Refineries/petro-products |
| 156 | MTNL | MAHANAGAR TELEPHONE NIGAM LTD. | Telecom services |
| 157 | NAHARSPING | NAHAR SPINNING MILLS LTD. | Textiles |
| 158 | NATPEROX | NATIONAL PEROXIDE LTD. | Commodity chemicals |
| 159 | NCC | NCC LTD. | Construction and engineering |
| 160 | NEPCMICON | NEPC INDIA LTD. | Heavy electrical equipment |
| 161 | NIITLTD | NIIT LTD. | IT training services |
| 162 | NIRLON | NIRLON LTD. | Misc.commercial services |
| 163 | OILCOUNTUB | OIL COUNTRY TUBULAR LTD. | Oil |
| 164 | ONGC | OIL AND NATURAL GAS CORPORATION LTD. | Oil and gas |
| 165 | ORIENTBANK | ORIENTAL BANK OF COMMERCE | Banks |
| 166 | ORIENTHOT | ORIENTAL HOTELS LTD. | Hotels |
| 167 | OSWALAGRO | OSWAL AGRO MILLS LTD. | Real estate |
| 168 | PANCM | PANYAM CEMENTS AND MINERAL INDS. | Cement and cement products |
| 169 | PARRYSUGAR | PARRYS SUGAR INDUSTRIES LTD. | Sugar |

(continued)

**Table 11.3** (continued)

| S.No. | ID | Name | Category |
|---|---|---|---|
| 170 | PDUMJEPULP | PUDUMJEE PULP AND PAPER MILLS LTD. | Paper and paper products |
| 171 | PEL | PIRAMAL ENTERPRISES LTD. | Pharmaceuticals |
| 172 | PENTAGRAPH | PENTAMEDIA GRAPHICS LTD. | Graphics |
| 173 | PHCAP | PH CAPITAL LTD. | Comm.trading and distribution |
| 174 | PIDILITIND | PIDILITE INDUSTRIES LTD. | Manufacturer |
| 175 | PILITA | PIL ITALICA LIFESTYLE LTD. | Plastic furniture manufacturers |
| 176 | PIXTRANS | PIX TRANSMISSIONS LTD. | Manufacturer of belts |
| 177 | PRAGBOS | PRAG BOSIMI SYNTHETICS LTD. | Textiles |
| 178 | PRIMESECU | PRIME SECURITIES LTD. | Security |
| 179 | PRISMCEM | PRISM CEMENT LTD.? | Cement and cement products |
| 180 | PUNJCOMMU | PUNJAB COMMUNICATIONS LTD. | Communication |
| 181 | RAIN | RAIN INDUSTRIES LTD. | Rain cements limited |
| 182 | RAJSREESUG | RAJSHREE SUGAR AND CHEMICAL LTD. | Sugar |
| 183 | RALLIS | RALLIS INDIA LIMITED-NITYA AGRO SERVICES | Agrochemicals supplier |
| 184 | RAMANEWS | SHREE RAMA NEWSPRINT LTD. | Newsprint and papers |
| 185 | RAMCOCEM | THE RAMCO CEMENTS LTD. | Cement and cement products |
| 186 | RAYMOND | RAYMOND GROUP | Fabrics |
| 187 | RELCAPITAL | RELIANCE CAPITAL LTD. | Finance (including NBFCs) |
| 188 | RSWM | R S W M Ltd. | Textiles |
| 189 | SAIL | STEEL AUTHORITY OF INDIA LTD. | Iron and steel/interm.products |
| 190 | SBI | STATE BANK OF INDIA | Banks |
| 191 | SBIN | STATE BANK OF INDIA | Banks |
| 192 | SPICEJET | SPICEJET LTD. | Airlines |
| 193 | SURYAROSNI | SURYA ROSHNI LTD. | Misc.commercial services |
| 194 | TITAN | TITAN COMPANY LTD. | Other apparels and accessories |
| 195 | TRENT | TRENT LTD. | Department stores |
| 196 | UFLEX | UFLEX LTD. | Containers and packaging |
| 197 | UMANGDAIR | UMANG DAIRIES LTD. | Packaged foods |
| 198 | VEDL | VEDANTA LTD. | Iron and steel/interm.products |
| 199 | WHIRLPOOL | WHIRLPOOL OF INDIA LTD. | Consumer electronics |

**Acknowledgements** This research was partially supported by the institute grant, IIM Ahmedabad. KS thanks University Grants Commission (Ministry of Human Research Development, Govt. of India) for her junior research fellowship. AC acknowledges financial support from the institutional research funding IUT (IUT39-1) of the Estonian Ministry of Education and Research, grant number BT/BI/03/004/2003(C) of Govt. of India, Ministry of Science and Technology, Department of Biotechnology, Bioinformatics division, and University of Potential Excellence-II grant (Project ID-47) of the Jawaharlal Nehru University, New Delhi, India.

## References

J.A. Bastos, J. Caiddo, Recurrence quantification analysis of global stock markets. Physica A **390**(7), 1315–1325 (2011)

W. Baumol, J. Benhabib, Chaos: significance, mechanism, and economic applications. J. Econ. Perspect. **3**, 77–105 (1989)

I. Borg, P. Groenen, *Modern Multidimensional Scaling: Theory and Applications* (Springer, New York, 2005)

J.-P. Bouchaud, M. Potters, *Theory of Financial Risk and Derivative Pricing: From Statistical Physics to Risk Management* (Cambridge University Press, Cambridge, 2009)

W. Brock, C. Sayers, Is the business cycle characterized by deterministic chaos? J. Monet. Econ. **22**, 71–90 (1988)

A. Chakraborti, An outlook on correlations in stock prices, in *Econophysics of Stock and Other Markets*, ed. by A. Chatterjee, B.K. Chakrabarti (Springer, Milan, 2006). Available at arXiv:physics/0605246

K. Guhathakurta, B, Bhattacharya, A. Roy Chowdhury, Using recurrence plot analysis to distinguish between endogenous and exogenous stock market crashes. Physica A **389**(9), 1874–1882 (2010)

D. Kenett, X. Huang, I. Vodenska, S. Havlin, H.E. Stanley, Partial correlation analysis: applications for financial markets. Quant. Finance **15**(4), 569–578 (2015)

C. Kuyyamudi, A.S. Chakrabarti, S. Sinha, Long-term evolution of the topological structure of interactions among stocks in the New York stock exchange 1925–2012, in *Econophysics and Data Driven Modelling of Market Dynamics*, ed. by F. Abergel et al. (Springer, Milan, 2015)

R.N. Mantegna, Hierarchical structure in financial markets. Eur. Phys. J. B **11**, 193 (1999)

R.N. Mantegna, H.E. Stanley, *Introduction to Econophysics* (Cambridge University Press, Cambridge, 1999)

N. Marwan, N. Wessel, U. Meyerfeldt, A. Schirdewan, J. Kurths, Recurrence plot based measures of complexity and its application to heart rate variability data. Phys. Rev. E, **66**, 026702 (2002)

N. Marwan, M.C. Romano, J. Thiel, J. Kurths, Recurrence plots for the analysis of complex systems. Phys. Rep. **438**(5–6), 237–329 (2007)

J.-P. Onnela, A. Chakraborti, K. Kaski, J. Kertesz, A. Kanto, Dynamics of market correlations: taxonomy and portfolio analysis. Phys. Rev. E **68**, 056110 (2003)

J.-P. Onnela, A. Chakraborti, K. Kaski, J. Kertesz, A. Kanto, Asset trees and asset graphs in financial markets. Phys. Scr. T **106**, 48 (2003)

R.K. Pan, S. Sinha, Collective behavior of stock price movements in an emerging market. Phys. Rev. E **76**, 046116 (2007)

R. Rammal, G. Toulouse, M.A. Virasoro, Ultrametricity for physicists. Rev. Mod. Phys. **58**, 765 (1986)

D. Sornette, *Why Stock Markets Crash?* (Princeton University Press, Princeton, 2004)

F. Takens, Detecting strange attractors in turbulence, in *Dynamicl Systems and Turbuence*, ed. by D. Rand, L. Young (Springer, Berlin, 1981), pp. 366–381

G. Tilak, T. Szell, R. Chicheportiche, A. Chakraborti, Study of statistical correlations in intraday and daily financial return time series, in *Econophysics of Systemic Risk and Network Dynamics*, ed. by F. Abergel et al. (Springer, Milan, 2012)

# Chapter 12
# The Divergence Rate of Share Price from Company Fundamentals: An Empirical Study at the Regional Level

**Michiko Miyano and Taisei Kaizoji**

**Abstract** We empirically analyze the divergence rate of share price from company fundamentals at the regional level. We use data from industrial companies publicly listed worldwide for the period 2004–2013. Based on ISO country codes, approximately 8,000 companies are divided into four regions: America, Asia, Europe, and the rest of the world. Following Kaizoji and Miyano (Stock market crash of 2008: an empirical study of the deviation of share prices from company fundamentals, Working paper, 2016b, arXiv:1607.03205: https://arxiv.org/abs/1607.03205), we develop a panel regression model for share price in which share price is the dependent variable and dividends per share, cash flow per share, and book value per share are explanatory variables. We identify the two-way fixed effects model as the best model for all four regions. To estimate individual company fundamentals for each year, we remove the time fixed effects from the theoretical value, that is, the fitted value in the regression model. We find that share prices significantly differ from company fundamentals in the years 2006 to 2008 in all regions, although the divergence rate differs by region.

## 12.1 Introduction

The mechanism of bubbles and crashes in stock markets is one of the most important research topics in finance and economics. Particularly, since the stock market crash of 1929 (Kindleberger 1978) and the global financial crisis of 2008 (Shiller and Robert 2015) triggered world economic crises, numerous attempts have been made by scholars to demonstrate the mechanism. Bubbles in stock markets are generally defined as the phenomenon in which stock prices diverge significantly from optimal stock prices which reflect a firm's fundamentals for the long term. Over the past

M. Miyano (✉) • T. Kaizoji
The Graduate School of Arts and Sciences, International Christian University, 3-10-2 Osawa, 181-8585, Mitaka, Tokyo, Japan
e-mail: g199006t@icu.ac.jp; kaizoji@icu.ac.jp



Y. Aruka, A. Kirman (eds.), *Economic Foundations for Social Complexity Science*, Evolutionary Economics and Social Complexity Science 9,
DOI 10.1007/978-981-10-5705-2_12

few decades, a number of hypotheses have been proposed to explain this deviation. One influential hypothesis is the noise-trader hypothesis proposed by Fisher Black (1986) and De Long et al. (1990a,b). (See also Kaizoji 2000 and Kaizoji et al 2015.) Since its introduction, a number of empirical studies and laboratory experiments have been performed on the noise-trader hypothesis (Jegadeesh and Titman 1993; DeBondt and Thaler 1985; Brunnermeier et al. 2004; Greenwood and Nagel 2008; Smith et al. 1988), and Haruvy et al (2007). After a great deal of effort, a belief that the noise-traders' behavior offers the key to understanding the occurrence and bursting of bubbles has slowly gained credibility. On the other hand, scholars who support the efficient market hypothesis proposed by Fama (1970) believe that speculative bubbles do not exist. To answer the question of whether bubbles actually exist, developing an econometric method to estimate a firm's fundamentals using real data would appear to be crucial (Shiller 1981; LeRoy et al. 1981).

Kaizoji and Miyano (2016b) used financial data from companies publicly listed worldwide to examine the question of whether the stock market crash of 2008 was an efficient response to financial shocks in line with fundamentals or was caused by investor panic. The aim of this paper is to investigate the extent to which share prices deviated from company fundamentals for the 10-year period 2004–2013 at the regional level, using the same framework as our previous studies. We use data from our earlier work and divide the 7,796 companies for which the data are available into four regions: America, Asia, Europe, and the rest of the world. The data for this regional study are as described in Miyano and Kaizoji (2016d). America includes North America, South America, and Central America. Asia includes Eastern Asia, Central Asia, and the Middle East. Europe includes Western Europe and Eastern Europe. The rest of the world includes Oceania and Africa.

In Fig. 12.1, we present the changes in the mean of the logarithmic share prices at the regional level. All regions show a decline in 2008 and then a gradual rise beginning in 2009, except for the rest of the world region, which shows a continued decline in 2009. It is clear that the declines evidenced in all regions in 2008 were caused by that year's global financial crisis.

To estimate the company fundamentals required for our study, we follow our previous studies and construct a panel regression model with share price as the dependent variable and three financial indicators – dividends per share, cash flow per share, and book value per share – as the explanatory variables. These financial indicators are representative variables commonly used to evaluate a company's business performance. The two-way fixed effects model was identified as the best panel regression model for all regions, based on the standard model selection tests for panel regression models.

The two-way fixed effects model has two fixed effects: the individual fixed effects that account for an individual company's heterogeneity, including such factors as the company's diversity of corporate governance and the quality of its employees, and the time fixed effects that indicate variables that fluctuate over time but are fixed across companies. The time fixed effects reflect various shocks, including financial shocks.

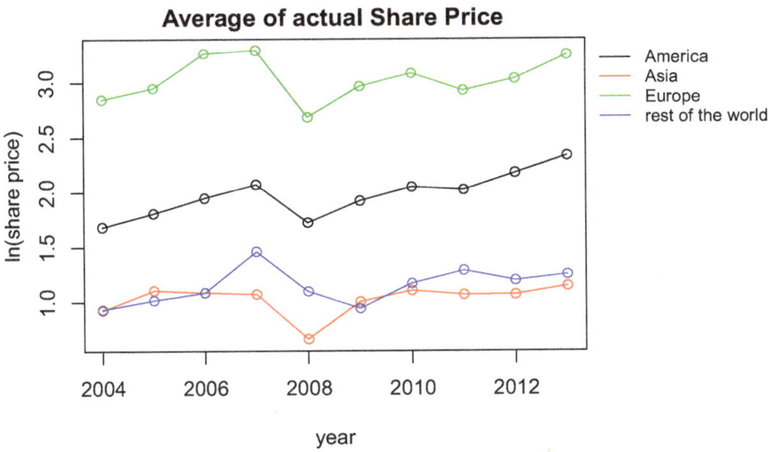


**Fig. 12.1** Changes in mean of share price

We define fundamentals as the theoretical value with the time fixed effects removed from the fitted value for the two-way fixed effects model. Thus, we can easily estimate company fundamentals using the two-way fixed effects model. We investigated the distributions of the divergence rate, which we defined as the logarithmic difference between share price and company fundamentals. We found that share prices deviated substantially from company fundamentals in the period 2006–2008 in all regions, just as our previous studies using comprehensive data for the world as a whole had found. The distributions of the divergence rate deviated in the positive direction in years 2006 and 2007 but shifted significantly from the positive side to the negative side in 2008. It is clear that share prices (on average) were overvalued against the fundamentals during the boom period from 2006 to 2007, while in 2008 they were significantly below the fundamentals. In addition, the distributions of the divergence rate for all regions were negatively skewed in 2008 but positively skewed in 2007. We found the same situation in every region – a bubble in 2006 and 2007 and a subsequent crash in 2008 – that our previous study had found in the world model, though there were some variations among the regions.

This paper is organized as follows: Sect. 12.2 describes the data used in this study; Sect. 12.3 discusses the panel data regression model for company fundamentals; Sect. 12.4 examines the divergence rate, that is, the deviation of share prices from company fundamentals; and Sect. 12.5 gives concluding remarks.

## 12.2 Data

The data source used here is the OSIRIS database provided by Bureau Van Dijk containing financial information on globally listed companies. In this study, we employ annual data for the period 2004–2013. Stock and financial data for a total of 7,796 industrial companies for which data were available over this 10-year period were extracted from the database. Using this data, we performed a panel data regression, with share price as the dependent variable and the three financial indicators – dividends per share, cash flow per share, and book value per share – as explanatory variables.

For analysis at the regional level, we divided the 7,796 companies selected into the four regions described above, using the ISO country code appropriate to the individual companies. The number of companies in each region was as follows: America, 1,886 companies; Asia, 4065 companies; Europe, 1436 companies; and the rest of the world, 409 companies.

## 12.3 Econometric Model for Company Fundamentals

In this section, we introduce a panel regression model to calculate company fundamentals. The panel regression model used in our previous model was confirmed to have high explanatory power with respect to share price. Therefore, we applied the same model for our regional data in this current study.

### 12.3.1 Panel Regression Model

Assuming the relationship between share price and the three financial indicators to be logarithmic and linear, the econometric model for our study can be written as

$$lnY_{it} = a + b_1 lnX_{1,it} + b_2 lnX_{2,it} + b_3 lnX_{3,it} + u_{it} \quad i = 1, \ldots, N; \quad t = 1, \ldots T \tag{12.1}$$

where $Y_{it}$ denotes the dependent variable (share price) for company $i$ in year $t$, $a$ denotes a constant, $X_{1,it}$ is the dividends per share of company $i$ in year $t$, $X_{2,it}$ is the cash flow per share of company $i$ in year $t$, $X_{3,it}$ is the book value per share of company $i$ in year $t$, and $u_{it}$ denotes error term.

We estimate the model in Eq. (12.1) using the panel least squares method. In the panel regression model, the error term, $u_{it}$, can be assumed to be divided into a pure disturbance term and an error term due to other factors. Assuming the two-way error component model with respect to error, the factors other than disturbance can be (i) factors due to unobservable individual effects and (ii) factors due to unobservable time effects. That is, the error term in Eq. (12.1) can be written as

$$u_{it} = \mu_i + \gamma_t + \epsilon_{it} \tag{12.2}$$

Here, $\mu_i$ denotes unobservable individual effects, $\gamma_t$ denotes unobservable time effects, and $\epsilon_{it}$ denotes pure disturbance. If both $\mu_i$ and $\gamma_t$ are equal to zero, Eq. (12.1) has only pure disturbance term, $\epsilon_{it}$, and is estimated using the pooled OLS method. If $\gamma_t$ is equal to zero, Eq. (12.2) becomes the one-way error component model written as

$$u_{it} = \mu_i + \epsilon_{it} \tag{12.3}$$

If neither $\mu_i$ nor $\gamma_t$ is equal to zero, Eq. (12.2) is the two-way error component model.

There are two estimation methods for estimating the error term in Eq. (12.2): the fixed effects estimation method and the random effects estimation method. In all, then, the possible estimation models include the pooled OLS, the individual fixed effects model, the time fixed effects model, the two-way fixed effects model, the individual random effects model, the time random effects model, and the two-way random effects model. We estimate five of the six models, excluding the two-way random effects model, by region.[1]

The model selection tests include the likelihood ratio test and the F-test for selection of the pooled OLS model vs the fixed effects model and the Hausman test for selection of the random effects model vs the fixed effects model. The selection test for the pooled OLS model vs the random effects model is based on the simple test proposed by Wooldridge and Jeffrey (2010).[2]

For all four regions, the two-way fixed effects model is identified as the best model among the six alternatives.

Substituting Eq. (12.2) for $u_{it}$ in Eq. (12.1), the two-way fixed effects model is written as

$$lnY_{it} = a + \mu_i + \gamma_t + b_1 lnX_{1,it} + b_2 lnX_{2,it} + b_3 lnX_{3,it} + \epsilon_{it} \tag{12.4}$$

where $a$ is a constant term common for all companies. $\mu_i$ denotes the individual fixed effects constant for time series. $\gamma_t$ denotes the time fixed effects constant for cross section. $\epsilon_{it}$ is the pure disturbance. The individual fixed effects, $\mu_i$, account for individual company heterogeneity, such factors as diversity in corporate governance or diversity in the quality of employees. The time fixed effects, $\gamma_t$, indicate variables that fluctuate over a period of time but are fixed across companies. The time fixed effects reflect various shocks, including financial shocks.

Table 12.1 shows the regional results for the two-way fixed effects model described in Eq. (12.1). The standard errors shown in Table 12.1 of the estimates are modified using the White cross-section method, since the residuals are

[1] We use the EViews software package to estimate the model. The two-way random effects model was unavailable since we use unbalanced panel data.

[2] Wooldridge and Jeffrey (2010, p. 299) proposes a method that uses residuals from pooled OLS and checks the existence of serial correlations.

**Table 12.1** Regional results of panel regression (two-way fixed effects model). Total observations presented in the table are unbalanced panel observations

| Region | | a | $b_1$ | $b_2$ | $b_3$ | $R^2$ | Total observations |
|---|---|---|---|---|---|---|---|
| America | Coefficient | 1.750 | 0.119 | 0.215 | 0.324 | 0.977 | 8,935 |
| | std. error | 0.044 | 0.017 | 0.020 | 0.019 | | |
| | *p*-value | 0.000 | 0.000 | 0.000 | 0.000 | | |
| Asia | Coefficient | 1.154 | 0.112 | 0.218 | 0.440 | 0.956 | 27,404 |
| | std. error | 0.052 | 0.010 | 0.029 | 0.041 | | |
| | *p*-value | 0.000 | 0.000 | 0.000 | 0.000 | | |
| Europe | Coefficient | 2.160 | 0.191 | 0.154 | 0.318 | 0.967 | 8,791 |
| | std. error | 0.048 | 0.018 | 0.020 | 0.020 | | |
| | *p*-value | 0.000 | 0.000 | 0.000 | 0.000 | | |
| Rest of the world | Coefficient | 1.546 | 0.189 | 0.207 | 0.416 | 0.953 | 2,028 |
| | std. error | 0.083 | 0.035 | 0.038 | 0.068 | | |
| | *p*-value | 0.000 | 0.000 | 0.000 | 0.000 | | |

heteroscedastic. The signs of the three coefficients are all positive, consistent with corporate value theory. The p-values of the coefficients are very close to zero, indicating statistical significance for all regions. In addition, the $R^2$ values are in the range 0.95–0.98, indicating that the estimated models explain the variation in share price quite well. The econometric model for share price, using dividends per share, cash flow per share, and book value per share as explanatory variables, fits the actual data quite well at the regional level as well as the world level.[3]

### 12.3.2 Theoretical Value and Fundamentals

Estimation of the two-way fixed effects model for share price, $ln\hat{Y}_{it}$, is written as

$$ln\hat{Y}_{it} = \hat{a} + \hat{\mu}_i + \hat{\gamma}_t + \hat{b}_1 lnX_{1,it} + \hat{b}_2 lnX_{2,it} + \hat{b}_3 lnX_{3,it} \qquad (12.5)$$

We call $\hat{Y}$ the theoretical value of share price. Figures 12.2, 12.3, 12.4, and 12.5 show the scatter diagrams of the theoretical value of share price plotted against actual share prices. Figure 12.2 suggests that the relationship between the theoretical value and share price is highly positive, even though a few observations deviate markedly from their theoretical values in the scatter diagram of Asia. We consider these few observations to be outliers in our study. Eliminating these observations, we performed a second panel regression for the Asia data. The two-way fixed effects model is again selected as the best model. The regression results are presented

[3] $R^2$ for the world model is 0.97, though detailed results are not reported here.

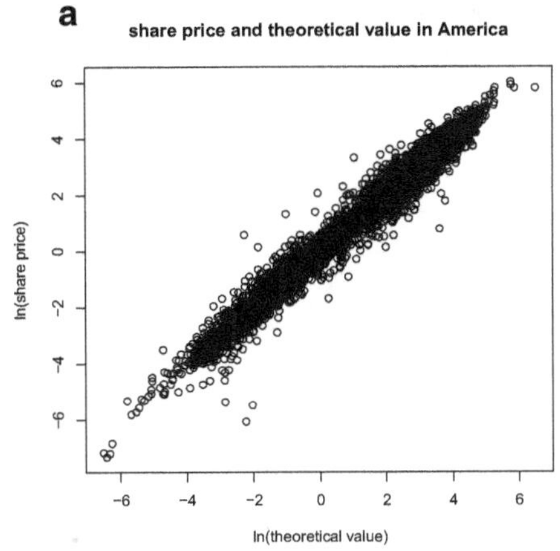


Fig. 12.2 Scatter diagram of America

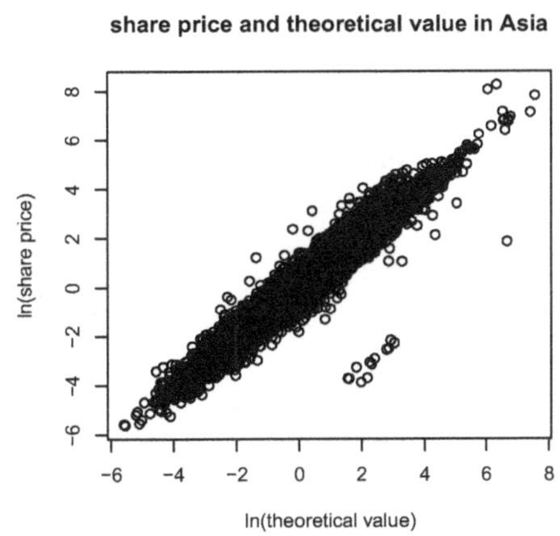


Fig. 12.3 Scatter diagram of Asia

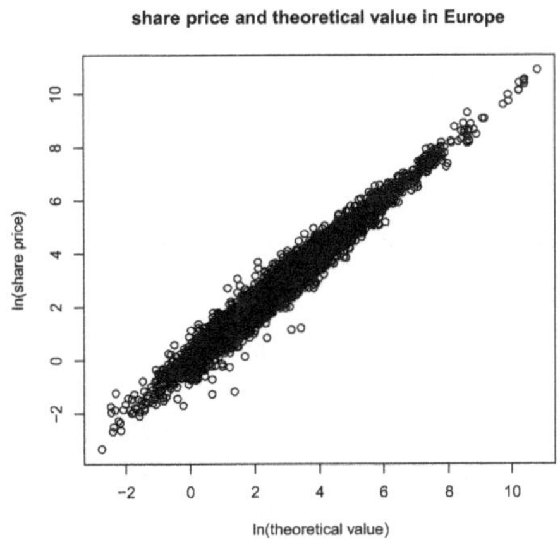


Fig. 12.4 Scatter diagram of Europe

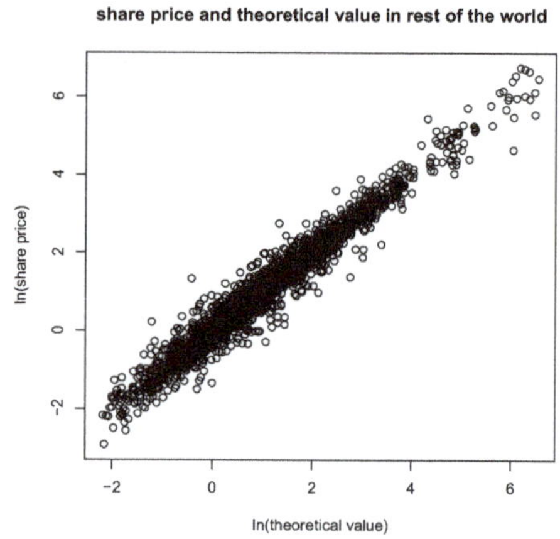


**Fig. 12.5** Scatter diagram of the rest of the world

**Table 12.2** Results for eliminating Asia observations that appear to be outliers

| Region | | a | $b_1$ | $b_2$ | $b_3$ | $R^2$ | Total observations |
|---|---|---|---|---|---|---|---|
| Asia | Coefficient | 1.159 | 0.112 | 0.219 | 0.440 | 0.963 | 27,394 |
| | std. error | 0.050 | 0.010 | 0.029 | 0.041 | | |
| | $p$-value | 0.000 | 0.000 | 0.000 | 0.000 | | |

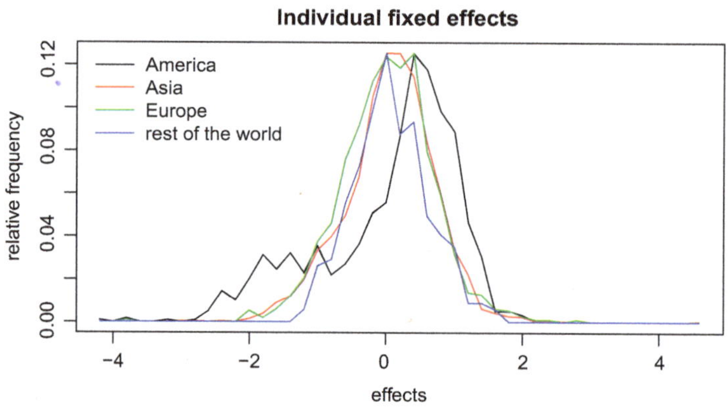


**Fig. 12.6** Distribution of individual fixed effects

in Table 12.2. There are few differences between the Asian results in Tables 12.1 and 12.2. The $R^2$ value is slightly improved.

Figure 12.6 shows the relative frequency distributions of the individual fixed effects, which are constant over time, by region. The distributions indicate a wide heterogeneity in unobservable company capabilities in all regions. As we can see

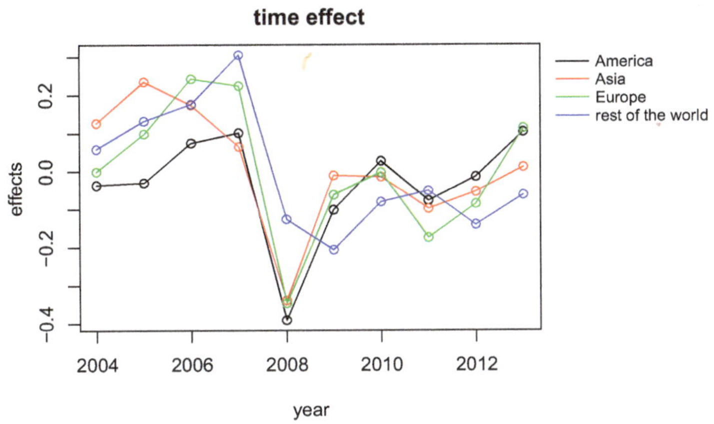


**Fig. 12.7** Change in time fixed effects

from Fig. 12.6, the distribution for the American region shows a somewhat thick and "lumpy" left-side tail that sets it apart from the other three regions.

Figure 12.7 shows the time fixed effects for each year by region. The time fixed effects are related to points in time (year) and the occurrence of events affecting stock markets. That is, the time fixed effects can be considered to be the results of temporal shocks to the stock market. Hence, the movements of time effects are similar to the changes in the mean of logarithmic share price shown in Fig. 12.1. All regions show positive effects for the years 2006 and 2007 and large negative effects in 2008 and subsequent periods.

To estimate the fundamentals of individual companies, we eliminate the time fixed effects from the theoretical value for the two-way fixed effects model, since the time fixed effects can be considered to be the results of temporal shocks to share price. We retained the individual fixed effects since these effects represent the individual company's unobservable heterogeneity as reflected in its share price. Therefore, we define the logarithmic form of a company fundamentals as

$$ln\tilde{Y}_{it} = \hat{a} + \hat{\mu}_i + \hat{b}_1 lnX_{1,it} + \hat{b}_2 lnX_{2,it} + \hat{b}_3 lnX_{3,it} \quad (12.6)$$

where $\tilde{Y}_{it}$ denotes the company fundamentals of company $i$ in year $t$. We utilize the fundamentals estimated here to investigate the divergence of share price from the company fundamentals. The estimates of the coefficients in Eq. (12.6), $\hat{a}$, $\hat{\mu}_i$, $\hat{b}_1$, $\hat{b}_2$, and $\hat{b}_3$, are constant over time. Therefore, if we can obtain values on dividends per share, cash flow per share, and book value per share for a company, we can easily estimate the company fundamentals using the model.

## 12.4 Divergence Rate of Share Price from Company Fundamentals

In this section we investigate the divergence rate of share price from company fundamentals using the model of company fundamentals ($\tilde{Y}$) during the 10-year period from 2004 through 2013. The divergence rate between share price and the company fundamentals is defined as

$$D_{it} = lnY_{it} - ln\tilde{Y}_{it} \tag{12.7}$$

where $Y_{it}$ denotes the share price of company $i$ in year $t$ and $\tilde{Y}_{it}$ denotes the fundamentals of company $i$ in year $t$. The divergence rate, $D_{it}$, for company $i$'s share prices is the rate of change between company $i$'s share price and company $i$'s fundamentals in year $t$. We calculate the divergence rate, $D_{it}$ for each company for each year. Table 12.3 shows the basic statistics for the divergence rates for each year over the 10-year period.[4] Figure 12.8 shows changes in the mean of the divergence rates. As indicated here, there are considerable variations in the divergence rate over time in all regions.

The mean divergence rate is positive for the years 2006 and 2007 in all regions. The means then fall sharply in 2008, indicating a negative divergence rate. In this year, share prices fall, on average, more than 30% below the fundamentals, except in the rest of the world, where the mean falls rather less than in the other regions. However, the situation reverses, and the mean for the rest of the world falls more than in the other regions in 2009, when the means in the other three regions show a rise. It is clear that the reason for the precipitous fall in the average divergence rate in 2008 was the global financial crisis of that year.

The contrasting movement of the divergence rate for the rest of the world (vs the other three regions) is principally due to the difference in closing dates for the companies that comprise the various regions. For our study, we used annual data based on company closing dates. While most of the companies set their closing date at the end of December, some had closing dates in other months, including March, June, and September. Given that the global financial crisis was triggered by the bankruptcy of Lehman Brothers on September 15, 2008, it can be said that those companies with closing dates prior to September 2008 were strongly affected by the global crisis, not in 2008, but rather in 2009. In the rest of the world, 74.8% of the companies set their closing dates before September 2008, while only 14.8% of companies in America, 4.9% in Asia, and 7.2% in Europe had such a closing date. This variation in closing dates also explains the different behavior of the rest of the world in Fig. 12.1, where share price means are shown, and in Fig. 12.7, where the time fixed effects are shown.

[4]The region shown in Table 12.3, (Asia)(in parenthesis), is for the case in which observations identified as outliers are excluded from the regression model.

**Table 12.3** Basic statistics for the divergence rates

| | Region | 2004 | 2005 | 2006 | 2007 | 2008 | 2009 | 2010 | 2011 | 2012 | 2013 |
|---|---|---|---|---|---|---|---|---|---|---|---|
| Mean | America | −0.04 | −0.03 | 0.08 | 010 | −0.39 | −0.10 | 0.03 | −0.08 | −0.01 | 0.10 |
| | Asia | 013 | 0.23 | 0.17 | 0.06 | −0.34 | −0.01 | −0.01 | −0.10 | −0.05 | 0.01 |
| | (Asia) | 0.13 | 0.24 | 0.19 | 0.09 | −0.33 | 0.00 | −0.01 | −0.09 | −0.05 | 0.02 |
| | Europe | 0.00 | 0.10 | 0.24 | 0.22 | −0.35 | −0.06 | 0.00 | −0.17 | −0.09 | 0.11 |
| | Rest of the world | 0.06 | 0.13 | 0.18 | 0.30 | −0.13 | −0.21 | −0.08 | −0.05 | −0.14 | −0.06 |
| Standard deviation | America | 0.38 | 0.37 | 0.30 | 0.32 | 0.43 | 0.25 | 0.25 | 0.27 | 0.29 | 0.34 |
| | Asia | 0.37 | 0.41 | 0.37 | 0.43 | 0.33 | 0.29 | 0.29 | 0.25 | 0.29 | 0.32 |
| | (Asia) | 0.37 | 0.40 | 0.29 | 0.33 | 0.33 | 0.28 | 0.29 | 0.24 | 0.27 | 0.31 |
| | Europe | 0.32 | 0.28 | 0.26 | 0.27 | 0.33 | 0.23 | 0.24 | 0.28 | 0.32 | 0.33 |
| | Rest of the world | 0.36 | 0.33 | 0.33 | 0.34 | 0.36 | 0.29 | 0.26 | 0.28 | 0.32 | 0.38 |
| Kurtosis | America | 8.46 | 14.05 | 10.30 | 3.64 | 6.71 | 2.02 | 1.93 | 3.17 | 3.32 | 2.96 |
| | Asia | 4.25 | 1.40 | 73.43 | 67.19 | 1.29 | 4.62 | 1.57 | 3.36 | 27.64 | 1.94 |
| | (Asia) | 3.93 | 1.08 | 4.88 | 1.42 | 0.93 | 4.23 | 1.37 | 2.17 | 1.73 | 1.69 |
| | Europe | 1.49 | 2.97 | 1.69 | 0.95 | 0.84 | 2.70 | 3.28 | 2.78 | 6.66 | 2.80 |
| | Rest of the world | 0.31 | 2.22 | 1.43 | 3.93 | 1.78 | 1.27 | 2.62 | 1.26 | 2.14 | 1.53 |
| Skewness | America | −0.73 | −1.43 | 0.01 | 0.64 | −0.45 | 0.47 | 0.29 | −0.69 | −0.20 | −0.03 |
| | Asia | 0.10 | −0.22 | −5.22 | −4.95 | −0.28 | 0.89 | 0.74 | 0.36 | −1.13 | 0.62 |
| | (Asia) | −0.05 | −0.34 | −0.22 | 0.64 | −0.43 | 0.75 | 0.67 | 0.08 | 0.37 | 0.49 |
| | Europe | 0.12 | −0.11 | −0.15 | 0.26 | −0.39 | −0.29 | 0.36 | −0.16 | −0.90 | −0.38 |
| | Rest of the world | −0.12 | 0.35 | −0.13 | 0.61 | −0.49 | −0.75 | −0.49 | 0.09 | −0.50 | −0.22 |
| Observations | America | 861 | 895 | 911 | 903 | 855 | 828 | 901 | 916 | 934 | 931 |
| | Asia | 2841 | 2861 | 2770 | 2833 | 2504 | 2560 | 2700 | 2723 | 2772 | 2843 |
| | (Asia) | 2841 | 2861 | 2765 | 2826 | 2504 | 2560 | 2700 | 2723 | 2771 | 2843 |
| | Europe | 911 | 921 | 932 | 963 | 818 | 790 | 866 | 876 | 861 | 853 |
| | Rest of the world | 194 | 207 | 209 | 215 | 206 | 186 | 208 | 204 | 203 | 196 |

We also investigate the distributions of the divergence rates of America, Asia, Europe, and the rest of the world. Figures 12.9,12.10, 12.11, 12.12, 12.13, 12.14, 12.15, and 12.16 show the distributions of the divergence rates by region. Part a of each figure presents the distribution of divergence rates for the period from 2006 through 2008; part b of each figure presents the distributions for the period from 2009 through 2013.

Regarding the distributions in the period 2006–2008, the distributions of the divergence rates in 2008 drastically shift toward the minus side from 2007 in America, Europe, and the rest of the world, while in Asia, the distributions start

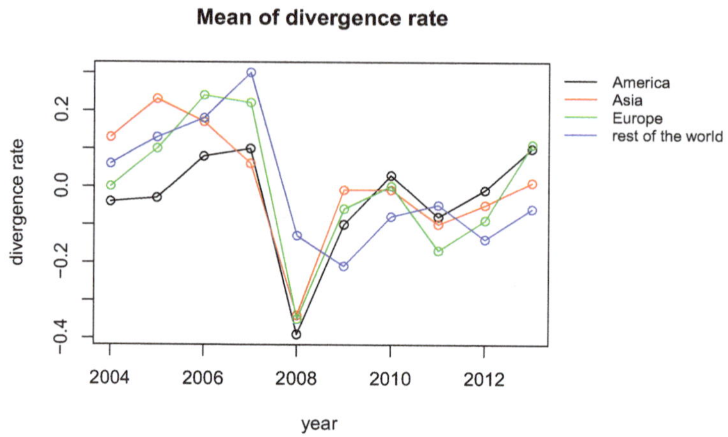


**Fig. 12.8** Change in mean of divergence rate

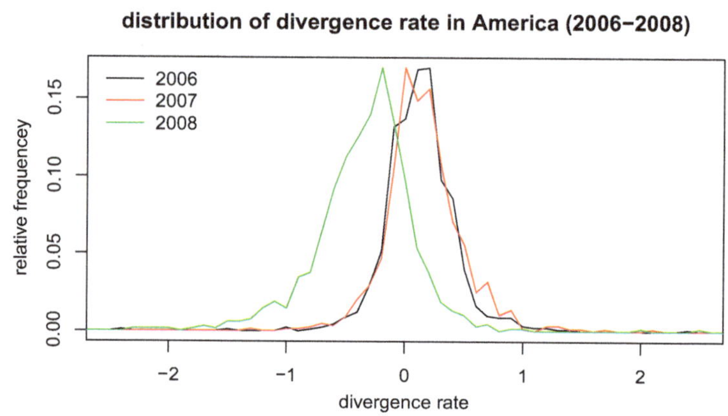


**Fig. 12.9** Distribution in America (2006–2008)

shifting toward the minus side in 2006 and continue to do so in 2007.[5] In America and Europe, there seems to be little difference between 2006 and 2007, while in the rest of the world, the distribution in 2007 shifts to the right side from 2006 but subsequently shifts considerably toward the minus side. On the other hand, the distributions of the divergence rate for the years 2009–2013 generally revert to the right side in all regions, although there are a few variations in shifting behavior by region.

In Table 12.3, we see changes in the kurtosis and skewness of the distributions of the divergence rates over the 10-year period examined in the study. As shown, in Asia, the kurtosis values in 2006, 2007, and 2012 are extremely large, and the

[5]This behavior is the same as the case eliminating the outliers in the Asia data.

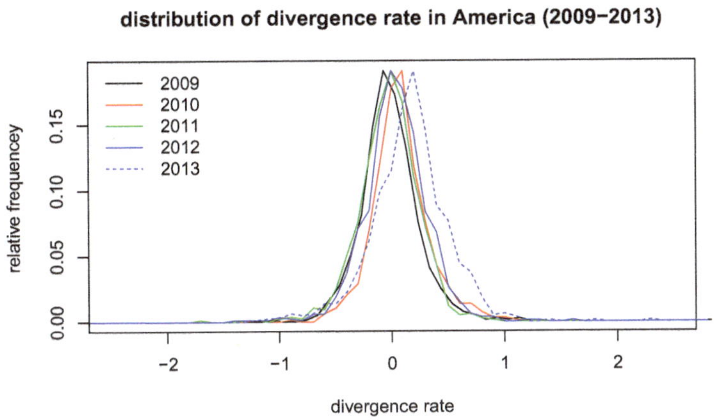


**Fig. 12.10** Distribution in America (2009–2013)

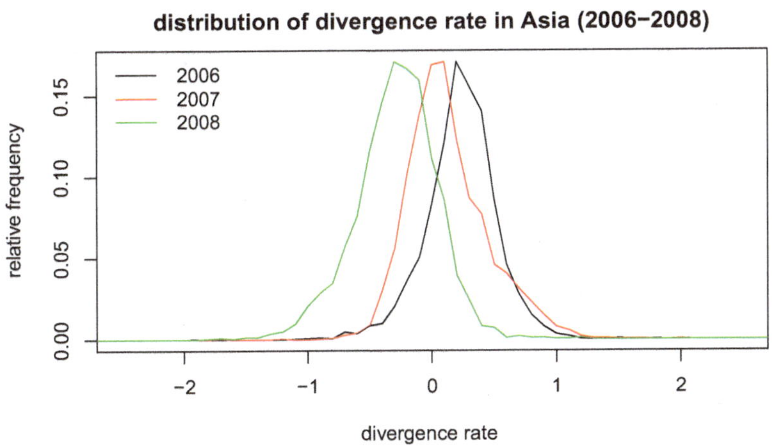


**Fig. 12.11** Distribution in Asia (2006–2008)

skewness values in 2006, 2007, and 2012 are likewise extremely large in absolute value. This clearly occurred due to the existence of a few outliers, as explained in Sect. 12.3. In Table 12.3, the lines labeled (Asia) ("Asia" in parentheses) give the results from the case in which the Asia outliers are eliminated. With the outliers removed from the Asia data, the kurtosis values in America are relatively large compared to the other regions, indicating "fatter" distributions than the normal distribution. The skewness values show a clear difference between 2007 and 2008. All regions show positive skewness in 2007 and negative skewness in 2008.

As reflected in the divergence rates by region over the 10-year period of the study, the global crisis in 2008 exerted a harmful influence on stock markets in all four regions. However, there is some variation in both the magnitude and pace of these influences among the regions. The distributions of the divergence rates in America

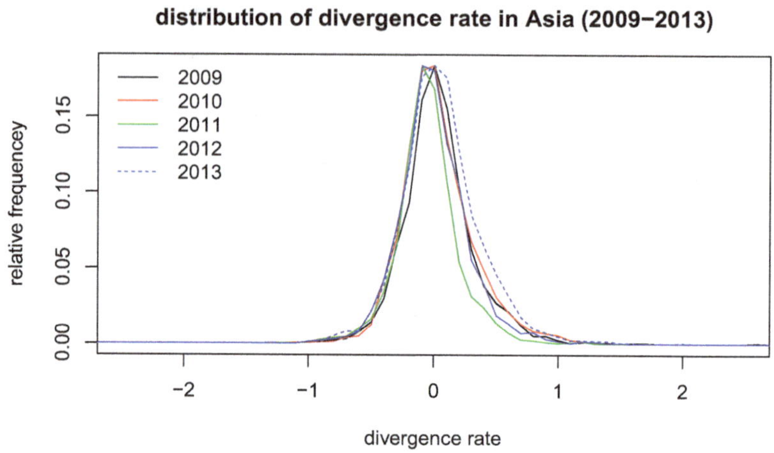


**Fig. 12.12** Distribution in Asia (2009–2013)

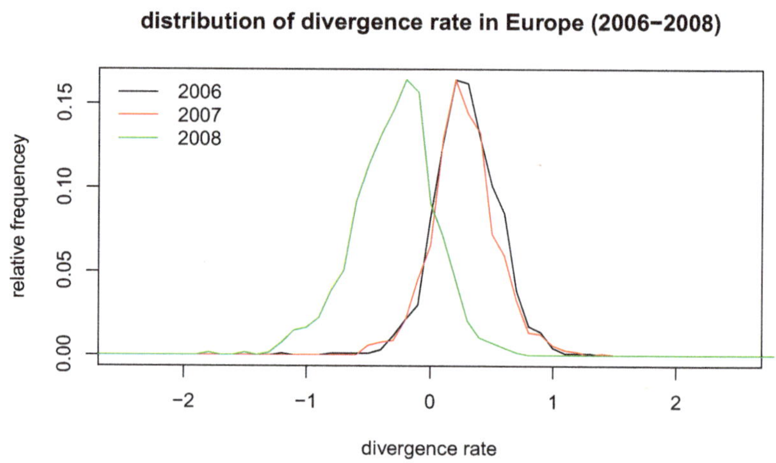


**Fig. 12.13** Distribution in Europe (2006–2008)

and Europe appear to exhibit similar behavior, while those in Asia and the rest of the world differ from one another and from those in America and Europe.

## 12.5 Concluding Remarks

In this study, we examined the divergence rate of share price from company fundamentals at the regional level. Using the same data as Kaizoji and Miyano (2016), we constructed regional data, dividing a total of 7,796 companies into four

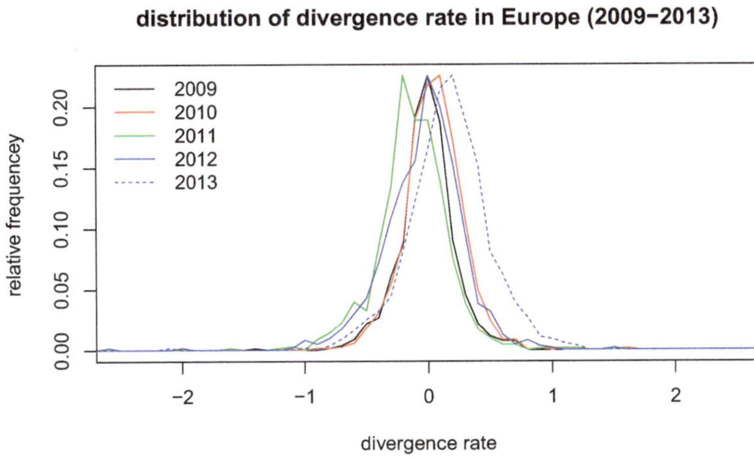


**Fig. 12.14** Distribution in Europe (2009–2013)

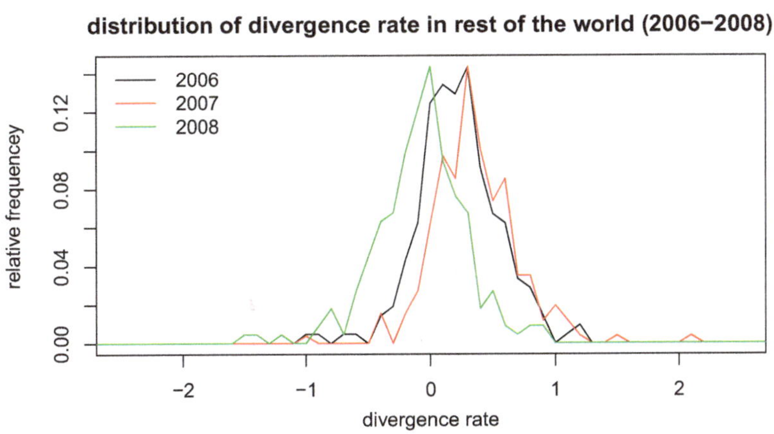


**Fig. 12.15** Distribution in rest of the world (2006–2008)

regions: America, Asia, Europe, and the rest of the world. We performed a panel regression model by region. Two-way fixed effects models for share price were selected as the best models in all four regions. As in our previous studies, the models were found to have quite a high power to explain share price using three financial indicators – dividends per share, cash flow per share, and book value per share. We estimated company fundamentals by eliminating time fixed effects, which are considered to be the result of temporal shocks to the stock market, from the theoretical values estimated for the two-way fixed effects model.

We then quantitatively investigated the extent to which share prices deviate from company fundamentals in the years 2004 to 2013, defining the divergence rate of the share price as the logarithmic difference between share price and company

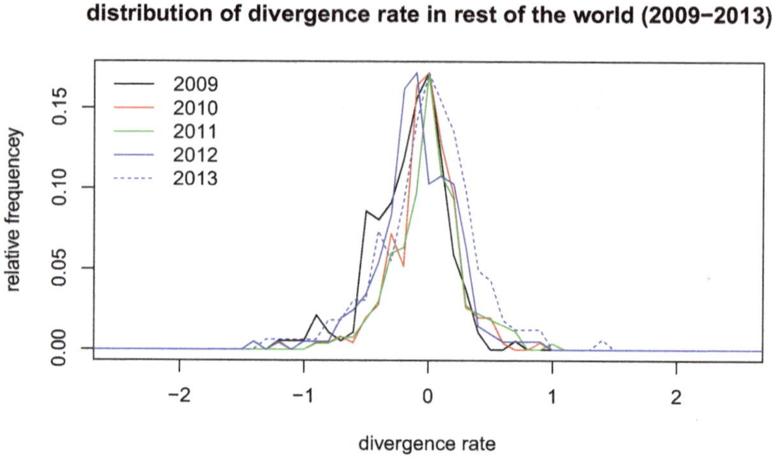


**Fig. 12.16** Distribution in rest of the world (2009–2013)

fundamentals. These values are positive in 2006 and 2007 in all regions but strongly negative in 2008 and in subsequent periods across all regions. Though there are some differences among regions, the movement of the divergence rates is similar to what we had found previously using world data. With respect to the differences in the divergence rates at the regional level, the magnitudes of the divergence rates differ somewhat from each other. On the other hand, the pattern of changes in the mean of the divergence rates is similar in all regions, especially in America and Europe. We found that the financial crisis of 2008 exerted a harmful influence on stock markets in all regions, with some variations in the magnitudes and pace of the impact among the regions.

**Acknowledgements** This research was supported by JSPS KAKENHI Grant Number 2538404, 2628089.

## References

F. Black, Noise. J. Financ. **41**(3), 529–543 (1986). Papers and Proceedings of the Forty-Fourth Annual Meeting of the America Finance Association, New York, 28–30 Dec 1985

M. Brunnermeier, N. Stefan, Hedge funds and the technology bubble. J. Financ. **59**(5), 2013–2040 (2004)

W.F.M. De Bondt, R. Thaler, Does the stock market overreact? J. Financ. **40**, 793–808 (1985)

J.B. De Long, A. Shleifer, L.H. Summers, R.J. Waldmann, Positive feedback investment strategies and destabilizing rational speculation. J. Financ. **45**(2), 379–395 (1990a)

J.B. De Long, A. Shleifer, L.H. Summers, R.J. Waldmann, Noise trader risk in financial markets. J. Polit Econ. **98**, 703–38 (1990b)

E. Fama, Efficient capital markets: a review of theory and empirical work. J. Financ. **25**(2), 383–417 (1970)

R. Greenwood, S. Nagel, Inexperienced investors and bubbles. J. Financ. Econ. **93**(2), 239–258 (2008)

E. Haruvy, Y. Lahavand, C. Noussair, Traders' expectations in asset markets: experimental evidence. Am. Econ. Rev. **97**(5), 1901–1920 (2007)

N. Jegadeesh, S. Titman, Returns to buying winners and selling losers: implications for stock market efficiency. J. Financ. **48**, 65–92 (1993)

T. Kaizoji, Speculative bubbles and crashes in stock markets: an interacting-agent model of speculative activity. Phys. A **287**, 493–506 (2000)

T. Kaizoji, M. Leiss, A. Saichev, D. Sornette, Super-exponential endogenous bubbles in an equilibrium model of fundamentalist and chartist traders. J. Econ. Behav. Organ. **112**, 289–310 (2015)

T. Kaizoji, M. Miyano, Why does power law for stock price hold? Chaos, Soliton Fractals **88**, 19–23 (2016a)

T. Kaizoji, M. Miyano, Stock market crash of 2008: an empirical study of the deviation of share prices from company fundamentals. Working paper. arXiv:1607.03205 (2016b), https://arxiv.org/abs/1607.03205

T. Kaizoji, M. Miyano, Zipf's law for company fundamentals and share price Working paper. arXiv:1702.00144 (2016c), http://arxiv.org/abs/1702.00144

T. Kaizoji, M. Miyano, Power law distribution for share price and financial indicators: analysis at the regional level. Published in the Proceedings of ECONOPHYS-2015 in the New Economic Windows series of Springe (2016d)

C. Kindleberger, *Manias, Panics, and Crashes: A History of Financial Crises* (Basic Books, New York, 1978)

S.F. LeRoy, R.D. Porter, The present value relation: tests based on implied variance bounds. Econometrica **49**, 555–574 (1981)

R.J. Shiller, Do stock prices move too much to be justified by subsequent changes in dividends? Am. Econ. Rev. **71**, 421–436 (1981)

R.J. Shiller, *Irrational Exuberance*, 3rd edn. (Princeton University Press, USA, 2015)

V.L. Smith, G.L. Suchanek, A.W. Williams, Bubbles, crashes and endogenous expectations in experimental spot asset markets. Econometrica **56**(5), 1119–1151 (1988)

J.M. Wooldridge, *Introductory Econometrics: A Modern Approach*, 4th edn. (South-Western Publication, Mason, 2008)

J.M. Wooldridge, *Econometric Analysis of Cross Section and Panel Data*, 2nd edn. (The MIT Press, London, 2010)

# Chapter 13
# Analyzing Relationships Among Financial Items of Banks' Balance Sheets

**Kunika Fukuda and Aki-Hiro Sato**

**Abstract** The bankruptcy of a big bank can have a chain reaction that causes a huge impact on other sectors, as demonstrated by the Lehman shock. Such an event is called systemic risk, and was the main subject of this study. In this report, the relationships among financial items of banks were analyzed using real data extracted from balance sheets, which were reported by banks listed in the first section of Tokyo Stock Exchange. Following the Cobb-Douglas production function approach, a power-law relationship between equity and debt for each bank become apparent. Furthermore, two variable regression analyses about the power-law relationship were conducted, and it was determined that the parameter estimate can be used as an index measuring macroeconomic condition. No correlation was found between the capital adequacy ratio and the return on equity, which are well-known indicators to characterize bank properties in a quantitative manner. This method makes it easier to understand macroeconomic conditions of banking sector from a comprehensive point of view.

## 13.1 Introduction

Lehman Brothers Holdings Inc., which was one of the largest American investment banks, made a failed investment in subprime mortgage loans and eventually went bankrupt on 15 September 2008. This led to a subsequent negative influence on financial markets worldwide, in a historic event called the Lehman Shock. Triggered by this big event, the Basel Committee introduced a new regulation, which was stronger than their previous one (Bank for international settelements 2015), and each individual country's central bank began discussing how to avoid a global financial crises, through remodeling the whole banking system.

---

K. Fukuda • A.-H. Sato (✉)
Department of Applied Mathematics and Physics, Graduate School of Informatics, Kyoto University, Kyoto, Japan
e-mail: aki@i.kyoto-u.ac.jp



Y. Aruka, A. Kirman (eds.), *Economic Foundations for Social Complexity Science*, Evolutionary Economics and Social Complexity Science 9,
DOI 10.1007/978-981-10-5705-2_13

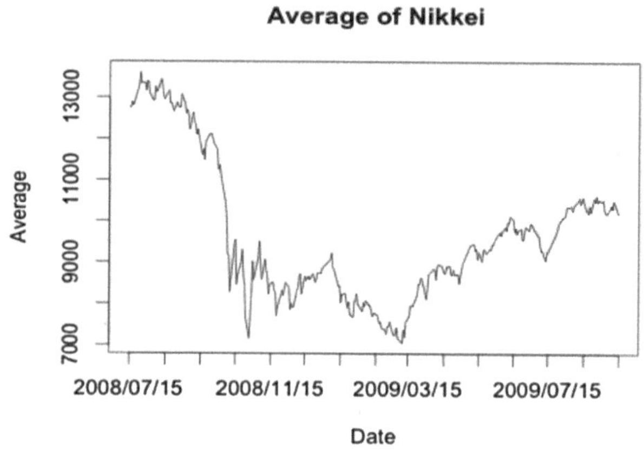


**Fig. 13.1** The fluctuation of the Nikkei Stock Average before and after the Lehman shock

Figure 13.1 shows the value over time of the Nikkei Stock Average before and after the Lehman Shock. The Nikkei Stock Average sharply dropped from the 13,000 JPY level to the 8,000 JPY level within 2 months following the event on 15 September 2008. This demonstrates that the bankruptcy of a big bank has a huge impact on other sectors, in a chain reaction (Lehman-shock chart 2015).

This indicates that traditional risk management methods, which focus on risk management for each bank individually, did not work well in the modern banking climate. Therefore, in later years, it was suggested that any regulation aimed at improving financial stability must give attention to systemic risk, which focuses losses occurring through the interaction of the banking and financial system, rather than to each individual institution (Kay 2012).

Previous studies investigated the systemic risk and estimated loss that would affect the whole banking system due to the bankruptcy of a one bank (Tasca and Battiston 2016; Iori et al. 2006; Gai and Kapadia 2010; Haldane and May 2011; Sato et al. 2015; Wei and Leon-Ledesma 2015). They divide market participants into two types, trend followers and contrarians. When the former type represents a large number of market participants, the price becomes unstable. When contrarians, however, become a majority of market participants, the price shows a mean recursive price fluctuation.

Additionally, contrarians tend to have a decreased capital adequacy ratio, while trend followers tend to have an increased ratio. Furthermore, in that simulation, sensitivity analysis for some elements over time was conducted in order to understand the relationship between assets and bankruptcy. It is necessary to know how the financial numeric value of real banks can be included in numerical simulation. When actual values extracted from bank's financial records are included, more realistic results are obtained. This was not done sufficiently in the previous study.

It is also important to note whether a financing numerical relationship is universal or whether it changes over time. Similarly, investigating this over a long term is crucial. This leads to the dilemma of how to use data to investigate systemic risk. It is important to generate a new index that can measure the systemic risk for all banks. Each bank can be studied and understood through the study of its data and financial trends. Therefore, problems and characteristics of each individual bank can be investigated and understood numerically. This must be helped by supportive corporate strategies.

## 13.2 Model

Financial statements were divided into assets and liabilities using the structured model of a bank (Sato et al. 2015), and the information about change of corporate value was then considered. Figure 13.2 shows the basic structured model of a bank's balance sheet.

Suppose $N$ banks have their own balance sheet. An assumption was made that when bank $i$ has cash $C_i(t)$ at time $t$ and holds $n_i(t)$ units of assets with a market price $S(t)$, an amount of risk assets $J_i(t)$ can be described as

$$J_i(t) = n_i(t)S(t). \tag{13.1}$$

If the bank buys the $V_i(t)$ units, the risk assets of within time difference $\Delta t$, then $n_i(t)$, and $C_i(t)$ are updated as

$$n_i(t) = n_i(t - \Delta t) + V_i(t), \tag{13.2}$$

$$C_i(t) = C_i(t - \Delta t) - V_i(t)S_i(t), \tag{13.3}$$

In addition, we assume that $W_{ij}(t)$ is denoted as borrowing amount money from bank $i$ to bank $j$ at time $t$. Thus, the credit of bank $i$, $K_i(t)$ is expressed as

$$K_i(t) = \sum_{j=1}^{N} W_{ij}(t), \tag{13.4}$$

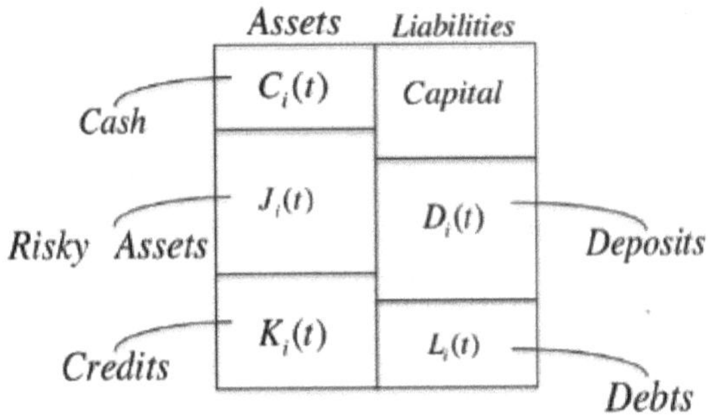


**Fig. 13.2** Fundamental model of bank's balance sheet

and the liability of bank $i$ at time $t$ $L_i(t)$ is described as

$$L_i(t) = \sum_{j=1}^{N} W_{ji}(t). \tag{13.5}$$

The deposit of the bank $i$ is assumed to be $D_i(t)$.

Return on equity (ROE) is commonly used as an index to quantify the profit rate of a bank. Using the abovementioned model, ROE can be defined as

$$\mathrm{ROE}_i(t) = \frac{C_i(t) - C_i(t-\Delta t) - \Lambda_D D_i(t)\Delta t - \Lambda_L L_i(t)\Delta t + \Lambda_K K_i(t)\Delta t}{C_i(t) + J_i(t) + K_i(t) - L_i(t) - D_i(t)} \times 100(\%), \tag{13.6}$$

where $\Lambda_D$, $\Lambda_L$, and $\Lambda_K$ are interest rates measured in units of time. This shows whether the capital dropped by a stockholder produces profit effectively.

Furthermore, capital adequacy ratio (CAR) is used as an index to quantify the soundness of a bank, and it can be defined as

$$\mathrm{CAR}_i(t) = \frac{C_i(t) + J_i(t) + K_i(t) - L_i(t) - D_i(t)}{J_i(t) + K_i(t)} \times 100(\%), \tag{13.7}$$

which serves to show the proportion of an amount of risky assets to an amount of equity. ROE is originally defined by using a ratio of net income by each term of the income statement, as follows:

$$\frac{\text{Equity}}{\text{Asset}}. \tag{13.8}$$

Recently, the Financial Services Agency announced that banks should increase the quantity of the owned capital after the introduction of Basel III. As for the Tier 1 ratio ($\approx$ capital adequacy ratio), 3.5% was the lowest standards in 2013. Phase in of the capital maintenance buffer is started, and the lowest standard of Tier 1 was set to raise up to 5.125% in 2016 and expected to become 7% in 2019. The capital adequacy ratio is generally defined as

$$\frac{\text{Common Equity Tier 1} + \text{Other Tier1} + \text{Tier 2}}{\text{Risky Asset}}, \tag{13.9}$$

where common equity Tier 1 means the capital which is the highest ability to absorb losses. In addition, from the viewpoint of reinforcement of the quality of the capital and of prevention of accumulation of the risk in the financial system, the capital possession of other financial institutions, intangible assets, and the deferred tax assets (such as goodwill) deduct from common equity Tier 1. Otherwise, inclusion profit (OCI), including the money of evaluation balance of other securities, it is counted as common equity Tier 1. Other Tier 1 refers to preferred stocks, and Tier 2 refers to subordinated debenture, subordinated loan, and general loan loss reserve

**Table 13.1** Correspondence of the item of a model and the balance sheet

| Parameter | Description |
|---|---|
| $J_i(t)$ | (Receivables under resale agreements) + (monetary receivables purchased) + (money held in trust) + (securities) + (foreign exchange) |
| $C_i(t)$ | Cash |
| $K_i(t)$ | (Loans receivable) + (call loans and bills bought) |
| $D_i(t)$ | Deposit |
| $L_i(t)$ | Borrowed money |

(the inclusion upper limit is 1.25% of credit risk assets) (Financial Services Agency, The Japanese Government 2015).

Qualitative studies about the banking system, such as the previous studies that used the agent simulation, have been well established. In this study, it was attempted to investigate the relationship among each items of the financial statements of the bank empirically, using real bank data.

Data about 74–78 banks listed in the first section of Tokyo Stock Exchange from 2006 to 2014 were collected and used to investigate the behavior of the bank and the relationship between some items of the balance sheet during that period. Table 13.1 shows notations of the statement used in the above model as well as the item(s) included in the real balance sheet.

However, we cannot completely unify $J_i$ because the expression of the balance sheet varies according to a company slightly and originally we do not define risk assets from the financing item of the balance sheet. $K_i$ is similar, too.

## 13.3 Data

The data used throughout this investigation were collected from the eol database (Pronexus Inc 2015). Balance sheet data about 74–78 banks listed in the first section of Tokyo Stock Exchange from 2006 to 2014 were extracted.[1] Income statements

[1]The banks corresponding to the security code include 7150 (THE SHIMANE BANK, LTD.; local bank), 7161 (Jimoto Holdings, Inc; local bank), 7167 (Ashikaga Holdings Co., Ltd.; local bank), 7173 (Tokyo TY Financial Group, Inc.; local bank), 7180 (Kyushu Financial Group, Inc.; –), 7182 (JAPAN POST BANK Co., Ltd.; –), 8301 (Bank of Japan; –), 8303 (Shinsei Bank, Limited; local bank), 8304 (Aozora Bank, Ltd.; local bank), 8306 (Mitsubishi UFJ Financial Group, Inc.; major bank), 8308 (Resona Holdings, Inc.; local bank), 8309 (Sumitomo Mitsui Trust Holdings, Inc.; local bank), 8316 (Sumitomo Mitsui Financial Group, Inc.; local bank), 8324 (The Daishi Bank, Ltd.; local bank), 8325 (The Hokuetsu Bank, Ltd.; local bank), 8327 (THE NISHI-NIPPON CITY BANK, LTD.; local bank), 8331 (The Chiba Bank, Ltd.; local bank), 8332 (The Bank of Yokohama, Ltd.; local bank), 8333 (The Joyo Bank, Ltd.; local bank), 8334 (The Gunma Bank, Ltd.; local bank), 8336 (The Musashino Bank, Ltd.; local bank), 8337 (The Chiba Kogyo Bank, Ltd.; local bank), 8338 (Tsukuba Bank, Ltd.; local bank), 8341 (The 77 Bank, Ltd.; local bank), 8342 (The Aomori Bank, Ltd.; local bank), 8343 (THE AKITA BANK, LTD.; local bank), 8344

were also downloaded from the Japanese Bankers Association website (Japanese Bankers Association 2015).

For example, some important financial items were found in a 2014 balance sheet for Sumitomo Mitsui Financial Group, Inc. The assets of the balance sheet express the purpose of the funds, while the debts and equities express their sources. Thus, the total of the debt and equity sections accords with the total of the assets section. In addition, the equity section was a part of the capital before 2007.

Applying the abovementioned model to the Sumitomo Mitsui Financial Group, Inc. at the end of 2014, the model sections become $J_i(t) = 33,041,825$; $C_i(t) = 32,991,113$; $K_i(t) = 69,475,923$; $D_i(t) = 94,331,925$, $L_i(t) = 7,020,841$. Table 13.2 shows values of model and the items of the balance sheet.

The assets of the balance sheet are divided into fluid assets and fixed assets. The fluid assets are those that can be immediately cashed, such as cash money, and the fixed assets are those that need time to be converted into cash, such as land and

(The Yamagata Bank, Ltd.; local bank), 8345 (The Bank of Iwate, Ltd.; local bank), 8346 (THE TOHO BANK, LTD.; local bank), 8349 (THE TOHOKU BANK, LTD.; local bank), 8350 (THE MICHINOKU BANK, LTD.; local bank), 8354 (Fukuoka Financial Group, Inc.; local bank), 8355 (THE SHIZUOKA BANK, LTD.; local bank), 8356 (The Juroku Bank, Ltd.; local bank), 8358 (Suruga Bank Ltd.; local bank), 8359 (The Hachijuni Bank, Ltd.; local bank), 8360 (The Yamanashi Chuo Bank, Ltd.; local bank), 8361 (The Ogaki Kyoritsu Bank, Ltd.; local bank), 8362 (The Fukui Bank, Ltd.; local bank), 8363 (The Hokkoku Bank, Ltd.; local bank), 8364 (THE SHIMIZU BANK, LTD.; local bank), 8365 (The Bank of Toyama, Ltd.; local bank), 8366 (THE SHIGA BANK, LTD.; local bank), 8367 (The Nanto Bank, Ltd.; local bank), 8368 (The Hyakugo Bank, Ltd.; local bank), 8369 (The Bank of Kyoto, Ltd.; local bank), 8370 (The Kiyo Bank, Ltd.; local bank), 8374 (The Mie Bank, Ltd.; local bank), 8377 (Hokuhoku Financial Group, Inc.; local bank), 8379 (The Hiroshima Bank, Ltd.; local bank), 8381 (The San-in Godo Bank, Ltd.; local bank), 8382 (The Chugoku Bank, Limited; local bank), 8383 (THE TOTTORI BANK, LTD.; local bank), 8385 (The Iyo Bank, Ltd.; local bank), 8386 (The Hyakujushi Bank, Ltd.; local bank), 8387 (The Shikoku Bank, Ltd.; local bank), 8388 (The Awa Bank, Ltd.; local bank), 8392 (THE OITA BANK, LTD.; local bank), 8393 (The Miyazaki Bank, Ltd.; local bank), 8395 (THE BANK OF SAGA LTD.; local bank), 8396 (The Eighteenth Bank, Limited; local bank), 8397 (The Bank of Okinawa, Ltd.; local bank), 8398 (The Chikuho Bank, Ltd.; local bank), 8399 (Bank of The Ryukyus, Limited; local bank), 8410 (Seven Bank, Ltd.; local bank), 8411 (Mizuho Financial Group, Inc.; Financial Conglomerate), 8416 (THE BANK OF KOCHI, LTD.; local bank), 8418 (Yamaguchi Financial Group, Inc.; local bank), 8521 (THE NAGANO BANK, LTD.; local bank), 8522 (The Bank of Nagoya, Ltd.; local bank), 8524 (North Pacific Bank, Ltd.; local bank), 8527 (The Aichi Bank, Ltd.; local bank), 8529 (The Daisan Bank, Ltd.; local bank), 8530 (The Chukyo Bank, Limited; local bank), 8536 (The Higashi-Nippon Bank, Limited; local bank), 8537 (THE TAIKO BANK, LTD.; local bank), 8540 (THE FUKUOKA CHUO BANK, LTD.; local bank), 8541 (The Ehime Bank, Ltd.; local bank), 8542 (TOMATO BANK, LTD.; local bank), 8543 (THE MINATO BANK, LTD.; local bank), 8544 (The Keiyo Bank, Ltd.; local bank), 8545 (Kansai Urban Banking Corporation; local bank), 8550 (THE TOCHIGI BANK, LTD.; local bank), 8551 (The Kita-Nippon Bank, Ltd.; local bank), 8554 (The Minami-Nippon Bank, Ltd.; local bank), 8558 (THE TOWA BANK, LTD.; local bank), 8559 (The Howa Bank, Ltd.; local bank), 8560 (The Miyazaki Taiyo Bank, Ltd.; local bank), 8562 (THE FUKUSHIMA BANK, LTD.; local bank), 8563 (THE DAITO BANK, LTD.; local bank), 8600 (TOMONY Holdings, Inc.; local bank), 8648 (Bank of America Corporation; major bank), 8710 (Citigroup Inc.; Financial Conglomerate), 8713 (FIDEA Holdings Co. Ltd.; local bank), and 8714 (Senshu Ikeda Holdings, Inc.; local bank). Income statements were also downloaded from the Japanese Bankers Association website.

**Table 13.2** Values of financial items of Sumitomo Mitsui Financial group in 2014 and 2015 (M(JPY))

| | 2014 | 2015 |
|---|---|---|
| $J_i(t)$ | 33,041,825 | 36,581,444 |
| $C_i(t)$ | 32,991,113 | 39,748,979 |
| $K_i(t)$ | 69,475,923 | 74,395,205 |
| $D_i(t)$ | 94,331,925 | 101,047,918 |
| $L_i(t)$ | 7,020,841 | 9,778,095 |

buildings. Debt can also be divided into fluid debt and fixed debt. Fluid debt is immediately cashable, such as borrowed money, and fixed debt do not need to be refunded immediately, such as corporate bonds.

## 13.4 Calculation and Method

The relationship between different items of the downloaded data was analyzed. The previous study (Fujino 2004) measured the effectiveness of local banks, using national bank financial statements data from 1994 to 2000. It considered whether production activity was effective or not and estimated a change in local financial institution effectiveness that depends on the geographic area and the type of industry. They used the Cobb-Douglas production function (Konishi 2004) to conduct imbalanced panel analysis varying in the number of the samples. The number of samples could vary if there was a merger or failure during the studied period. They assumed that:

$$\ln Q_{\mathrm{it}} = \alpha_0 + \alpha_1 \ln Y_{1_{\mathrm{it}}} + \alpha_2 \ln Y_{2_{\mathrm{it}}} + \alpha_3 \ln Y_{3_{\mathrm{it}}} + \alpha_4 \ln Y_{4_{it}} + v_{\mathrm{it}} - u_{\mathrm{it}}, \quad (13.10)$$

where explanatory variables:

$Q_{\mathrm{it}} = X_{1_{\mathrm{it}}} + X_{2_{\mathrm{it}}} + X_{3_{\mathrm{it}}}$: Product,
$X_{1_{\mathrm{it}}}$: Net profit on loans = interest on loans and discounts − written off of loans − discreteness allowance for loan losses,
$X_{2_{\mathrm{it}}}$: Interest on deposits with banks,
$X_{3_{\mathrm{it}}}$: Gain on trading account securities transactions = interest and dividends on securities + gains on sales of bonds + gains on redemption of bonds + gain on sales of stocks and other securities − loss on sales of bonds − loss on redemption of bonds − loss on devaluation of bonds,

Explained variables:

$Y_{1_{\mathrm{it}}}, Y_{2_{\mathrm{it}}}, Y_{3_{\mathrm{it}}}, Y_{4_{\mathrm{it}}}$: Injection element,
$Y_{1_{\mathrm{it}}}$: Labor (Regular post number),
$Y_{2_{\mathrm{it}}}$: Debt (Financing balance = deposit savings + negotiable certificates of deposit + borrowed money),
$Y_{3_{\mathrm{it}}}$: Capital (Capital stock),
$Y_{4_{\mathrm{it}}}$: Bank premises and real estate, (Total of bank premises and real estate),

where $Q_{\mathrm{it}}$ is a production at period $t$ of the financial institution of the $i$ joint, and $\alpha$ is the coefficient vector that needs to be estimated. Because of multicollinearity for Eq. (13.10), two variable regression analyses were adopted in this paper. Among the items in this study, the following relationships should be focused on:

- Relationships between equity and debt
  - Slope: Risk preference
  - Intercept: Relations with the leverage ratio
- Relationships between CAR and ROE

It is thought that the degree of slope of the regression line in the relationship between equity and debt shows risk preference, where the intercept is related to a leverage ratio. Also important is analyzing the relationships between CAR, thought of as the quantity of stock, and ROE, the quantity of flow. The analyses conducted for 2006–2014 with annual resolution are shown. $\mathrm{CAR}_i(t)$ and $\mathrm{ROE}_i(t)$ are calculated by Eqs. (13.7) and (13.6).

### 13.4.1 Relationships Between Equity and Debt

We show the cross-sectional relations of the equity and debt from 2006 to 2014 in the Fig. 13.3.

The cross-sectional relationship between equity and debt from 2006 to 2014 in depicted in Fig. 13.3, which shows the double-logarithmic plots between equity and debt for 74–78 banks. From the $t$-test, the power-law relations are adopted between equity and a debt in all years.

$$(\mathrm{Debt}) = A \times (\mathrm{Equity})^{\alpha}, \tag{13.11}$$

where $A$ is a positive constant and $\alpha(> 0)$ is an exponent. The deviation from the regression line in each year shows risk preference of the bank. For example, if a slope has a large value, then the debt is relatively larger in terms of equity. Similarly, if a slope has a low value, then the debt is relatively small in terms of equity. The values of $A$ and $\alpha$ can be estimated by a regression analysis. Two methods of ordinary least squares (OLS) and reduced major axis (RMA) in this report as a method of regression.

Ordinary least squares (OLS) and reduced major axis (RMA) regressions are explained as method (regression analysis) that can fittingly perform the relations of an explanation variable and the cover explanation variable using a linear function (Sato 2014).

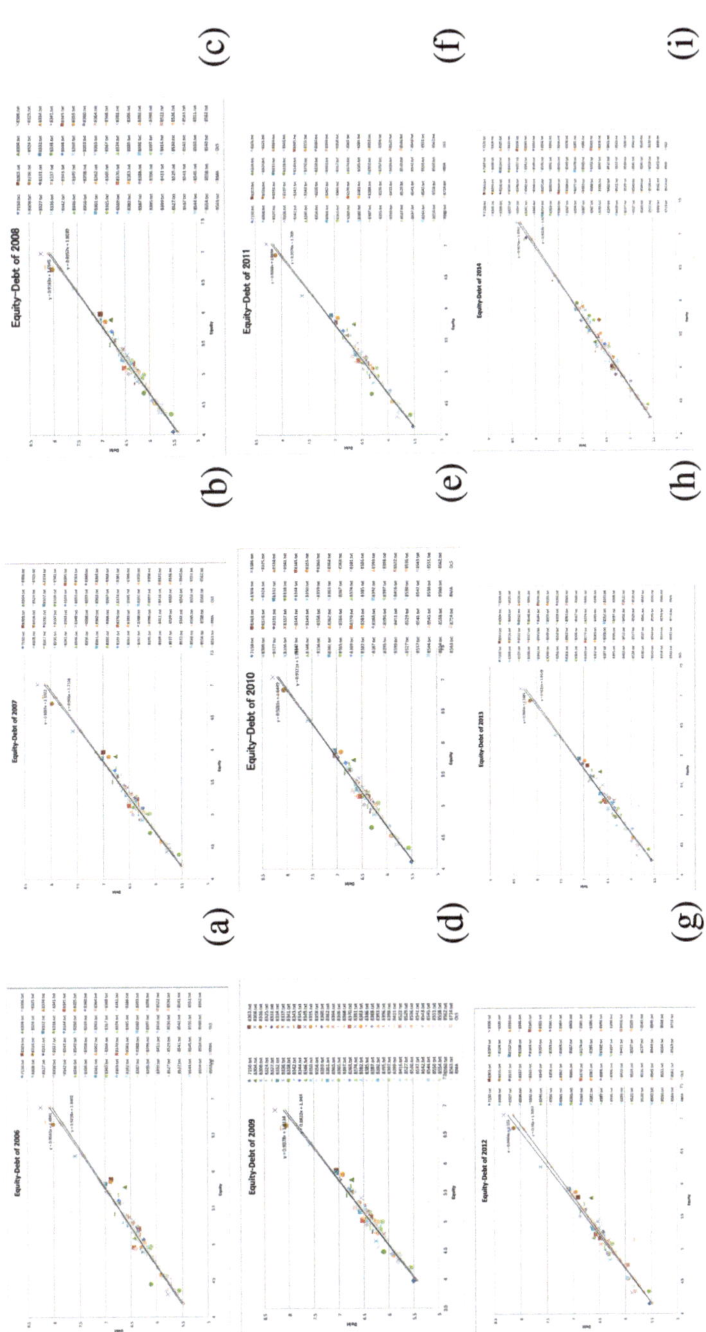


**Fig. 13.3** Double-logarithmic plots of equity and debt for 74–78 banks from 2006 (a) to 2014 (i)

A square error for $y = ax + b$ of OLS is defined as

$$E(a,b) = \sum_{i=1}^{N} (y_i - ax_i - b)^2. \tag{13.12}$$

Minimizing this square error, $a$ and $b$ are given as solutions of the normal equations:

$$\frac{\partial E}{\partial \alpha} = 0, \ \frac{\partial E}{\partial b} = 0. \tag{13.13}$$

The solutions are expressed as

$$\widehat{a} = \frac{\mathrm{Cov}(X,Y)}{\mathrm{Var}(X)}, \ \widehat{b} = \mathrm{E}(Y) - \widehat{a}\mathrm{E}(X). \tag{13.14}$$

Even if RMA reverses $x$ and $y$ and estimates a coefficient, we can obtain the same value. On the other hand, OLS reverses $x$ and $y$, a different coefficient is estimated.

Consider the area of a triangle consisting of the line $y = ax + b$ and the $i$-th data point $(x_i, y_i)$. The area of this triangle is calculated as

$$\frac{1}{2}|x_i - \frac{y_i - b}{a}||ax_i + b - y_i| = \frac{1}{2}\frac{(ax_i + b - y_i)^2}{|a|}. \tag{13.15}$$

Therefore, the total area of the triangles computed from $N$ data points, which is an objective function, is calculated as

$$f(a,b) = \frac{1}{2}\sum_{i=1}^{N} \frac{(ax_i + b - y_i)^2}{|a|}. \tag{13.16}$$

For $a$ >0, minimizing $f(a,b)$ in terms of $(\widehat{a}, \widehat{b})$ implies

$$\frac{\partial f}{\partial \alpha} = \frac{1}{2a^2}\sum_{i=1}^{N} [a \times 2(ax_i + b - y_i)x_i - (ax_i + b - y_i)^2] = 0, \tag{13.17}$$

$$\frac{\partial f}{\partial b} = \frac{1}{2}\sum_{i=1}^{N} \frac{2(ax_i + b - y_i)}{a} = 0. \tag{13.18}$$

From Eq. (13.18), we have

$$\widehat{b} = \frac{\sum_{i=1}^{N} y_i}{N} - \widehat{a}\frac{\sum_{i=1}^{N} x_i}{N}. \tag{13.19}$$

Inserting Eq. (13.18) into Eq. (13.17), we obtain

$$\widehat{a^2}\left[\sum_{i=1}^{N}(x_i)^2 - \frac{\left(\sum_{i=1}^{N} x_i\right)^2}{N}\right] - \left[\sum_{i=1}^{N}(y_i)^2 - \frac{\left(\sum_{i=1}^{N} x_i\right)^2}{N}\right] = 0. \tag{13.20}$$

Thus, since we impose $a \geq 0$, we have

$$\widehat{a} = \sqrt{\frac{\sum_{i=1}^{N}(y_i)^2 - \frac{\left(\sum_{i=1}^{N} x_i\right)^2}{N}}{\sum_{i=1}^{N}(x_i)^2 - \frac{\left(\sum_{i=1}^{N} x_i\right)^2}{N}}}. \tag{13.21}$$

For $a \leq 0$, we obtain the solution in the same manner. Consequently, since we impose $a \leq 0$, we get

$$\widehat{a} = -\sqrt{\frac{\sum_{i=1}^{N}(y_i)^2 - \frac{\left(\sum_{i=1}^{N} x_i\right)^2}{N}}{\sum_{i=1}^{N}(x_i)^2 - \frac{\left(\sum_{i=1}^{N} x_i\right)^2}{N}}}. \tag{13.22}$$

Thus, they are expressed as

$$\widehat{a} = \pm\sqrt{\frac{\sum_{i=1}^{N}(y_i)^2 - \frac{\left(\sum_{i=1}^{N} x_i\right)^2}{N}}{\sum_{i=1}^{N}(x_i)^2 - \frac{\left(\sum_{i=1}^{N} x_i\right)^2}{N}}}. \tag{13.23}$$

Therefore, the sign of $\widehat{a}$ should be chosen according to the sign of the second derivative of $f(a, b)$ in terms of $a$. Thus, the sign of $\widehat{a}$ is equivalent to the sign of $\mathrm{Cov}[X, Y]$. We can also obtain $\widehat{b}$ from Eqs. (13.19) and (13.23). The coefficients of determination $r^2 = \frac{\mathrm{Cov}(X, Y)^2}{\mathrm{Var}(X)\mathrm{Var}(Y)}$ and errors are calculated as

$$\sigma_a = \sqrt{\frac{\mathrm{MSE}}{N\mathrm{Var}[X]}}, \tag{13.24}$$

$$\sigma_b = \sqrt{\mathrm{MSE}\left(\frac{1}{N} + \frac{\mathrm{E}[X]^2}{N\mathrm{Var}[X]}\right)}, \tag{13.25}$$

where the mean square error MSE is computed as

$$\mathrm{MSE} = \frac{1}{N-2}\sum_{i=1}^{N}\left(y_i - \widehat{a}x_i - \widehat{b}\right)^2 = (\mathrm{Var}[Y] - \widehat{a}\mathrm{Cov}[X,Y])\,\frac{2N}{N-2}. \tag{13.26}$$

An analysis method for OLS and RMA and the estimated error are explained.

### 13.4.2 Reduced Major Axis Regression: Slope

The value of slope of RMA may change annually, and it may be related to a macroeconomic factor. The value was calculated for each year to investigate this quantitatively. Table 13.3 shows the value of slope RMA.

In addition, Fig. 13.4 shows that the time series of slope of the regression line is obtained from RMA regression.

According to Fig. 13.4 and Table 13.3, the slope appears to decrease in 2006 and increase in 2010. This is due to the fact that the slope expresses a ratio between equity and debt, and the economy was brisk before the Lehman shock, and banks generally took a lot of debt for equity. Moreover, the slope is at its lowest point in 2009, before recovering to the pre-Lehman Shock after 2012, when monetary alleviation measures were applied, such as the reduction of the official discount rate of the Bank of Japan or the purchase of governmental bonds. The value of the slope of RMA regression suggests a change in the trend of the world or the Japanese economy. Table 13.4 shows a chronological list of the main event, both in Japan and worldwide.

**Table 13.3** Slope obtained with RMA regression

| Year | Slope | Estimated error | T-test by OLS: $p$ value |
|---|---|---|---|
| 2006 | 0.9560 | 0.033999487 | 0.00987 |
| 2007 | 0.9288 | 0.030409938 | 0.00893 |
| 2008 | 0.9168 | 0.029382901 | 0.00951 |
| 2009 | 0.9077 | 0.030695075 | 0.00923 |
| 2010 | 0.9285 | 0.027436006 | 0.00887 |
| 2011 | 0.9267 | 0.028286922 | 0.00875 |
| 2012 | 0.9463 | 0.032464705 | 0.00849 |
| 2013 | 0.9484 | 0.031653672 | 0.00945 |
| 2014 | 0.9572 | 0.029351073 | 0.00396 |

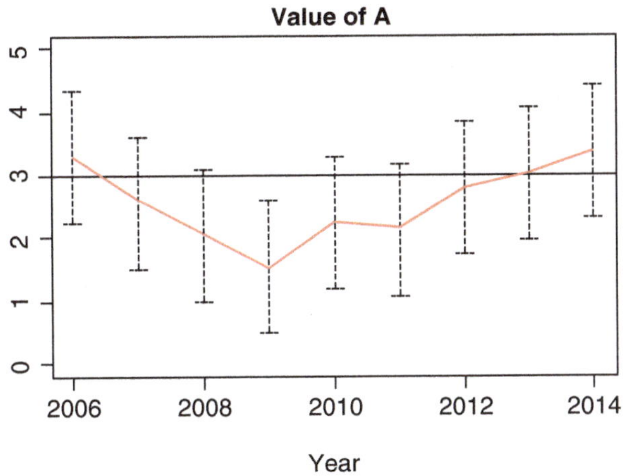


**Fig. 13.4** Time series of slope of the RMA from 2006 to 2014

**Table 13.4** Major economic events worldwide or Japan

| | | |
|---|---|---|
| 2006 | USA | Most of subprime loan fall down |
| Aug-07 | FRA | BNP Paribas shock |
| Mar-08 | USA | Bear Stearns shock |
| Sep-08 | USA | Lehman shock |
| Mar-09 | JPN | The Nikkei Average is lowest after the bubble |
| Oct-10 | JPN | Announcement of the comprehensive monetary easing policy |
| Mar-11 | JPN | In response to the Great East Japan Earthquake, strengthening of monetary easing |
| Apr-13 | JPN | Different dimension monetary easing (Kuroda bazooka) |
| Oct-14 | USA | The end of the quantitative easing policy by FRB |

## *13.4.3 Reduced Major Axis Regression: Constant A*

Constant $A$ of RMA may also be annually, and the value may be related to a leverage ratio. A was calculated for each year by RMA regression, to investigate this quantitatively. The leverage ratio is defined as

$$\text{Leverage} = \frac{\text{Tier 1}}{\text{exposure amount}} \geq 3(\%). \tag{13.27}$$

Leverage ratio is the non-risk-based index that is not regulated by the risk weight, and it complements the capital adequacy ratio, which is a risk-based index. The purpose is to control the heaping up of leverage in the banking department (Suzuki 2015). The exposure amount is calculated as the sum of four points:

1. on balance,
2. derivative transactions,

**Table 13.5** *A* from RMA regression, the reciprocal percentage and the estimated errors from 2006 to 2014

| Year | $A$ | $1/A\times$ (%) | Estimated error |
|---|---|---|---|
| 2006 | 30.47 | 3.280 | 1.04503426 |
| 2007 | 39.08 | 2.558 | 1.041145413 |
| 2008 | 49.54 | 2.018 | 1.04068293 |
| 2009 | 65.01 | 1.538 | 1.043144602 |
| 2010 | 44.14 | 2.265 | 1.037190841 |
| 2011 | 46.75 | 2.138 | 1.038835674 |
| 2012 | 35.72 | 2.798 | 1.042588331 |
| 2013 | 32.96 | 3.032 | 1.042176255 |
| 2014 | 29.97 | 3.336 | 1.038340723 |

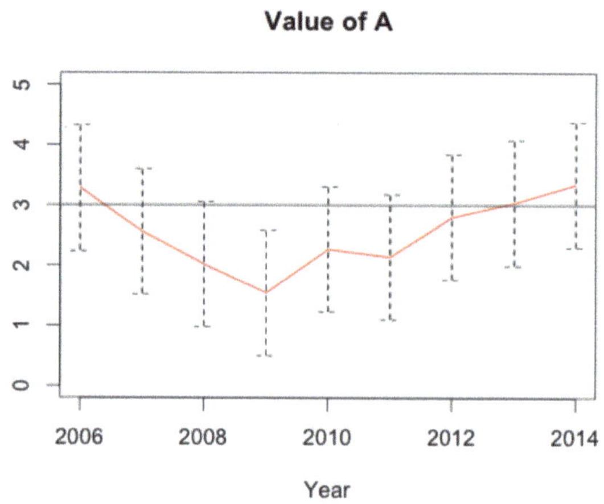


**Fig. 13.5** Time series of *A* from 2006 to 2014

3. securities financial dealing such as repurchase transaction (SFT),
4. off-balance. Some suppositions are necessary to estimate Tier 1 and the amount of exposure from disclosures about the configuration of capital.

Thus, we cannot estimate leverage ratio from a balance sheet. However, if there is a bank with one JPY as limiting a value in a certain cross section, considering the total debt is useful as one of the indicators. An ideal liability *A* can be calculated as

$$D_i + L_i = A\Delta_i{}^{\alpha}, \tag{13.28}$$

$$A = \lim_{\Delta_i \to 1} (D_i + L_i), \tag{13.29}$$

where $\Delta_i$ is an ideal equity. Table 13.5 shows *A* from RMA regression, the reciprocal percentage and the estimated errors.

The debt of an ideal bank, having equity of one yen, was thus around 30 JPY in 2006 but increased up to 65 JPY plot to the Lehman Shock in 2009. Figure 13.5 shows the time series of A of the regression line derived from RMA.

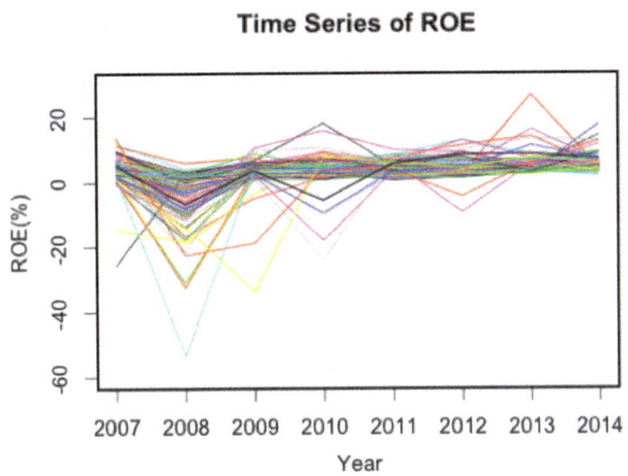


**Fig. 13.6** Time series of the Return on Equity (ROE) for 78 banks from 2007 to 2014

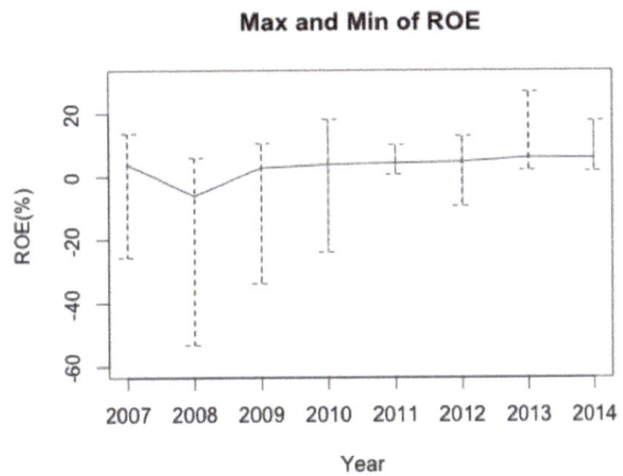


**Fig. 13.7** Mean, the maximum, and the minimum of the ROE of each year of 78 banks

According to the Basel III accords, the leverage ratio that the Basel Committee quires is more than 3.

## *13.4.4 Return on Equity*

ROE will be considered in this section. Figure 13.6 shows the time series of the ROE for 78 banks from 2007 to 2014.

Seventy-four of seventy-eight banks raised profit in 2007, but only 6 of 78 raised their ROE in 2008. Only four banks turned profit in 2 years (Seven Bank, Ltd, Bank of the Ryukyus, Limited, the Yamanashi Chuo Bank, Ltd, and the San-in Godo Bank, Ltd). After that, the value of ROE raised gradually in all companies, and the ROE took a positive value in all companies in 2013. Figure 13.7 shows the mean, the maximum, and the minimum of the ROE for each year for the 78 banks.

Only 5.12% of the banks had negative ROE in 2007, but this increased to 57.69% in 2008, when the mean of ROE became −6.056%. This implies that economic deterioration by Lehman Shock, when the subprime mortgages problem caused the export industry in particular to cool down, had adverse effects on the performance of banks.

### *13.4.5 Capital Adequacy Ratio*

In this section, we consider CAR. Figure 13.8 shows time series of CAR for the 68 banks from 2006 to 2014.

Aozora Bank (code 8304) where CAR was maximum in 2008 (60.755%) started to reduced CAR from about 2009 and fell to 33.75% in 2010. CAR of Aozora Bank was pulled out Mizuho Financial Group (code 8411) in 2013. Figure 13.9 shows the mean, the maximum and the minimum of CAR of 68 banks.

Aozora Bank (code 8304) had a maximum of 60.755% CAR in 2008, but its CAR started to decline starting in 2009, falling to 33.75% in 2010. The CAR of Aozora Bank was pulled out Mizuho Financial Group (code 8411) in 2013. Figure 13.9 shows the mean, the maximum, and the minimum of CAR of 68 banks.

The CAR mean was 11.22% in 2006, but it fell down to 8.39% in 2009, when 44 banks out of the 68 were in danger of not meeting the Financial Services Agency standard (8%). Note that the mean CAR in 2013 was higher than it was in 2006.

Here, the temporal difference of Capital Adequacy Ratio is defined as

$$\Delta_{\mathrm{CAR}} = \mathrm{CAR}(t + \Delta t) - \mathrm{CAR}(t). \tag{13.30}$$

If the last year of ROE is negative, $\Delta_{\mathrm{CAR}}$ has a strong tendency to be negative, so that CAR decreases. When ROE is almost 0, $\Delta_{\mathrm{CAR}}$ tends to be negative. Additionally, there is a tendency toward $\Delta_{\mathrm{CAR}} \geq 0$ if ROE$\geq$0.

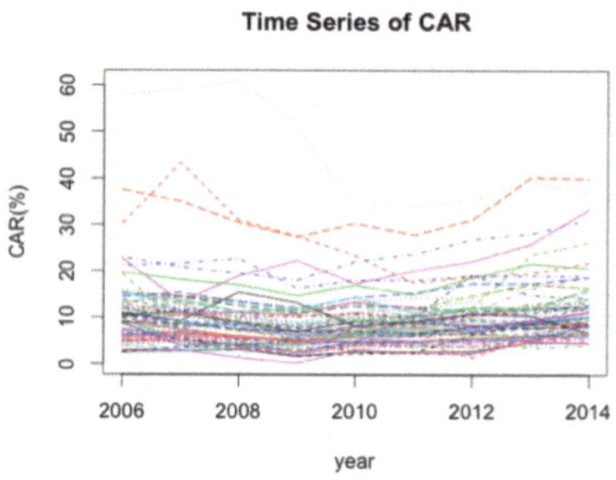


**Fig. 13.8** Time series of Capital Adequacy Ratio for the 68 banks from 2006 to 2014

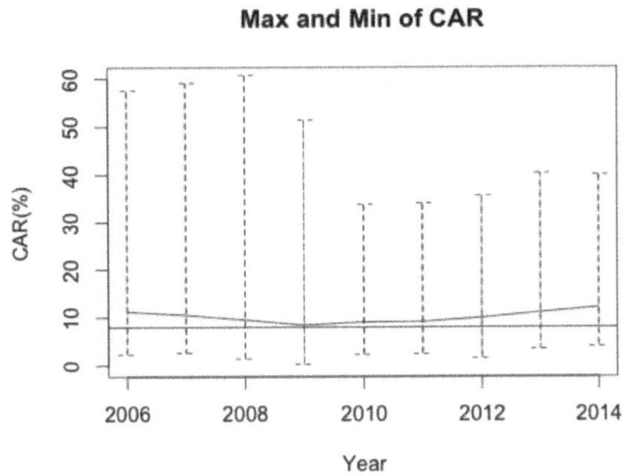


**Fig. 13.9** Mean, the maximum and the minimum of Capital Adequacy Ratio of 68 banks

## 13.5 Discussion

This study confirmed the existence of a power-law relationship between equity and debt. This result shows that the approach of the production function is applicable to the banking industry, which lacks a constant technical system such as the one employed by the manufacturing industry. Furthermore, the simulations conducted by previous studies (Sato et al. 2015) were repeated using actual values extracted from real financial data, which is very advanced.

According to flow of funds statistics reported by the Bank of Japan (2015), the contribution ratio funds of finance business takes a negative value around 2009. Figure 13.10 shows the contribution ratio of the breakdown of the private financial institution loans compared with the previous year. The value of the slope diminishes in 2009, as shown in Table 13.6, which means that equity decreased compared to debt during this period. In fact, banks suffered from a lack of cash and monetary supply was shortage after the Lehman shock. The result is consistent with the intuitive understanding.

Focusing on individual banks, points are almost concentrated from 2006 to 2011 on the regression line, but one bank which deviated greatly to the equity side improved from 2012 to 2014. That was Seven Bank, coded 8410. In 2012, Seven Bank had an equity of 138,045 million yen and a debt of 674,486 million yen. According to the RMA regression line, a debt of 2,614,267 million yen was the average taken during that period. This makes it clear that the business model of Seven Bank is clearly different from the other banks. Seven Bank's business specialized in the use of the ATM, receiving 95% of its profits from ATM fees, which accounts for only about 20% of the profits of other banks. Moreover, ATM fees are the bases of income from other banks for Seven Bank, which uses its ATMs to also benefit from the sales of 7-Eleven. Thus, Seven Bank was able to reduce its amount of deposit possession.

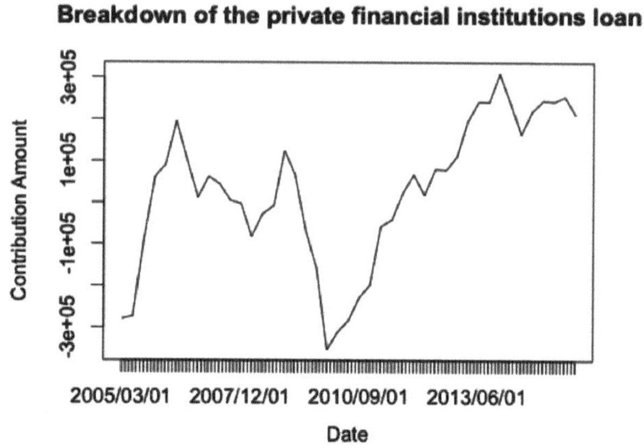


**Fig. 13.10** Contribution amount of the private financial institution's loan

**Table 13.6** Comparison between disclosure matter and model

| | CAR | Denominator | Numerator | |
|---|---|---|---|---|
| Disclosure | 17.74% | 65,364,586 | 9,011,926 (Tier 1) | 2,620,476 (Tier 2) |
| Model | 33.31% | 102,517,748 | 34,156,095 | |

The value of slope $\alpha$ of RMA regression changes in terms of time and is related through a macroeconomy factor. As a result of having greatly reduced equity in 2008, according to the flow of funds statistics, the finance business (bank) increases internal reservation assets by reducing outflow (loans) of the funds from the finance business in 2009.

If the economy was stable before the Lehman Shock, $\alpha \geq 0.95$ is desirable. In addition, the estimated error of the value of the slope obtained from RMA regression is around plus or minus 0.03 every year (See Table 13.3). Since this range is $1\sigma$, it is normally distributed in this range at 68.27%. So, for example, it is hard to think that it would be $0.9077 + 0.0306 = 0.9383$ in 2009 and $0.9560 - 0.0339 = 0.9221$ in 2006. Therefore, a value of slope $\alpha$ of RMA regression may have a correlation with a real event in Japan and worldwide with around 0.03 ranges.

In addition, constant $A$ of RMA regression varies in time and has been suggested to have a high association with the leverage ratio of the bank. It is difficult to find the amount of exposure exactly from the financing item of the balance sheet. We have four points:

1. The amount of on-balance assets is given mainly in (total assets – Tier 1).
2. The derivatives trading does not appear in a balance sheet because when a bank made a contract of derivatives, it becomes worthless.
3. Repurchase transaction is to trade securities with cash as a security, and the item of security of securities borrowed is not available on a balance sheet (Japan Securities Dealers Association 2016).
4. The off-balance trade is business not to be included on a balance sheet such as swap and options, financial futures, and the futures foreign exchange.

Thus, we cannot calculate the amount of exposure from the balance sheet. The amount of accurate exposure should be read from disclosures about the constitution of the leverage ratio of the companies. Three points are to be considered for constant $A$ becoming smaller:

1. A debt is relatively bigger in terms of equity relatively.
2. Equity is relatively in terms of a debt relatively.
3. A value of slope of RMA is big. Namely, the dispersion of the debt has a bigger slope than the dispersion of equity.

Taking the constant $A$ minimum in 2009 shows that the deflection width of the debt became bigger than an average year as a result of the Lehman Shock. In addition, a value of constant $A$ of RMA regression may have a correlation with an event in the real world and Japan with around 1 range, because estimated error is approximately one year-round. We propose that constant $A$ of RMA regression may be used as an indicator to measure the degree of systemic risk.

North Pacific Bank, Ltd. (code 8524) was the best bank with the largest value of ROE in 2013. This is due to it acquiring first preferred shares that cost 30 billion JPN, and then amortizing them, allowing the ROE to rise as the price per one rose. Conversely, the Minami-Nippon Bank, Ltd. (code 8554) was the worst bank with smallest ROE value in 2008. Since operation costs increased, the amount of loan loss reserve transfer and stock amortization were nearly 20 billion JPN, and thus no profit was made. A bank which did raise its profit rate from 2007 to 2008 was THE SHIMANE BANK, LTD. (code 7150), which had increased expenses of the amount of loan loss reserve transfer in 2007.

Aozora Bank, Ltd. (code 8304) was the best bank with the largest value of CAR out of the 74 banks, because it had the smallest deposit amount. The Bank of Iwate, Ltd. (code 8345) was the worst bank with the smallest value of CAR out of the 74 banks, because it was the most careful about loans which meant that its total assets did not mature. This bank thus had a small CAR, but also a stable management. In addition, the capital adequacy ratio officially reported by Sumitomo Mitsui Financial Group, Inc. was 17.79%, along with the disclosure about the constitution of its owned capital. In that case, CAR according to the model is calculated 33.31.

The model defines CAR according to its concept. The denominator was found to be approximately 1.5 times larger than those of disclosure, and the numerator is calculated approximately three times larger than that of the disclosure values; when this tendency is considered from a concept of CAR, it seems that the gap is caused by the numerator of the model being much larger than those of disclosure. In the debt section, in the case of a major bank, account deposit, borrowing money, negotiable deposit, call money, and payables under repurchase agreements were not taken into account, and therefore a value with a large error was obtained. CAR appears to be 15.53% when these variables are put in the calculation of the model. Thus, it will be necessary to change a model type according to individual bank characteristics (city bank, local bank, the second local bank) to calculate CAR in a more accurate model.

Using the scatter plots, no correlation was found between the CAR and the ROE, and thus the bank management could be considered stable even if profit ratio was

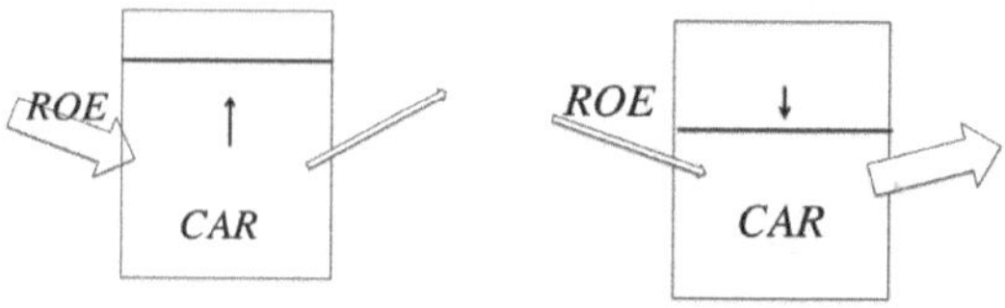


**Fig. 13.11** A pictorial illustration of flow and stock variables

not always good, and vice versa. However, we can make a connection that CAR and ROE by introducing $\Delta_{CAR}$.

Since CAR based on stock variables saved equity until last year, ROE is calculated on the basis of flow variables that add equity up to the previous year. Figure 13.11 shows the pictorial illustration of amount of flow and stock variables.

## 13.6 Conclusion

In this study, we obtained the following results:

- The relations of equity and the debt may change on time, and it was confirmed to be a power-law relationship.
- The approach of the production function is applicable to banking.
- Value of slope $\alpha$ of RMA regression changes over time and is related to macroeconomic factors. Furthermore, we proposed that $\alpha$ showed correlations with events observed in the real world.
- Constant $A$ of RMA regression changes over time, and it is suggested that its value is related to the leverage ratio of the bank and that this constant $A$ can be used as an indicator to measure the degree of systemic risk.
- ROE and CAR according to the model reflected an event in the world.
- ROE and CAR do not have the correlation, but a correlation between them can be made by introducing the difference of one year of CAR.
- The proposed method is usable in the analysis from a macroeconomic viewpoint of the banking system.
- Properties of individual banks were described and analyzed.

The following problems should be addressed in the future:

- The analysis method of this report might be applicable to industrial sectors other than the banking. CRD association makes use of data in management among medium and small-sized businesses. The risk identification of various sectors of the Japanese industry becomes possible by calculating this proposed technique, according to the type of industry using data of this CRD.
- The analysis method of this report might be applicable to industrial sectors other than the banking. CRD association makes use of data in management among medium- and small-sized businesses. The risk identification of various sectors of the Japanese industry becomes possible by calculating this proposed technique, according to the type of industry using data of this CRD.

**Acknowledgements** This work was financially supported by a Grants-in-Aid for Scientific Research (KAKENHI) (B) (#26282089).

## References

Bank for International Settelements, [Online] Available http://www.bis.org/publ/bcbs189.htm. Accessed 24 Dec 2015

Bank of Japan, [Online] Available https://www.boj.or.jp/statistics/sj/sjexp.pdf. Accessed 24 Dec 2015

Financial Services Agency, The Japanese Government, [Online] Available http://www.fsa.go.jp/policy/basel_ii/. Accessed 24 Dec 2015

T. Fujino, Effectiveness analysis of the local bank -empirical analysis by the probabilistic frontier production function- (In Japanese only), monthly report of Shinkin Central Bank Research Institute, Mar 2004

P. Gai, S. Kapadia, Contagion in financial networks. Proc. R. Soc. A **466**, 2401–2423 (2010)

A.G. Haldane, R.M. May, Systemic risk in banking ecosystems. Nature **469**, 351–355 (2011)

G. Iori, S. Jafarey, F.G. Padilla, Systemic risk on the interbank market. J. Econ. Behav. Organ. **61**, 525–542 (2006)

Japanese Bankers Association, [Online] Available http://www.zenginkyo.or.jp/abstract/stats/year2-02/account2014-terminal/. Accessed Dec 24 2015

Japan Securities Dealers Association, [Online] Available http://www.jsda.or.jp/katsudou/kisoku/index.html. Accessed Jan 27 2016

J. Kay, Understanding the bigger picture. FST J. **9**, 5–6 (2012)

Y. Konishi, Nonparametric statistical analysis of the production function (In Japanese only). Appl. Stat. **33**(2), 157–179 (2004)

Lehman-shock chart, [Online] Available http://tyoikabu.web.fc2.com/lehman.html. Accessed 24 Dec 2015

Pronexus Inc., [Online] Available http://eoldb.jp/EolDb/Result/CompanyHitlist.aspx. Accessed 24 Dec 2015

A.-H. Sato, *Applied Data-Centric Social Sciences* (Springer, Tokyo, 2014), pp. 88–93

A.-H. Sato, P. Tasca, T. Isogai, Dynamic interaction between asset prices and bank behavior: a systemic risk perspective [Online]. Available http://arxiv.org/abs/1504.07152. Accessed 24 Dec 2015

T. Suzuki, What is a leverage ratio? (In Japanese only) [Online] Available http://www.dir.co.jp/research/report/finance/basel3/20141208_009229.html. Accessed 24 Dec 2015

P. Tasca, S. Battiston, Market procyclicality and systemic risk. Quant. Finan. **16**, 1219–1235 (2016)

M. Wei, M. Leon-Ledesma, G. Marcelli, S. Spurgeon, Dynamic network model of banking system stability, in *21st Computing in Economics and Finance Conference – CEF 2015*, Taipei, 20–22 June 2015

FSC
www.fsc.org
MIX
Papier | Fördert
gute Waldnutzung
FSC® C083411

Kolibri 360 GmbH
Ferdinand-Jühlke-Straße 7
99095 Erfurt, Deutschland
produktsicherheit@kolibri360.de